Sanjay Rana
Jayant Sharma

Frontiers of Geographic Information Technology

Sanjay Rana
Jayant Sharma
Editors

Frontiers of Geographic Information Technology

With 107 Figures, 14 in Color, and 11 Tables

Springer

Editors

Sanjay Rana

Centre for Transport Studies
Department of Civil and
Environmental Engineering
University College London
London WC1E 6BT
UK

E-mail: s.rana@ucl.ac.uk

Jayant Sharma

Oracle Corporation
1 Oracle Drive
Nashua NH 03062
USA

E-mail: Jayant.Sharma@oracle.com

Library of Congress Control Number: 2005930053

ISBN-10 3-540-25685-7 Springer Berlin Heidelberg New York
ISBN-13 978-3-540-25685-4 Springer Berlin Heidelberg New York

Springer-Verlag is a part of Springer Science + Business Media
springer.com

Printed in the Netherlands

Cover design: Erich Kirchner, Heidelberg
Typesetting: SPI Publisher Sevices
Production: Almas Schimmel
Printing: Krips bv, Meppel
Binding: Stürtz AG, Würzburg

Printed on acid free paper 30/3141/as - 5 4 3 2 1 0

Table of Contents

List of Contributors

Satyaprakash Centre for Spatial Database Management and Solutions, G - 4, Sector - 39, Noida - 201 301 (UP), India

Michael Batty Centre for Advanced Spatial Analysis, University College London, 1-19 Torrington Place, London WC1E 6BT, UK

Alastair R. Beresford Computer Laboratory, University of Cambridge, 15 JJ Thomson Ave, Cambridge CB3 0FD, UK

Yaser Bishr Image Matters LLC, 201 Loudoun Street SW, Leesburg, VA 20175, USA

Liping Di 6301 Ivy Lane, Suite 620, Greenbelt, MD 20770, USA

Reginald G. Golledge Department of Geography and Research Unit on Spatial Cognition and Choice, Elison Hall 3611, University of California Santa Barbara, Santa Barbara, CA 93106, USA

David Gillingwater Centre for Transport Studies, Department of Civil and Building Engineering, Loughborough University, Loughborough LE11 3TU, UK

Ravi Gupta Centre for Spatial Database Management and Solutions, G - 4, Sector - 39, Noida - 201 301 (UP), India

Chris Hinde Department of Computer Science, Loughborough University, Loughborough LE11 3TU, UK

Daniel Holweg Fraunhofer Institute for Computer Graphics Datenverarbeitung, Fraunhoferstr. 5, 64283 Darmstadt, Germany

Ayanna Howard NASA Jet Propulsion Laboratory, California Institute of Technology, Pasadena, California, 91109, USA

R. Daniel Jacobson Department of Geography, Faculty of Social Sciences, University of Calgary, Earth Sciences 356, 2500 University Drive N.W., Calgary, Alberta T2N 1N4, Canada

Ursula Kretschmer Fraunhofer Institute for Computer Graphics Research, Fraunhoferstr. 5, 64283 Darmstadt, Germany

Sanjay Rana Centre for Transport Studies, Department of Civil and Environmental Engineering, University College London, London WC1E 6BT, UK

Matthew T. Rice Department of Geography and Research Unit on Spatial Cognition and Choice, Elison Hall 3611, University of California Santa Barbara, Santa Barbara, CA 93106, USA

Jayant Sharma Oracle Corporation, 1 Oracle Drive, Nashua NH 03062, USA

Jantien Stoter ITC, Department of Geo-Information Processing, PO Box 6, 7500 AA Enschede, The Netherlands

C. Vincent Tao Geospatial Information and Communication Technology (GeoICT) Lab, Department of Earth and Space Science and Engineering, York University, 4700 Keele Street, Toronto, ON, Canada M3J 1P3

Edward Tunstell NASA Jet Propulsion Laboratory, California Institute of Technology, Pasadena, California, 91109, USA

Ayon K. Tarafdar Centre for Spatial Database Management and Solutions, G - 4, Sector - 39, Noida - 201 301 (UP), India

Sabine Timpf Geographic Information Science Center, Department of Geography, University of Zurich-Ircehel, Winterthurerstr. 190 CH-8057, Zurich, Switzerland

Chaowei (Phil) Yang Earth Systems and GeoInformation Sciences, School of Computational Sciences, Center for Earth Observing and Space Research, George Mason University, 4400 University Drive, Fairfax, VA, 22030-4444, USA

Yinjie Yang Centre for Computational Intelligence, De Montfort University, The Gateway, Leicester LE1 9BH, UK

Sisi Zlatanova Delft University of Technology, OTB Research Institute for Housing, Urban and Mobility Studies, Jaffalaan 9, 2628 BX Delft, PO Box 5030, 2600 GA Delft, The Netherlands

Acknowledgements

It is always exciting and great responsibility to prepare a book which lays a claim to such an ambitious title. It is exciting because one can let the imagination rather than practical nitty gritty to guide the thoughts. It is also a great responsibility as most information is required to be accurate and representative of the entire spectrum of the presented topic.

Geographic Information Technologies are now an integral part of many decision making processes in modern society ranging from urban planning to military ventures. The computing technologies also seem to have reached a state of maturity whereby we have sufficient experience of the past and the awareness of the challenges facing us in the future. Hence, a book which presents an account of the prominent ideas that might be going to occupy future intellectual and commercial ventures is timely. However, just like any other multi-party efforts of this kind it wasn't always plain sailing from the beginning. It was felt by Sanjay from the beginning that this book requires a co-editor in order to remain balanced. Sanjay first floated the idea of book in the mid- 2001 but it fizzled out quickly due to the lack of enthusiasm from the early collaborators. However later that year, Simon Doyle showed a bit more interest and wrote a one-page proposal with Sanjay briefly outlining the objectives of the book. Sanjay would like to thank Simon for those constructive brainstorming sessions. Sadly, this partnership couldn't last long due to Simon's busy corporate schedule. Sanjay then threw the idea around again eventually teaming up with Jayant in 2002 and the formal manuscript preparation started in mid 2003. The next big task was to collect some authoritative chapters which could provide sufficient depth and outlook so as to come up to the rather ambitious book title. However, being an edited book has the disadvantage that it doesn't appeal to big names due to the rather low impact in research funding scenario. Therefore, following several apologies and excellent discussions from mentors such as Alan MacEachren, the first set of 11 chapters were commissioned. Since then 3 more chapters were added to broaden the scope of the book. We are grateful to all the authors for responding to the call despite the hectic schedules and sticking to the deadline.

We are also grateful to Kurt Buehler for writing the foreword. The support and patience of Springer-Verlag book editor Wolfgang Engel and assistants Susanna Pohl and Helen Rachner were also fundamental to keep the project alive and on track. We would like to thank the referees of the chapters, namely Allan Doyle, Claire Ellul, Lars Bodum, Laxmi Subramanian, and Young-Hoon Kim who helped us in evaluating the manuscripts and spared us time to work on the overall book layout. Many of the colour images have been taken from various personal and official websites. We are thankful to all the organisations and individuals for their generosity. We thank David Unwin and Ian Fieggen for their constructive comments at various stages of the book.

Foreword

Emerging from the collective experiences of millions of people who rely on geospatial information and technology everyday, knowingly or not, are many challenges and issues for the community of GIScience and GeoComputation researchers. There are enough problems to go around and, thankfully, there are also enough intelligent people around (or in training) to solve the issue and challenges that we face as a society in our quest to make decisions better, understanding deeper, profits greater, and our environment more sustainable. Jayant and Sanjay have assembled an astounding array of knowledgeable authors to present the broad frontier of geographic information technology that is emerging to address these many issues and challenges. They have correctly identified that the current state of geographic information technology consists of geographic information (GI), GI hardware, and GI software and that the current state has been heavily influenced by the IT industry and by the needs of society over time. This is as it should be. In fact, while reading the chapters of this book, I suggest that you keep a mental map of the relationship of the topics being discussed with this "data, hardware, software" classification and with the additional classification of technologies into those of data collection and handling (to include information creation and management), analysis (to include modeling, simulation, and knowledge generation and synthesis), and visualization (to include all forms of presentation with the goal of understanding). This helped me to organize my thoughts and to make the book a truly exciting reading experience.

When reading the chapter on autonomous systems control, my mental map yielded applications of automatic information generation using sensor systems mounted on visually impaired humans (right here on earth) as discussed in the chapter on multimodal interfaces and I could foresee a future where safe and effective wayfinding was made possible. When reading the chapters on distributed GI services and the geospatial grid, I kept flashing over to the chapters on agent technology and the semantic grid and to the issues presented there and to the promise of solutions emerging that will make "distributed geospatial semantic web grid services" a reality. My mind then went to the issues of privacy, solved by the application of appropriate IT technologies, and then to the inversion of the GeoICT pyramid and I visualized a society where local data is collected, stored, documented, and semantically described by local people and local government into a 3D, ontology-capable DBMS. I then thought of how this new database was exposed on the distributed geospatial semantic web grid (using standards-based interfaces and ontologies) where it was analyzed using agent-based models developed by "higher" levels of Government and effectively combined with data resources of many other local entities to provide a solution of a regionally-serving business leader trying to decide where to site his next facility to maximize his profit or by a state-level decision makers trying to decide how to more effectively manage their ever-more-precious land resources. Finally, I could see where augmented reality could help these decision makers interpret the results of analyses by presenting them with a picture of the future overlain on the present.

Combining the ideas in the book to form a picture of the future is the job of every reader, but beyond that, it is the responsibility of each of us to grasp one of the unexplored or less understood topics and further refine and contribute to the advancement therein for society, for science, and/or for profit. This book wonderfully sets the stage for all of us in the GIScience and GeoComputation domains and provides us with many possibilities for advancement. Please join me and the editors and authors in this volume in helping to expand the frontiers of geographic information technology.

Kurt Buehler
President, Image Matters LLC
Co-founder Open Geospatial Consortium

Conventions

1. Internet References in the chapters are indicated by an index shown in superscript next to the text e.g., Intel [125] is the Internet reference no. 125.

2. The format used in references is as follows:

   Journal Article:

   [Author (s)], [Year]. [Article Title], [*Journal Name*], [Volume No.], [Page Nos.].

   Abowd, G.D., Atkenson, C.G., Hong, J., Long, S., Kooper, R., Pinkerton, M., 1997. Cyberguide: A Mobile Context-Aware Tour Guide, *Wireless Networks*, **3**, 421-433.

   Book Chapter:

   [Author (s)], [Year]. [Chapter Title], In: [Editor (s)], [*Book Title*], [Publisher], [City, State, Country], [Page Nos.].

   Blades, M., 1991. Wayfinding Theory and Research: The Need for a New Approach, In: D.M. Mark, and A.U. Frank (Eds.), *Cognitive and Linguistic Aspects of Geographic Space*, Kluwer Academic Publishers, Dordrecht, The Netherlands, 137-165.

   Conference Proceedings:

   [Author (s)], [Year]. [Chapter Title], In Proceedings [*Conference Title*], [City, State, Country].

   Bly, S., 1982. Presenting Information in Sound, In Proceedings *Human Factors in Computer Systems*, Gaithersburg, MD, USA.

   Book:

   [Author (s)], [Year]. [*Book Title*], [Publisher], [City, State, Country].

   Bonabeau, E., Dorigo, M., and Theraulaz, G., 1999. *Swarm Intelligence: From Natural to Artificial Systems*, Oxford University Press, NY, USA.

Chapter 1

Geographic Information Technologies – An Overview

Sanjay Rana and Jayant Sharma

The ever-expanding growth of international commercial ventures and the awareness of resultant enormous profits have made us realise the potentials of our planet and its people. The driving force behind this sea change from the iron-curtain post WWII-era is simply the increasing accumulation and easy accessibility of *geographic information* and its utilisation. A simple and yet so universally relevant piece of information, *geographic information* essentially describes geographical entities and phenomena. Geographical phenomena involve living beings and their habitats, the host celestial body, and its non-living matter. Please note that our rather nebulous definition deviates from the similar *Geospatial Data*, which is specifically defined as "*information that identifies the geographic location and characteristics of natural or constructed features and boundaries of earth*" (Federal Geographic Data Committee, quoted by National Research Council 2003). For example, information such as *location* of a house, *population density* patterns in a city, *distribution* of an endangered plant/animal species, *spatial variation* of financial wealth in a country, weather *pattern*, *topography* of ice caps on the Mars North Pole, patterns of shopper *movements* in a shopping mall and our *store preferences*, are all examples of *geographic information* (GI). In other words, arguably all kinds of conceivable information have a direct or indirect geographic context. GI technology is the set of ideas and tools that lead to the collection and implementation of GI information. While the direct links to a geographic reference are obviously useful, developing indirect geographic reference allows sophisticated spatial analysis. For example, while *travel time* is not strictly a *geographic information* (National Research Council 2003), its application to a transport network can help in solving the shortest-route problem. Thus, a recent remark from Jack Dangermond, the founder of the company ESRI which sells the most GI systems (GIS), that "*The application of GIS is limited only by the imagination of those who use it*" is a succinct characterisation of this potential. After all, it was the mapping of the extents of the earth and its environment which allowed us to travel to far off destinations, record our land parcels, build cities and so on. Once we conquered the earth, the focus shifted to mapping people as part of socio-economic surveys relevant for the essential fabric of human societies i.e., social services, transport networks, financial institutions, natural resources and so on. It should therefore come as no surprise that the existing 5-billion USD global geospatial technology industry is predicted to rise to 30-billion USD by 2005 (Gewin 2004). This is a massive increase from the predicted 2-billion USD for 2004 (Daratech 2004).

Figure 1.1 shows the stock prices of two key publicly traded GIS companies, namely Intergraph and Leica GeoSystems. The figure clearly shows that while the NASDAQ index took a tumble after the dot com bust and 9/11, these two major GIS companies generally outperformed NASDAQ index many times over. Perhaps, this sheltered life and service-oriented aspect of the GIS industry has helped in producing sustained profits. This trend is likely to continue in the near future because of renewed demands on geospatial technologies particularly in the vast and untapped natural resources of developing countries such as India, China and other parts of world such as Middle-East and Russia. The aim of this book is to present a number of geospatial technologies that are getting a lot of attention from the research community and also from the key funding agencies across the globe. At the same time, we are aware that these chapters most likely represent only a fraction of the geospatial technologies currently being developed in research centres. These unreported "black technologies", whether they are variants/combinations of the proposed ideas or entirely novel, are perhaps too secretive or commercially sensitive to be made available to the general public at this date. One important point to note here that the individual technologies seldom provide the complete solution, therefore another aim of the book is to present ideas that would be used together to create the next generation GIS.

However, this is not the first such effort to compile the state of the art of the broader disciplines of GI Systems and Science and predict their trends (or scenarios). Previous examples include essays by Wegener and Masser (1996), Longley et al. (1998, 1999), Goodchild (1999a), Openshaw (1999), US National Research Council (1997, 2003), and Sui (2004). One of the main differences between this book and earlier works is the attitude of the predictions. While, we try to present an exhaustive review of existing GI technologies, we avoid being cursorily alarmist or merely a compilation of *all* ideas. This book demonstrates examples of working prototypes and proposes a commercially viable and challenging scenario. In any case, we believe that despite the claims by economists, planners and thinkers, the "*future is anything but what we imagine*" though it is a good idea to keep an eye on short-term trends (Wegener and Masser 1996).

This chapter presents a broad classification of the various components of GI technology and their historical development. The underlying motive behind such a classification is to show the wide extent and heritage of the modern GI technologies. This approach is similar to the discussion of the computers in geography by Maguire (1989) albeit we extend our classification to non-computational aspects of the technology. The chapter concludes with a summary of the other chapters in the book.

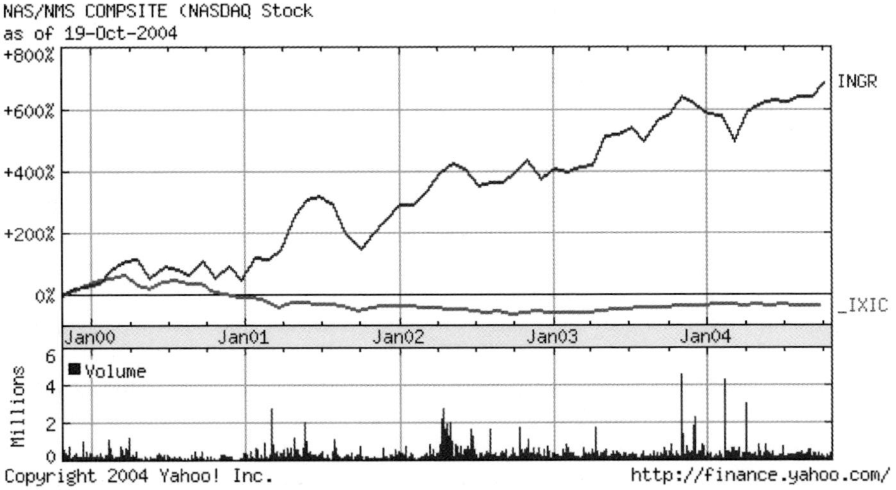

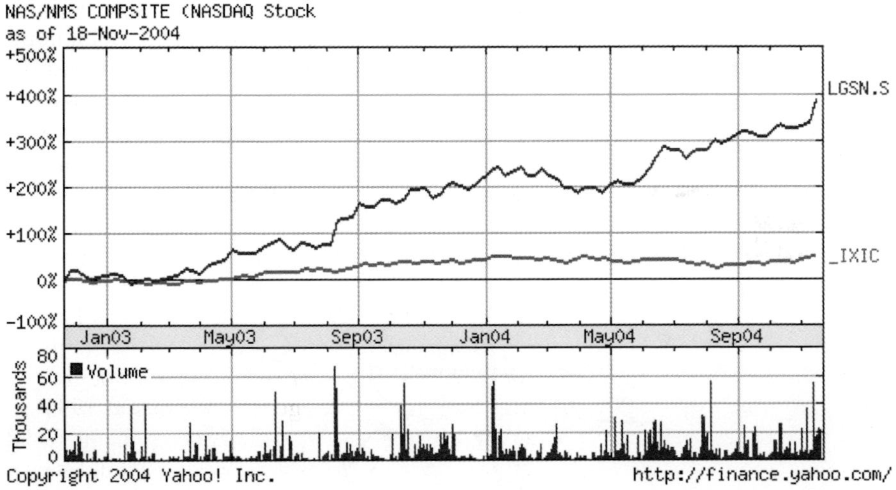

Figure 1.1. Performance of the Intergraph (top) and Leica Geosystems (bottom) stock prices in comparison to the NASDAQ index in the recent years. Reproduced with permission of Yahoo! Inc. © 2005 by Yahoo! Inc. YAHOO! and the YAHOO! logo are trademarks of Yahoo! Inc.

1.1 A brief history of GI Technology

We have been mapping our surroundings for more than 4000 years, from the surveyors in ancient Babylonia and Egypt who manually measured the land parcels to the latest laser-guided systems used by modern surveyors. Clearly, such a significant jump in the GI technology is an outcome of the accumulation of many hundreds of years of human innovation. This section traces back the origins of the current technology into different lineages. Figure 1.2 shows the three main elements of modern GI technology, namely GI software, geographic information and GI hardware, and how they relate to each other. These three elements drive changes in each other and are themselves influenced by the changes in the IT industry and society's demands. The relationships suggest that an organisation that influences all the three elements of the GI technology triangle is likely to benefit most. A good example is Leica Geosystems, which manufactures geospatial data collection sensors and sells the software systems to exploit the geospatial data. However, in our view, the scope of the GI technology is still rather limited and is significantly dependent upon the moods of the IT industry and the society in large. For example, the unfortunate incident of 9/11 has fuelled national governments interest in regularly updating geospatial data such as satellite images and demographic datasets for anti-terrorism purposes. What used to be only a piece of useful technology, mostly used to deliver and maintain a record of the natural resources has now become a front-line tool. Another example, of the increased relevance of GI technology is the online Foot and Mouth Disease information service provided by

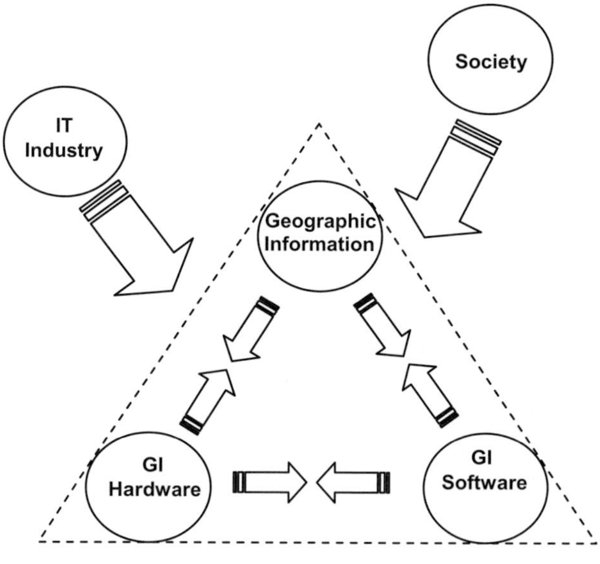

Figure 1.2. GI technology triangle and the external factors that influences its internal elements, namely geographic information, GI software and GI hardware.

Department for Environment, Food, and Rural Affairs, UK. During the epidemic of Foot and Mouth Disease in 2002-2003, the Multi-Agency Geographic Information for the Countryside (MAGIC) website[1] (Figure 1.3), provided daily reports of the occurrence of new cases thus informing the public and other parts of the government. Such practical examples of GI technology boosted the relevance of online mapping systems, especially amongst government organisations.

Figure 1.2 also highlights the fact that GI technology generally adopts the available software environments and hardware platforms. At the same time, there are instances where the IT industry benefited from developments in GI technology. For example, computer graphics researchers use various terrain data structures in games for faster animations. Nonetheless, this is still limited compared to the importance of computer games industry that has led to the mass production and innovation of hardware graphics cards, and related software libraries, dedicated for computer animations.

During the collection of literature on the history of three elements of GI technology triangle, it was realised that GI software e.g., GIS, spatial databases, have been widely documented in several summaries (e.g., Schuurman 2004) as well extensive texts (e.g., Foresman 1998). Hence, this chapter does not present an account of the history of GI software. The reader is requested to follow the suggested references. However, it is worth highlighting an important point as seen by the critics (Unwin 1998) of these texts. While it is generally accepted that the origin of the GI software lies with Roger Tomlinson's Canadian Geographic Information System (CGIS), some authors like to highlight the theoretical concepts that perhaps inspired researchers in the first place. For instance, Schuurman (2004)

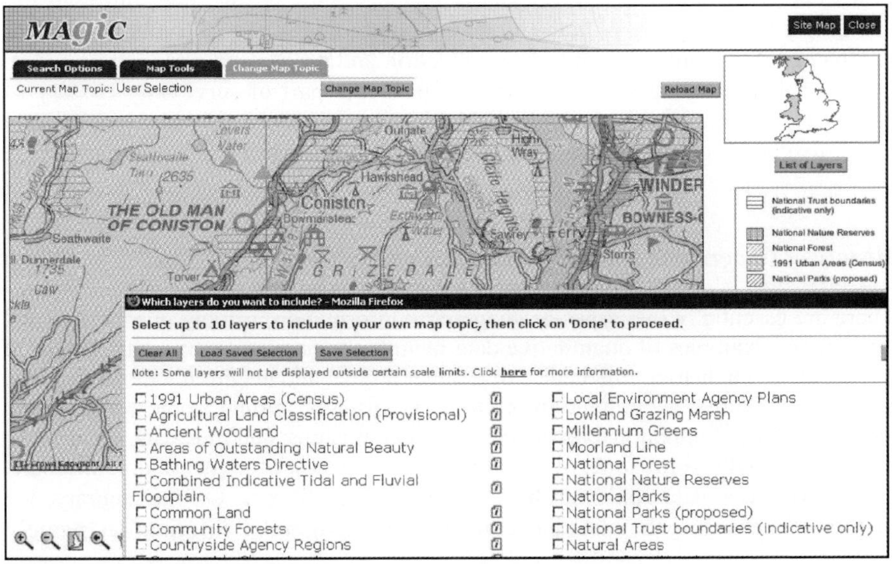

Figure 1.3. Mapping interface of Multi-Agency Geographic Information for the Countryside (MAGIC), which was used for Foot and Mouth Disease Mapping.

credits Ian Mcharg for proposing the idea of layer overlay, which characterises GIS. Other authors (Unwin 1998) question the over emphasis given to certain individuals, organisations, and sequence of events.

The history of GI Hardware and Data however presented exciting information, new to us and possibly our readers. Hence we present a broad overview here intended to cover the most important events.

1.1.1 History of GI Hardware

Advances in tool sophistication mark the milestones of a technological discipline. GI technology has similarly evolved from the stone engraving, paper parchments to Augmented Reality and location based systems. Generally, there are three types of GI hardware, depending upon their role, namely **data collection tools**, **data visualisation tools**, and **data analysis tools**. These tools typically represent the life cycle and organisation of most GI Systems. We need data collection tools such as Global Positioning System (GPS), and handheld Personal Data Assistant (PDA) to collect spatial information, which is then visualised using a data visualisation tool, typically a computer monitor or map, at the same time processed by a sophisticated data analysis tool such as a high-end workstation.

The history of GI hardware shares its heritage with the history of astronomy, geometry, surveying, geography, mathematics, computing, and several other disciplines. Thus, it is difficult to untangle the various lineages and present an exhaustive classification. Broadly, the history of GI hardware can be encapsulated into two threads, namely surveying and computing. These threads in turn inherit from several other disciplines. History of surveying is as vast as history of human settlements. A detailed treatment is beyond the scope of this section thus the following paragraphs will merely glance through some of the main milestones in the history of surveying. With the advent of computing in the 20th century, many of the manual surveying and other data collection and processing were automated so much so that computers are now an indispensable part of surveying. History of computing is comparatively simpler to the history of surveying, as it is really only about six decades old and involves primarily electronics and product design.

History of Geographic Information collection tools

There are essentially two types of geographic information, namely quantitative and qualitative. Examples of quantitative data include the population of a city, length of a road, area of a district, reflectance of a mineral type, and height of a mountain. The types of qualitative geographic information is virtually infinite, some common examples include shopping habits, topological relationships, and metadata. Qualitative data is essentially some form of description or extension of a quantifiable measure. Thus, qualitative data could not have a well defined history. On the contrary, the history of measuring the most basic quantitative geographic information, namely *location* (*longitude*, *latitude*, *elevation*) is fairly well documented especially in terms of the improvement in accuracy and sophistication. The ability to record location information leads to the concept of *length*, which is a fundamental geometrical measure. In ancient times, time was also equivalent to distance e.g., the journeys were often described in terms of days and nights rather than distance. Another important measure, which evolved from length (distance) and played a significant role, is *angle*. The ability to measure angular separation was crucial to the development of trigonometry and cartography. It allowed mariners to travel around the globe by measuring angular separation between stars, and surveyors to map the heights and shapes of the land. In fact, the history of geographic information collection tools is primarily the history of tools for collecting location and angular information.

In the beginning, location information was collected to solve measurement problems. The earliest records of length measures come from the Egyptians and they were based on the length of human body parts. Around 3000 BC, Egyptians developed the cubit, which was the length of an arm from the bent elbow to the extended middle finger plus the width of the palm of the hand of the Pharaoh at that time (Figure 1.4). Smaller lengths were obtained by an unequal subdivision of the length of the cubit[2]. In summary, the human limbs were the first geographic information collection tools used for measuring. This practise continued with later civilisations (2500 – 1700 BC), namely Babylonians, Harrappans, Greeks, Romans, although of course, varied genetic make up of the different societies meant that the size measurements changed considerably. The Harappan civilisation had a well developed decimal system of measuring fractions. Precision in numbering systems suggests sophistication in measurement techniques, which is reflected in the architecture of Harappan cities. Ancient Greeks used a number of tools made of twisted rushes (e.g., *schoinion*), wooden rods (e.g., *kalamos*) and iron chains (e.g., *halysis*). The Romans adopted the Greek measurement system and added longer length measures such as a mile, which was supposed to be 1000 paces. Romans measured longer distances by leapfrogging a pair of rods. In a different method, Alexander employed a staff called bematistai, whose job was to count the number of paces marched and record the places. The Roman system of measurements then spread

Figure 1.4. The Royal Cubit Master

across rest of the Europe during the Roman empire, subsequently local measurement systems such as the *perch*, *rod*, *furlong*, *fathom* (distance from fingertip to fingertip of outstretched arms), and *ell* used by the Angles, Saxons, Jutes were replaced.

At this stage, it is important to introduce the developments that had taken place by this era in the field of trigonometry as it relates to the development of devices used for measuring angles. One of the precursors to the idea of angles particularly in the western civilisations is the awareness of the circle resulting from its occurrence in nature (the moon, the eye, etc.). It is generally believed that the Babylonians first used the angular subdivision of a circle into 360 parts in their sexagesimal (base 60) numeration systems (Maor 1998). The term *degree*, now used routinely to describe angles, has an immediate origin from the Latin word *de gradus* but is actually derived from the Greek word μοιρα (moira). Babylonians recorded the angular distances (this also means that Babylonians were conscious of the sphericity of earth) between sun, stars, planets, and occurrences of eclipses in the sky using a simple device called *gnomon*, which was simply a wooden stick standing vertically on the ground. The length of the shadow cast by the *gnomon* indicated the hour. After centuries of "technology transfers" between different continents, the *gnomon* later transformed into the *sun dial*. *Gnomon* was used widely by the Greeks and in one noteworthy incidence Greek mathematician Thales of Miletus, accurately measured the height of the pyramid by comparing the length of the shadows cast by the *pyramid* and the *gnomon* (Maor 1998; Figure 1.7).

The Egyptians understood how the angles separations could be used in planar geometry to create regular shapes, which they used effectively in building the pyramids and make cadastres by the 2nd century BC. The Egyptian levelling tools included the *merkhet* (literally meaning "the instrument of knowing") and *A-frame level*. *Merkhet* "consists of a split-palm leaf used as the 'backsight' and a plumb-line for the 'fore sight'[3]" (Lewis 2001; Figure 1.5). *Merkhet* was used by two observers facing one another. Each one held a merkhet, and they sat lined up so that one could see the North Star just above the other's head. The second person then sat very still as the first one watched and recorded when stars passed by the top of the other's head, by his ear, shoulder, etc. Astronomical positions were thus recorded so people could predict how the stars moved[4]. In surveying, *merkhet* was used to align pyramid basements. The *A*-frame level is a right-angled isosceles triangle with a cross bar shaped tool, generally made up of wood and bronze parts, and a plumb line hung from the apex (Lewis 2001; Figure 1.6).

The Romans significantly advanced the technology behind the tools of the Egyptians and Greeks. For example, two common surveying tools used by the Romans called the *groma* (Figure 1.8) and *dioptra* (Figure 1.9) were essentially improved versions of the Egyptian and Greek ones. In particular, the *dioptra* was fairly sophisticated enough to allow levelling, angle measurement, and distance measurement, which are essentially the characteristics of the modern *theodolites*.

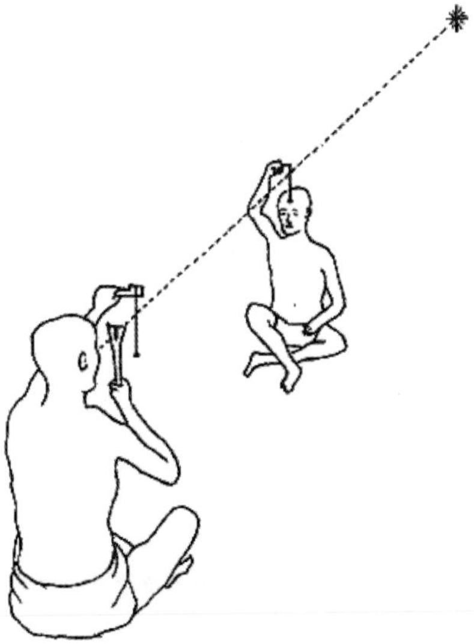

Figure 1.5. Merkhet in use.

Figure 1.6. A small ***A***-frame level being used to peg out home garden.

The *chorobates* was a relatively simpler and effective levelling tool, especially useful for vertical measurements (Figure 1.10).

Although most of the ancient surveying tools were relatively simple, the associated science was fundamental enough to last modern times. The Greeks derived many geometric and trigonometric functions, which were fundamental in calculating basic location information such as height and triangulation. Thales of Miletus showed how the principle of similar triangles could be used to compute distances

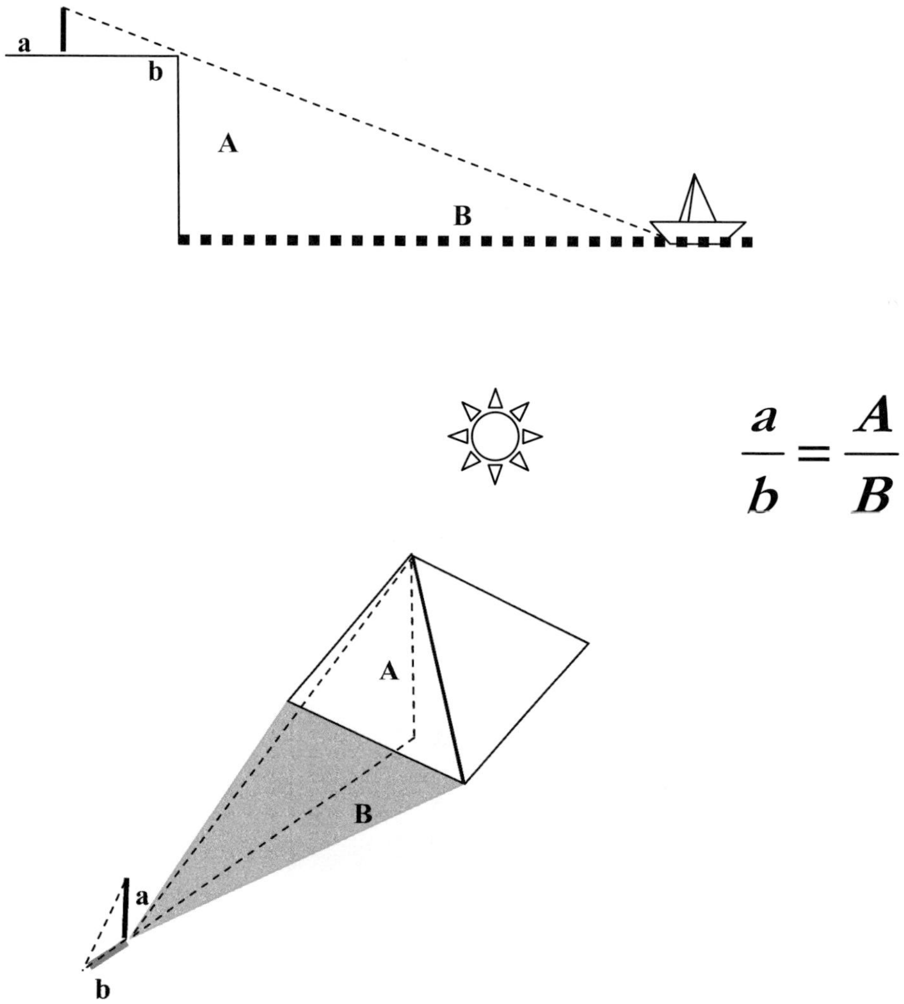

Figure 1.7. Application of *gnomons* for finding out distance and heights. Principle of similar right-angled triangles used by Thales of Miletus to calculate the distance to ships out in the sea (modified after Lewis 2001) and height of the Pyramid (may not be similar to Thales's original arrangement).

to ships out in the sea or compute the heights of tall structures (e.g., pyramids) (Figure 1.7). Hipparchus of Nicaea, Ptolemy of Alexandria and Pythagoras of Samos produced trigonometric tables and functions, and concepts for representing the earth into a grid of longitude and latitude. Ptolemy produced the seminal book *Geography*, in which he used the map projection based on Hipparchus's grid system to present the round earth onto a flat paper. Over in Southeast Asia, Vedic mathematician Aryabhatta derived the sine (called *ardh-jya* in Sanskrit) angle tables (ca. 400 AD; Maor 1998) and the ancient Chinese text *Zhoubi suanjing* shows an independent development of the device *gnomon* and an equivalent of Pythagoras theorem, the *Gougu rule*[5]. This leads on to another aspect of angular measurements, namely celestial navigation, which ultimately influenced geodesy and cartography.

It is unclear as to when exactly humans were able to navigate in the seas using celestial navigation. There is some evidence that people successfully navigated in sea e.g., the Cretans sailed in open galleys across the Mediterranean as early as 600 BC. Rod Cardoza in his essay "Evolution of Sextant" quotes a Sanskrit Mu'allim of 434 AD[6] "*He knows the course of the stars, both regular, accidental, and abnormal, of good and bad weather: he distinguishes regions of the ocean by the fish, the colour of the sea, the nature of the bottom, the birds of the mountains, and other indications. And the only aids he possesseth are his memory, helped by a pilot book, and a sounding lead or staff*".

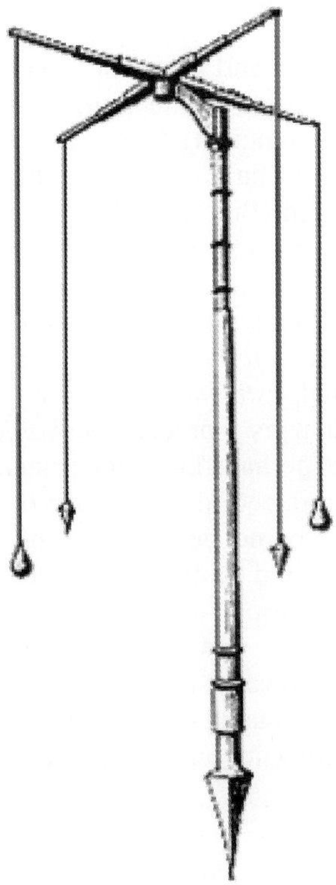

Figure 1.8. Groma

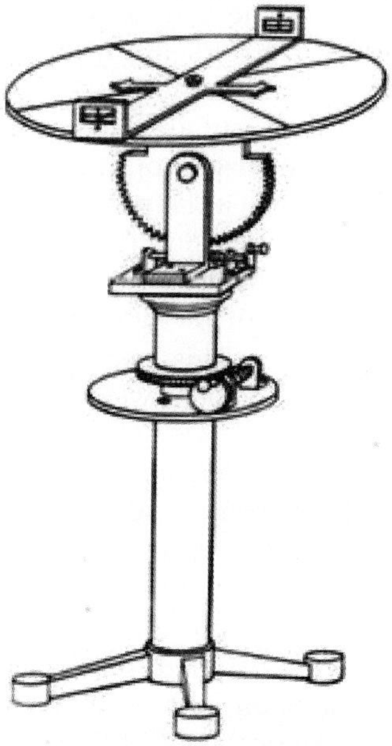

Figure 1.9. Dioptra

Peter Ilfand, a well known collector and writer on sextants have compiled an exhaustive review of the history of tools and techniques of celestial navigation (Ilfand 1999). According to Ilfand[7], (ancient) "*Arabs knew all about this technique*" (celestial navigation). Further, they built a device called "*kamal*" made up of a cord attached to a wooden transom (Figure 1.11). "*Before leaving homeport, the navigator would tie a knot in the cord so that, by holding it in his teeth, he could sight Polaris along the top of the transom and the horizon along the bottom. To return to homeport, he would sail north or south as needed to bring Polaris to the altitude he'd observed when he left home, then sail down the latitude.*" In contrast to the rather simple kamal, *astrolabe* (Figure 1.12) also credited to the ancients Arabs for making it popular, is a precise *mechanical* instrument for celestial navigation. While the Greeks perhaps knew the principles of these instruments as early as 2 BC[8], they were introduced in the Europe only in the 11th century via the Islamic Spain. An *astrolabe* could be used to find out the time during the day/night, time for sunset/sunrise, and finding directions. An *astrolabe* consists of a graduated circle in degrees with a pair of rotating arms (called *alidade*) in the middle. In order to use the *astrolabe*, one of the arms is set at horizon i.e., 0 degrees and the other arm is sighted towards the celestial body e.g., a star. The angular elevation is then read off the graduated circle as the separation between the two arms. Astrolabe evolved into a simpler and easy to use *quadrant* (Figure 1.13),

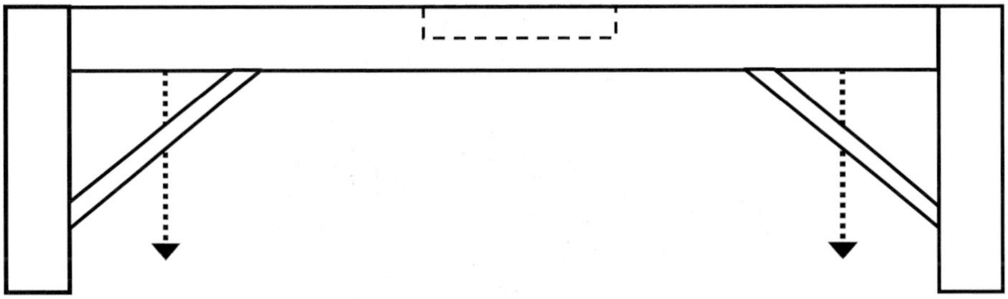

Figure 1.10. Chorobates

Figure 1.11. Kamal, reproduced by Peter Ilfand.

which was basically a graduated quarter-circle made up of wood or bronze. The gradations were in whole degrees and spanned up to 90 degrees. An even simpler and stable instrument for measuring angular elevation, which was around at this time, is the *cross-staff* (Figure 1.14). The *cross-staff* resembles a Christian cross and consists of long horizontal stick and a sliding vertical stick. The angular elevation of a star (commonly the North Star) above the horizon was calculated by keeping the horizontal stick at eye level such that the bottom edge of vertical stick touches the horizon and then sliding the vertical stick till the star can be sighted over the upper edge of the vertical stick. A number of variants of *quadrants*, *astrolabes* and *cross-staff* were also developed to make them stable, sea-worthy and precise. The culmination of all these various instrument is the *sextant* (Figure 1.15), which was probably the most influential and widely used geographic information collection tools, especially for maritime navigation/charts until the arrival of inertial guidance and satellite navigation. The scale of sextant is divided into 6 intervals hence

Figure 1.12. 15th Century Astrolabe.

the term *sextant*. The crucial innovation in sextant is the use of mirrors, prisms, telescope mechanisms for sighting the celestial objects. The use of optics made *sextant* a compact and robust instrument.

For surveying, it is also essential to be able to measure horizontal or azimuthal angles. The discovery of the magnetic properties of the lodestones as early as 200 BC in China was precursor to the modern day common tool for finding the azimuth, *compass*, which revolutionised navigation as the sailors didn't have to depend upon the celestial navigation alone. In the early years of compass, it was merely used for *geomancy* (Figure 1.16). However by 12th- century its application for navigation was well known to both the Chinese and European navigators. Now with the discovery of compass, the time for two instruments, which brought together the principles of compass, sextant, horizon and vertical levelling tools had naturally arrived. However it took several iterations for this to happen. At first, Victorian astronomer and mathematician, Leonard Digges invented the *theodolitus* (now known as *theodolite*) around 1571. *Theodolitus* (Figure 1.17) had a circular plate divided into 360 degrees, and a pivoting *alidade* with sight vanes at either end. In 1720s, vertical arcs were added to the construction. In 1831, American William J. Young enhanced the *theodolite* by attaching a compass and fixing the telescope so that it can move vertically, and his version came to be known as *surveyor's transit* (Figure 1.18). The basic *theodolite* construction has unchanged although advanced digital versions of the individual components have replaced the mechanical ones (Figure 1.19).

Figure 1.13. Mariner's Brass quadrant.

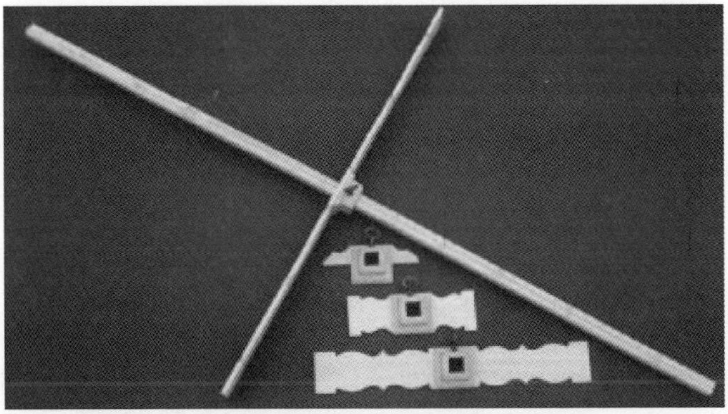

Figure 1.14. Cross-staff.

With the discovery of balloons, aeroplanes and wide spread use of the imaging technology principally the camera and telescope, the science and technology of geographic information collection tools changed from personal and regional scale to public and global scale. History of camera and telescope are prohibitively long and at places uncertain, to be presented satisfactorily in this section. But, there is little doubt that they influenced the development of aerial photography, remote sensing and field surveying in general. The following paragraphs present a short and simple version of the history of aerial photography and remote sensing.

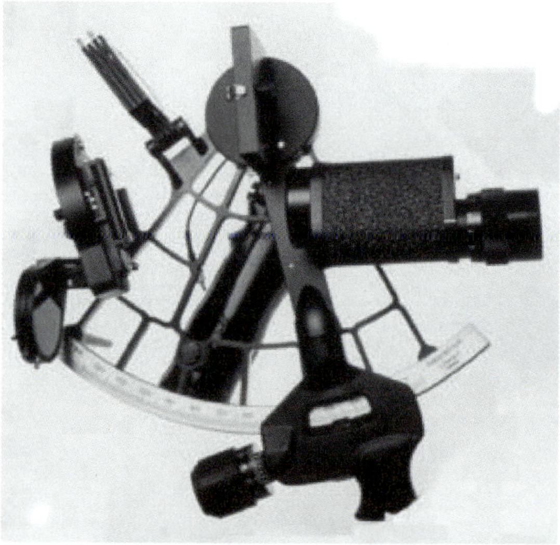

Figure 1.15. Sextant.

The first known example of aerial photography is by the French photographer Gaspar Felix Tournachon also known as "Nadar" (Figure 1.20a). In 1858 he photographed the village of Petit-Becetre from a balloon hoisted to a height of 80 meters. This event marked the beginning of aerial photography from rather amusing platforms such as balloons (Figure 12.0b), pigeon breasts (Figures 1.20c,d), kites (Figures 1.20e,f), and finally in 1909 to the more widely used platform, the aeroplane (Figures 1.20g,h) by Wilbur Wright.

Figure 1.16. Ancient Chinese compass for Geomancy

Figure 1.17. Common theodolite with telescope.

Figure 1.18. Surveyor's transit.

For a long time, high-altitude aeroplanes and rockets remained the main platform for aerial photographs, especially for orthophotographs. 1960s brought the launch of a series of Earth Observation satellites starting with the TIROS1 (Figures 1.20 i,j) meteorological satellite. The launch of Landsat-1 in 1972

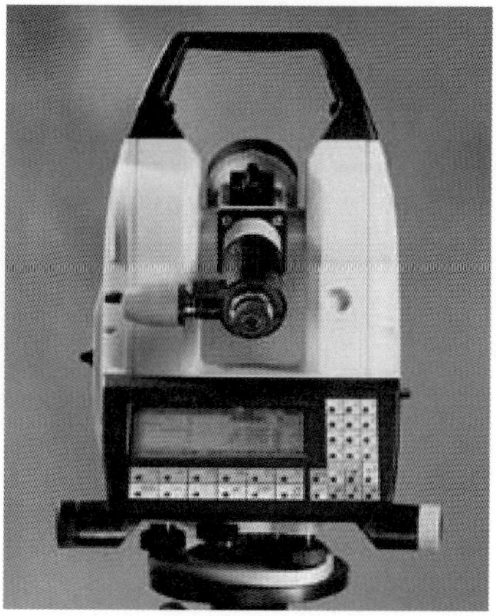

Figure 1.19. Leica TM51000A Industrial Theodolite

started the first routine multi-spectral remote sensing of earth so much so that today, there are several hundred satellites (many of whom are now defunct) of various types rotating around the earth, some of whom with spatial resolutions in millimetres.

Of all these satellites, the 24 global positioning satellites, orbiting earth at about 20,000 KM, are now indispensable parts of the GI technology. These satellites make up the Global Positioning System (GPS) developed by the US Department of Defence in 1973. As the name suggests, these GPS satellites are used primarily for accurate locational and time information i.e., (x,y,z,t). This information is useful and fundamental to multiple purposes e.g., GPS are now being used for land mapping, navigation systems (ranging from cars to air craft carriers), earthquake monitoring, guided missile systems, and tracking animals (and humans). Figure 1.21 shows the GPS and three typical outdoor GPS receivers.

In GPS, a GPS receiver (e.g., a handheld device) works out its location by first finding its distance from each of the four or more GPS satellites and then triangulating the distance to a precise location. The distance to a satellite is calculated by measuring the time delay between the time the signal was sent by the satellite and when it was received on the ground. In order for the GPS to work, the GPS satellite has to know its exact location and current time, which is achieved with constant satellite path monitoring (and manipulation if necessary) and synchronisation with atomic clocks.

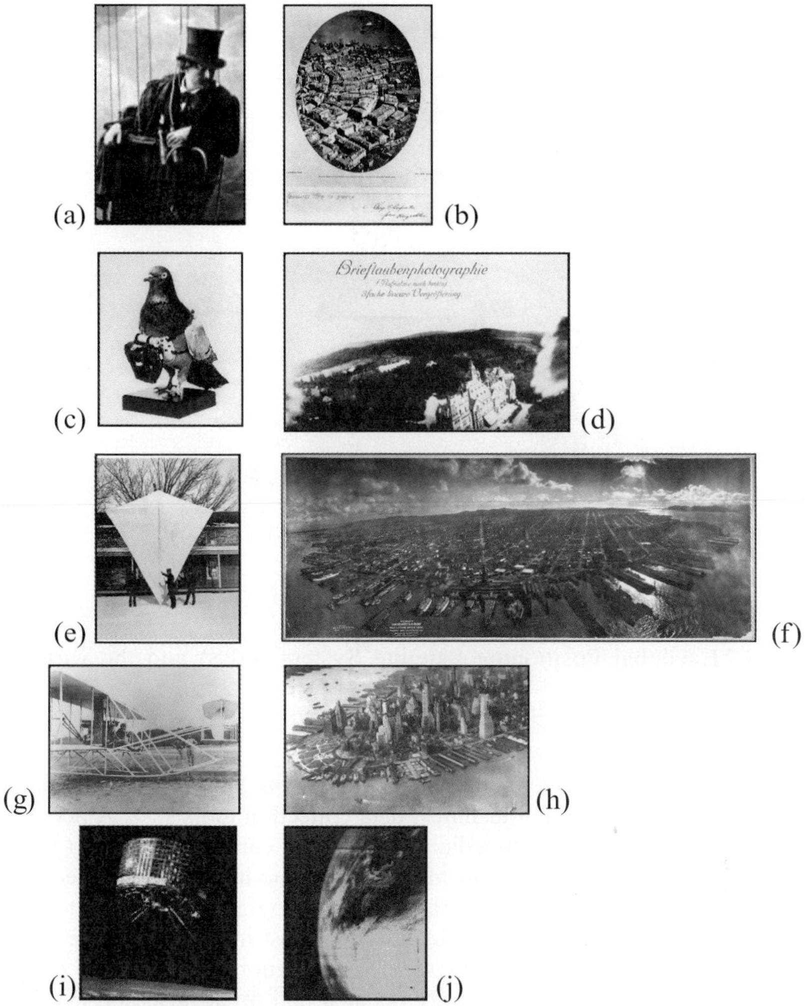

Figure 1.20. Various remote sensing platforms and their outputs spatial data.

(a) French Portrait Photographer Nadar in a balloon.
(b) Balloon view of Boston taken by James Wallace Black in 1960.
(c) Miniature pigeon camera patented by Dr. Julius Neubronner.
(d) Castle of Kronberg.
(e) Massive photo Kites used around the late 1890s.
(f) A Panoramic photography of the devastation of San Francisco in 1906 photographed by George Lawrence using several kite cameras.
(g) Lt. Frank P. Lahm and Orville qualify Army's first aircraft, Fort Myer, Virginia.
(h) Lower Manhattan, New York in the 1930's.
(i) An Artist impression of TIROS satellite.
(j) The first TV image taken by TIROS1 on April 1 1960.

Figure 1.21. Global Positioning Satellite System (GPS) and the various GPS receivers.

In addition to the conventional engineering tools for geographic information collections, there are also non-technical data collection tools such as personal and online questionnaire and multimedia tools such as camera, and audio recorders. The history of non-technological forms of geographic information collection tools is dealt in more detail in the section on history of geographic information.

Geographic information is undoubtedly the most important and generally the costliest part of the technology. Geographic information could become out of date quickly and thus need to be constantly updated and often raise non-trivial issues such as data versioning, standards, and semantics. Now when the geographic information collection technology has somewhat matured, it is these technical issues (and not technological) that will take over the GI technology agenda. This book contains several chapters that present the history and current research on these themes.

History of Geographic Information visualisation tools

The history of geographic information visualisation tools is a history of our awareness towards geographical representations, human psychology and demands on geography. The term Geovisualisation will be used from now on to loosely describe the discipline of geographic information visualisation. Geovisualisation is a fairly vast field and its scope varies from conventional cartographic forms e.g., maps, a statistical chart, to multi-modal visualisation interfaces such as voice controlled and touch-sensitive computer displays. Perhaps, a geographic map is a suitable type of geographic information that can be used to trace the history of geographic information visualisation tools because it is common between the ancient geographers to modern day Geovisualisation researchers.

Without any doubts, the earliest form of map representation was either a sketch on the face of rock cave or on a layer of sand, perhaps describing the lay of the land with its animals. The history of geographic information visualisation tools is unknown until the discovery of natural colours (such as plant and animal extracts) and print mediums such as wooden boards, tree barks/papyrus, cloth, papers and metal plates. Before the onset of the digital era in the late 20th century paper, and for some specialised purposes other materials such as cloth and leather sheets, were the dominant media for drawing maps.

Cartographers used careful aesthetic concepts of drawing shapes and salient colours to make the maps. The choice of symbol and colour remained the focus of discussions in Geovisualisation research for a long time (Bertin 1967). The description of the principle of camera obscura by Leonardo Da Vinci in the early 16th century followed by the discovery of silver chloride's sensitivity to sunlight in the 19th century introduced a new form of Geovisualisation tool-photography. Another significant scientific development, which took place in almost parallel to photography (and influenced Geovisualisation greatly), was the concept of binocular vision and stereoscopy.

After two centuries of evolution of ideas emanating from the 16th century drawings of Giovanni Battista della Porta on binocular vision and the Jesuit Francois d'Aguillion formulation of principles of "stéréoscopique", in 1838 Sir Charles Wheatstone proposed the design of first functional *stereoscope*[9]. The roots of virtual reality Geovisualisation were taking grounds as far as back in the early 19th century.

The next major leap in the technology of Geovisualisation tools happened with the invention of computers with plotters and displays in the 1950s. The discovery of this particular GI technology was fundamental to the core premises of GIS, namely the ability to visualise and correlate multiple layers of geographic information and perform large spatial analyses. The ability to draw and erase an image quickly on a display also paved the way for animation in cartography (also referred as dynamic mapping) that previously only existed in cartoon animations and motion photography. While the current technology have come a long way in performance, design, portability, and above all costs, the fundamental principles remain more or less the same when the above events took place. Figure 1.22 shows a rough timeline of events that link the various Geovisualisation tools.

The above summary of Geovisualisation tools is an abridged version of events at many places as bulk of historical developments actually took place in the science behind understanding and utilising the tools. In addition, because of a strong interdisciplinary nature of Geovisualisation, only a separate book can possibly summarise all the developments thoroughly. For example, there is a vast literature on cartography involving the principles of map making, projection, coordinate systems, non-geographic mapping i.e., thematic mapping, colour classification, cartographic variables, scientific visualisation, Virtual Reality modelling, multimodal interfaces and so on. A number of textbooks have been written on these topics e.g., MacEachren (1994a, 1995), Robinson et al. (1995), and Monmonier (1996) that provide an in depth treatment.

History of Geographic Information analysis tools

Geographic information analyses range from simple map overlays for studying contour patterns to sophisticated software-hardware systems for numerical modelling. The stage of analysis comes after data collection hence traditionally most data analyses have been done using laboratory based tools although some modern portable computers are robust and lightweight enough to be equally useful in field conditions. There is however, a slight overlap between what can be regarded as data collection tool and data analysis tool. In many cases, a tool can serve the purpose of both creating and analysing data. For example, an ordinary ruler can be used both to make and analyse contour maps. Typically, such a multipurpose tool has the characteristic that it can only analyse the data if it is in the format it is capable of producing itself. This logic applies to most geometrical (e.g., protractor, degree scale) and drafting (architect's T- bar) tools. Arguably, the output of an analysis is also some form of data. However, in the latter case, the output geographic information is more interpretation and less numbers. Despite the subtle overlap, it can be suggested that computers, as the geographic information analysis tools, have defined the horizons of GI analysis. This observation links back to the proposal mentioned at the beginning of the chapter that the state of GI hardware pretty much dictates the type and accuracy of the possible GI analyses. Thus, a brief overview (which is by no means an exhaustive and strictly chronological) of the developments of computing is relevant.

As discussed in the section on history of geographic information collection tools, the earliest records of geographic information collection come from the Babylonians and Egyptians. The datasets were mostly land surveys, astronomical charts and census datasets. The analysis of these datasets was done manually with the help of tables and charts. It was also mostly number crunching. Such a manual comparisons based analytical approach remained in use for a long time until the development of the mechanical and automated means for calculations. One of the earliest known forms of a numerical calculation tool is the *abacus*. The oldest surviving form of abacus is the stone *Salamis tablet* (ca. 300 BC) (Figure 1.23a) used by the Babylonians, which doesn't appear to be "industrial" and portable enough to be used routinely. During the middle ages (by ca. 1400 AD), the well known form of *abacus* was developed in Japan, China and Russia which used wooden boards and beads and have remained till this day, mainly for a bit of fun (Figure 1.23b). The first calculating *machine* was invented by Blaise Pascal in 1642 although it is also believed that Leonardo Da Vinci and William Schikard had created designs of a similar machine long before Pascal. Pascal's machine was essentially a metal board with 8 dials manipulated by a stylus (Figure 1.23c). This was followed by several types of such calculation

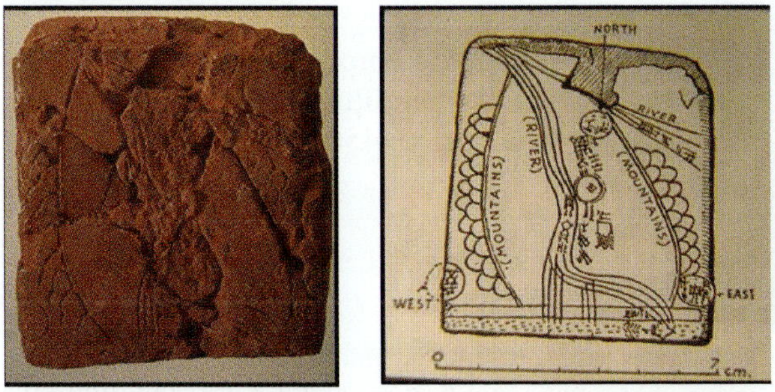

(a) Clay Tablet map from Ga-Sur (Ancient Nazu, near Kirkuk in modern day Iraq).
Age: 2500 BC

(b) Map of warring states carved on a wooden board, China.
Age: ca. 299 BC

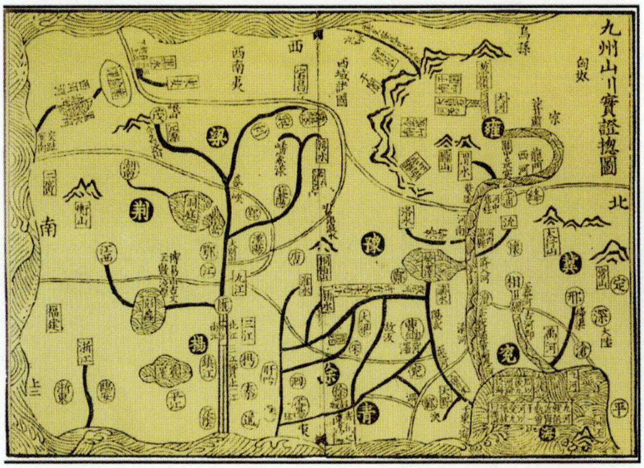

(c) Earliest hand printed paper map, China.
Age: 1177

(d) Reproduction of Gutenberg-era Printing Press.
Age: 1450

(e) Box Camera by Fox Talbot.
Age: 1835

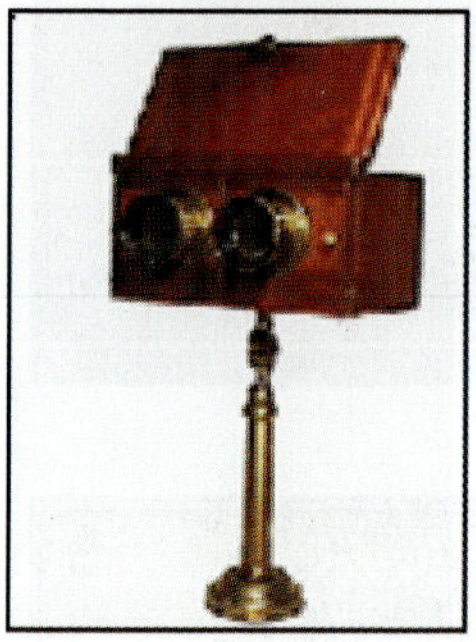

(f) Stereoscope by Sir Charles Wheatstone, which is now preserved at Science Museum London.
Age: 1838

(g) Whirlwhind Computer with vectorscope, developed at MIT.
Age: ca. 1951

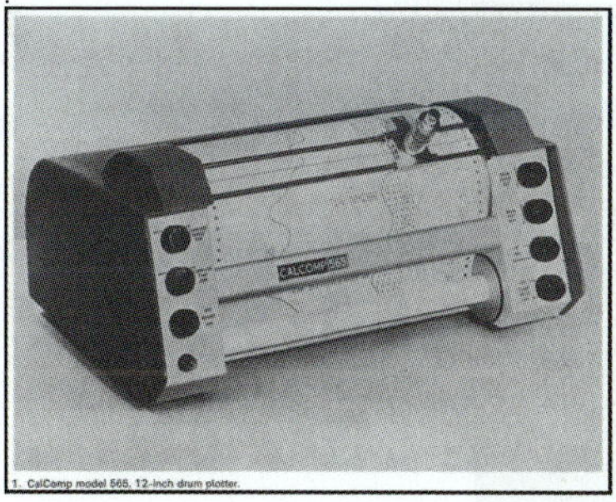

1. CalComp model 565, 12-inch drum plotter.

(h) Numerical controlled digital drafting machines, APT II (Automated Programming Tools) - MIT CalComp 565 drum plotter.
Age: 1958

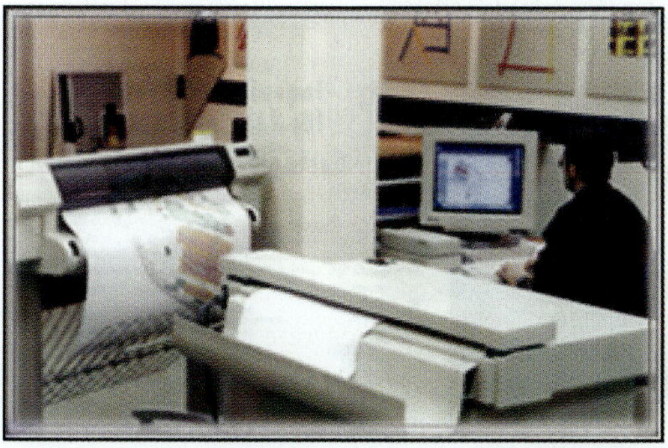

(i) Automated Mapping Systems.
Age: Post 1970

(j) Immersive virtual reality being used at VR Media Laboratory, University of Aalborg, Denmark.
Age: Post 1990

(k) Portable Augmented Reality Geovisualisation, GEIST project at Fraunhofer Institute.
Age: Post 1990

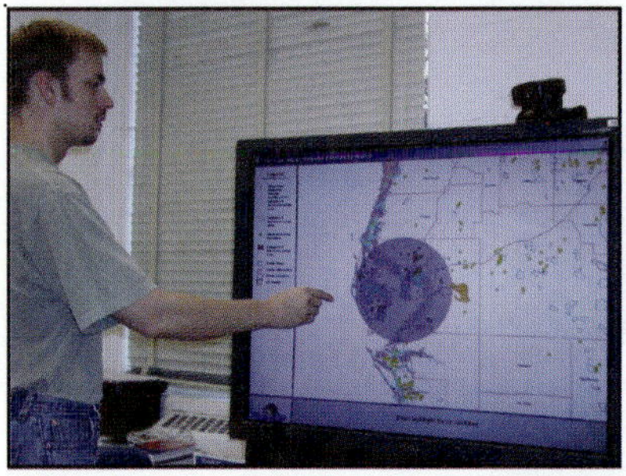

(l) Multi-modal interfaces, GeoVista Group, Pennsylvania State University.
Age: Post 1990

Figure 1.22. A timeline of events that links the various Geovisualisation technologies.

devices but these lacked fundamental qualities of an ideal analytical device, namely memory and hence programmability. In a series of developments, starting from 1820's Charles Babbage, developed the first prototype of modern computer called the *difference engine* (Figure 1.23d), which could take inputs from punch cards (run-time, read-only memory) and perform relatively higher order computation involving polynomials. These arithmetic engines were mechanical and worked by means of cranks and gears. One of the significant deliverables of the Babbage era was the punch card, instrumental in the 1890 US census, which used electro-mechanical tabulators for data input. This event also marked the origin of the computing giant IBM, whose co-founder Hermann Hollerith was the major contractor of these tabulator machines. Even today IBM makes a substantial chunk of revenues by processing government back-office jobs and data mining.

The successful development of *electronic valve* (also called *electronic tube*) by Lee De Forest in 1906 hailed the dominance of electronics in computing and also the dawn of 20th century's race for computing supremacy. Hence it deserves a separate paragraph. Electric valves could be switched on and off thus corresponded to the 1s and 0s (called bits) of the Boolean logic. The significant event in computers based on electronic valves took place near the beginning of the WWII when John V. Atanasoff and Clifford Berry developed a 16-bit adder device based on electronic valves. Atanasoff and Berry improved their model to compute simultaneous linear equations, which they aptly named ABC for "Atanasoff-Berry Computer".

(a) The Salamis Tablet: The oldest counting board, made of marble, has a large crack across the middle.

(b) A modern abacus by Lee Kai-chen.

(c) Pascaline : Pascal's Arithmetic Machine.

(d) Parts of Babbage's Difference Engine, assembled after Babbage's death by his son.

(e) Colossus.

(f) IBM 610 Auto-Point Computer, the first personal computer.

(g) Modern desktop (left) and handheld computers (right) by Hewlett-Packard.

Figure 1.23. Some milestones in the computing hardware.

The potential uses of ABC type device for cracking enemy encryptions were instantly recognised by the allied forces. In 1943, *Harvard Mark 1* (HM1) and *Colossus* (Figure 1.23e), significantly improved versions of ABC with card readers, paper tape readers, typewriters, memory and above all programmability were developed in US and UK respectively. Colossus was in fact used to crack the Lorenz code to understand the functioning of the *Enigma* machine. Some noteworthy successors of HM1 and colossus were the *ENIAC* (1946), *UNIVAC* (1951) and *EDVAC* (1952), which started the wide-scale commercialisation of computers.

The famous phrase "I think there is a world market for maybe five computers." by Thomas Watson in 1943, then chairman of IBM was outdated in just 8 years. But in some ways it was also true because an invention that eventually brought the demise of electronic valves based computers (so called "First Generation" computers) had already taken place.

First, in 1947 William B. Shockley, John Bardeen and Walter H. Brattain developed the transistor at Bell Laboratories and then in 1958-59, Jack St Clair Kilby at Texas Instruments and Robert Noyce separately developed the Integrated Circuit. These inventions gave birth to the "Second-, Third-, Fourth- generation" computers based on silicon transistors and integrated circuits differing essentially by a decreasing size of transistors and increasing computational power of the microprocessors (Figures 1.23f,g).

The development of computer has been well documented therefore it will serve little purpose to dwell further on the miniaturisation and performance of computing technology. Starting from the humble and hugely oversized beginnings of light valves, computers are now literally everywhere. In fact, for certain processes e.g., satellite data processing; the entire system could be fully automated from receiving the data from satellite, processing it to the right format and analysing the desired information. NASA's Mars rovers are perhaps one of the finest working examples of automated geo-processing.

It is quite ironical that many crises have been the focal point and causes of several technological innovations. *Necessity is* indeed *the mother of invention.* During the Cold War days of 1960s, the research on secure and robust computer networks by the US Department of Defence produced the *Advanced Research Projects Agency Network* (ARPANET). ARPANET was one of the earliest secure and wide area computer networks, and it started off with just four remote computers (Figure 1.24). ARPANET's achievements were crucial to highlight the field of *distributed computing* and birth of the two giant computer network ideas, namely *Internet* and *World Wide Web* (WWW).

Today, the extent and enthusiasm of *Internet* and *WWW* have reached feverish pitch. Internet and WWW offers something for everyone; governments can save money and time by providing e-governance, customers get significant discounts with online purchase as the online sellers pass on the savings from avoiding overhead costs, big universities can increase the revenues by providing distance learning and so on. In GI technology, internet played a key role in attracting attention to distributed computing to wider public. The idea of *distributed geo-computing* has been brewing for some time. The early ideas in early–mid 1990s, focussed mainly around experimenting the capabilities of *parallel processing* in computer networks to share

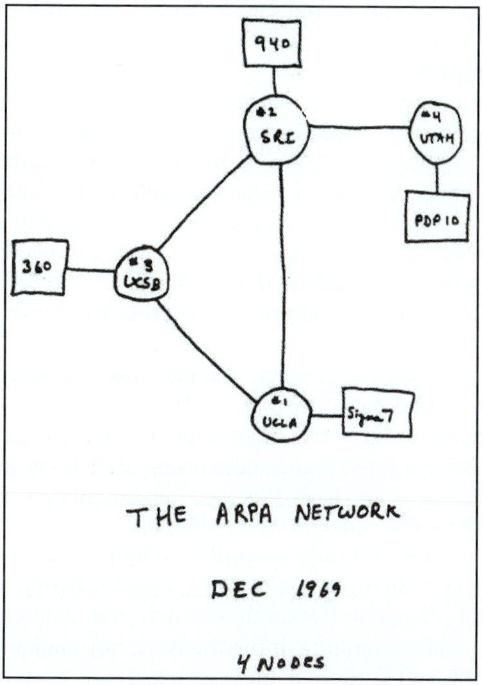

Figure 1.24. Famous sketch by Alex McKenzie of ARPA network, then only made up of 4 nodes.

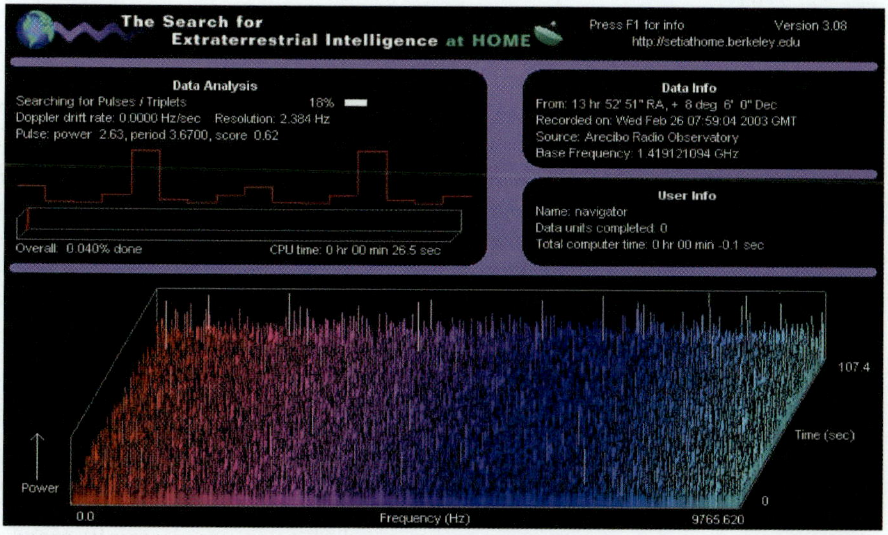

Figure 1.25. SETI@home screen saver.

the computation load of large spatial analyses e.g., viewshed computation. However, these ideas remain limited to large institutions with access to *super computer* facilities and such large networks.

In 1985, Miron Livny realised that the vast number of networked computers in large organisations (e.g., universities) sat idle for a lot of the time. He developed the *Condor* software system[10] to utilise the CPU of these idle processors for scientific computing. Another popular example of Internet computing is the project SETI@home[11] screen saver (Figure 1.25), which downloads a chunk of data from the SETI servers in USA, uses the client computer to process it and sends back the processed data back to the SETI server. The SETI servers delegate the processing and distribute data amongst several computers around the world in this way.

In the case of GIS, researchers were mainly using the Internet and *WWW* for online-mapping and spatial data distribution. However, this is set to change with the national initiatives by several international research funding agencies to promote the latest concept in computer networking called the *grid computing* networks. The *grid computing* networks are essentially a form of *parallel, heterogeneous, distributed computing*, whereby new types of networking protocols are being used to not only share the computational load but also to link up with data sensors and other resources into a single network. Since in UK alone, the Engineering and Physical Research Council has announced multi-million pounds support on grid computing infrastructure, big names from the industry such HP, Oracle, IBM, and GIS companies such as Laser Scan have started to take an interest. Thus grid computing is here to stay for some time and it might just be possible that a researcher based in UK can do an experiment using a software model based in the Europe and data based in the USA.

These are clearly exciting possibilities, of course only once we have passed the hurdles of data standards, licensing, privacy and above all human indifference. Therefore, a lot is at stake in these recent initiatives as it is uncommon to see such a massive financial support for research and development of infrastructure.

1.1.2 History of Geographic Information

In general, the geographic information remains the most costly and demanding part of GI technology but *information is power*. Therefore, the flow of data has relatively been more restricted compared to the flow of hardware technologies. In addition to the obvious strategic reasons for this irregular flow, accurate data collection requires training and an attitude to details. These qualities are not always transferable. While one could perhaps construct the history of geographic information by following the history of geographic information collection tools, we decided to dedicate a small separate section to highlight some influential individuals and events, otherwise remain unnoticed in the massive history of GI technology.

The earliest well documented geographic information were the cadastral, astronomical, civil engineering and census types of information by the ancient Babylonians and Egyptians. In Asia, at this time the Chinese recorded similar information but also a great deal also on physiography of natural landscapes (e.g., mountains, rivers, vegetation) for the simple reason that they were set in a different geographical setting. In ancient Greece and Roman cultures, census played a crucial role in their respective democracies but the Greeks particularly excelled in cartography and map projections. The medieval period observed little progress in the original geographic information concepts explorers and remained so until the late 18th – early 19th century however the knowledge of geographic information concepts spread far a field by the Arab and European explorers.

The late 18th -19th century period was truly the age of rediscovery of many old ideas and development of new GI technologies. In 1790, the "Board of Ordnance", the original name of the modern "Ordnance Survey" (OS) was formed, which conducted extensive mapping surveys primarily for military purposes. In 1832, Thomas F. Colby then superintendent of "Board of Ordnance" ordered a geological survey of Devon (UK). The success of the survey led to the formation of Geological Ordinance Survey, now known as the British Geological Survey (BGS), in 1835. The counter part of BGS in the USA, the United States Geological Survey (USGS) was established in 1879, which was inspired from an earlier "Coast Survey" established in 1807. The establishment of these mapping agencies started off the era of regular and widespread collection of administrative and natural resources mapping. This century also started off the collection of socio-economic geographic information. The first modern census in Britain took place in 1801, which consisted of just five questions and counted 10 million people (the population of London in 2001 was roughly 7 million) living in two million households. Figure 1.26 shows an example of a census return from the 1800s of a very famous person – Queen Victoria.

The invention of aeroplanes in the early 20th century marked the beginning of remotely sensed geographic information. At first, aerial photographs were used for reconnaissance but later they found their most important and wide use in forest mapping. Then came, satellite imagery initially only in visible wavelengths and later in multi-spectral bands. The research for technologies for air raid early warning systems during WWII brought the *Radio Detection and Ranging* (RADAR), which worked on the principle of emitting radiowave and collecting the reflected echoes. The time difference between the transmission and collection of radio waves could be used to calculate the distance to approaching aircraft. In remote sensing, the range finding aspect of RADAR technology was put to use to create high resolution digital elevation models (DEMs). The latest in the series of airborne height datasets collected from airborne platforms is the Light Detection and Ranging (LIDAR) datasets. The basic principle of LIDAR is same as RADAR except that LIDAR employs short bursts of lasers lights instead of radiowave. Because of very rapid bursts of lasers, LIDAR based DEMs achieve millimetre level spatial resolution (Figure 1.27).

Finally, until the last 3-4 decades, it was almost impossible to assign a sense of collection time and surroundings to the field data records as they could only be at best shown with a set of static images and observations. However, now the advancement in digital electronics and computer software has made it possible to not only capture streaming video, sounds, images but also to embed them within the Geovisualisation interfaces.

While the above summary provides a brief overview of the key events related to the history of the various geographic information types, it is also interesting to note

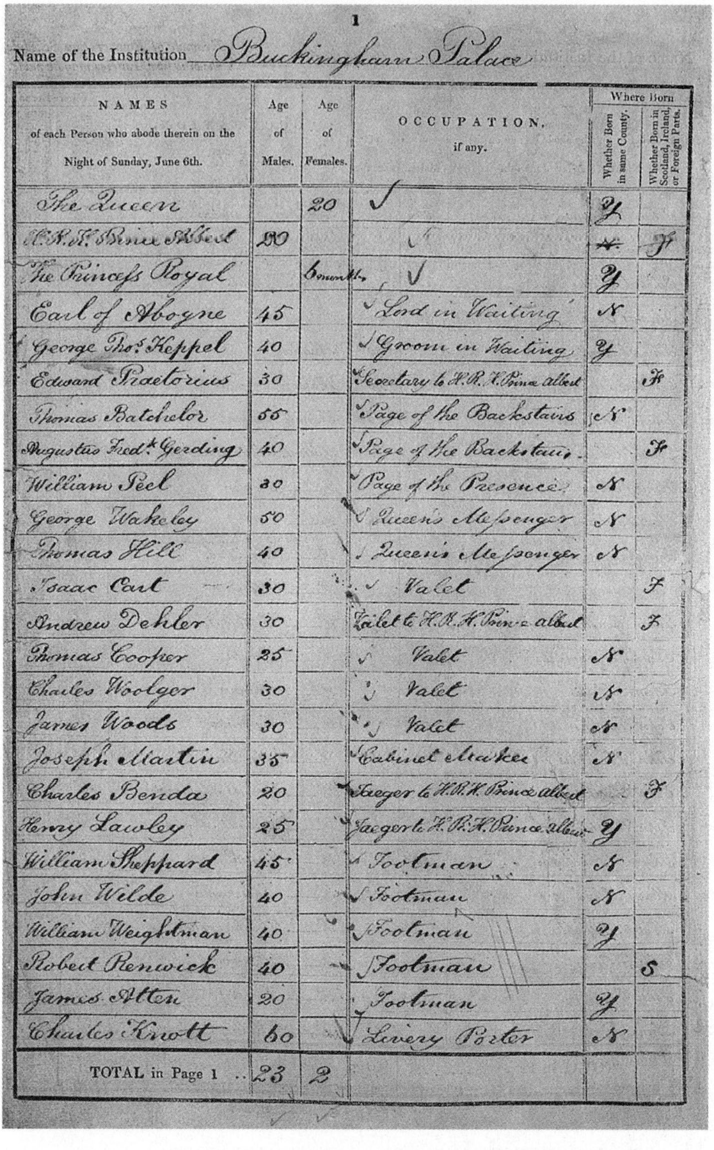

1

Name of the Institution Buckingham Palace

NAMES of each Person who abode therein on the Night of Sunday, June 6th.	Age of Males.	Age of Females.	OCCUPATION, if any.	Where Born: Whether Born in same County.	Whether Born in Scotland, Ireland, or Foreign Parts.
The Queen		20	✓	Y	
H.R.H. Prince Albert	20		✓	N	F
The Princess Royal		6 month	✓	Y	
Earl of Aboyne	45		Lord in Waiting	N	
George Thos Keppel	40		Groom in Waiting	Y	
Edward Praetorius	30		Secretary to H.R.H. Prince Albert		F
Thomas Batchelor	55		Page of the Backstairs	N	
Augustus Fredk Gerding	40		Page of the Backstairs		F
William Peel	30		Page of the Presence	N	
George Wakeley	50		Queen's Messenger	N	
Thomas Hill	40		Queen's Messenger	N	
Isaac Cart	30		Valet		F
Andrew Dehler	30		Valet to H.R.H. Prince Albert		F
Thomas Cooper	25		Valet	N	
Charles Woolger	30		Valet	N	
James Woods	30		Valet	N	
Joseph Martin	35		Cabinet Maker	N	
Charles Benda	20		Jaeger to H.R.H. Prince Albert		F
Henry Lawley	25		Jaeger to H.R.H. Prince Albert	Y	
William Sheppard	45		Footman	N	
John Wilde	40		Footman	N	
William Weightman	40		Footman	Y	
Robert Renwick	40		Footman		S
James Atten	20		Footman	Y	
Charles Knott	60		Livery Porter	N	
TOTAL in Page 1	23	2			

Figure 1.26. Census return by Queen Victoria in the 1800s.

the history of geographic information formats. Clearly, the bulk of the geographic information format history is based around the digital era. In order to trace some key developments in the history of geographic information formats it is convenient to follow the classical divisions in geographic information, namely the vector and raster data types as these data types evolved fairly independently. The geographic information formats were influenced by the data structures in computer science therefore a thorough review of geographic information format requires a treatment

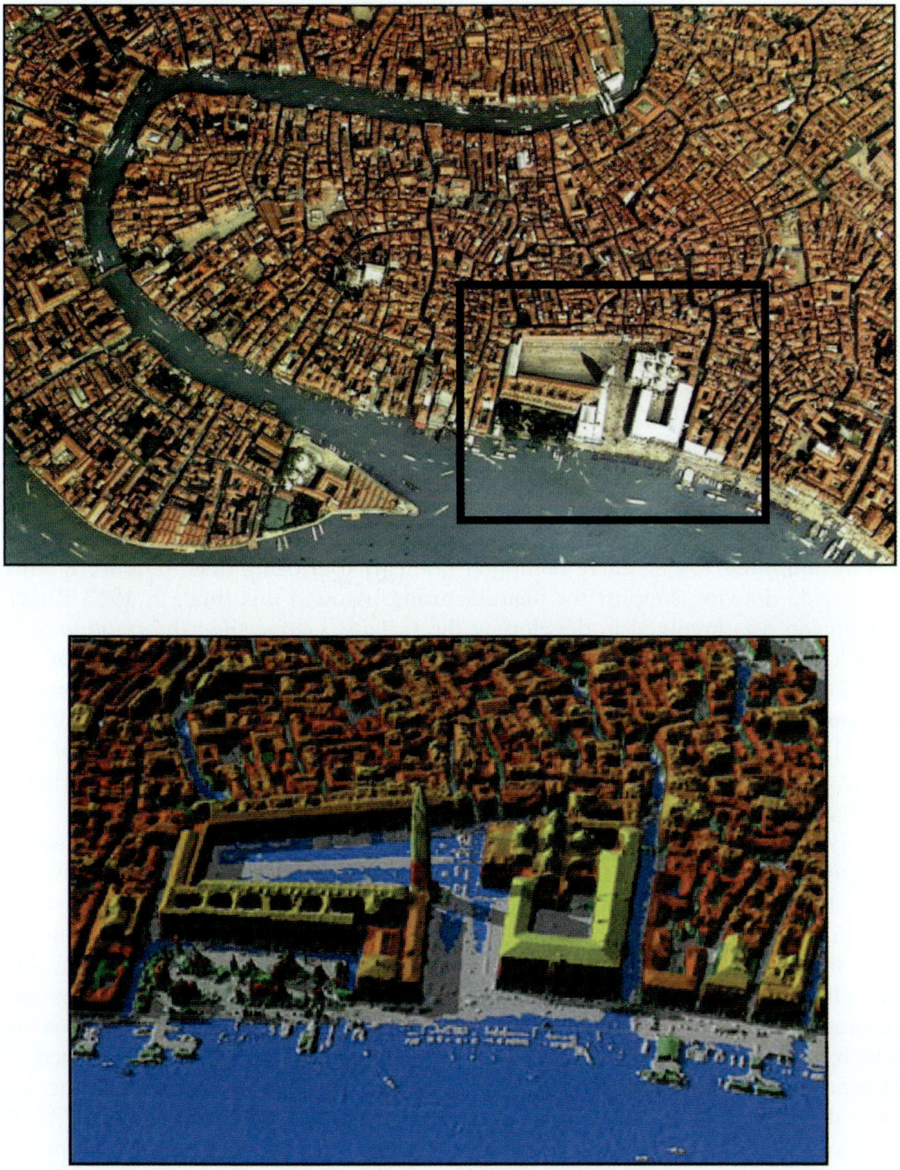

Figure 1.27. Aerial photograph (top) showing urban built-up in Venice around San Marco (inset box) and its LIDAR based reconstruction (bottom).

of data structures however that deserves a separate book. The section is based on the article "GIS Data formats" at the geocommunity website[12].

There are mainly three types information stored in a geographic information, namely the *geographic information*, *attribute information* and *display information*. The geographic information format types differ in the way each store (or sometimes don't store) these separate information. The *geographic information* as the name suggests includes the geometrical data about geographic features, *attribute information* provides non-graphic information associated with the geographic features, and finally the *display information* describes how to show the geographic features on the screen. At present, there are bewildering numbers of commercial and non-commercial geographic information formats being used around the world. The following descriptions mention only the most widely used types.

The principles of vector data formats have been around for a longer time than the raster data formats for the reason that the vector data format relate well to the construction of general graph data structures and hence to the fundamental geospatial relations such as topology, attributes, attribute linkages, database structures, and display information. The earliest vector formats to store spatial data structures were the ones used for computer-aided design (CAD). In the late 1950s and most 1960s, early computer controlled drafting machines were built to provide drawing support for manufacturing. Around this time, in 1963 Roger Tomlinson was involved in developing the *Canada Geographic Information Systems* (*CGIS*) for the Canadian Land Inventory project. Several aspects of the *CGIS* data formats were the building blocks of all geographic information formats to come in future. For example, the CGIS data format had coding of object boundaries by arc, with pointers to left and right area objects, relationships between arcs and areas were coded in the database, data were separated into attribute and locational files with "descriptor dataset" and "image dataset".

By 1970, *CGIS* was in trouble due to various reasons and but the GIS scene was thriving across the border in US, with several individuals involved in developing new GIS. In 1966, Howard Fisher started the *Synagraphic Mapping System* (*SYMAP*) at the Northwestern Technology Institute (Chicago) and completed it at the Harvard Laboratory for Computer Graphics and Spatial Analysis. Although, SYMAP itself was fairly simplistic in its input (punch cards) and output (characters on line printers) it transpired into several improved versions four of which, namely *CALFORM*, *SYMVU*, *POLYVRT* and *ODYSSEY* (all developed at Harvard Lab. in roughly the same chronological order) firmly established the future of vector data formats. Since each polygon was stored separately, *SYMAP* polygons often had duplicate records for internal boundaries. In order to remove double-encoding, *CALFORM* allowed the storage of internal boundaries as a table of point locations and a set of polygons represented as sequences of point IDs. *SYMVU* generated the 3D perspective views of *SYMVU* output. The *POLYVRT* and *ODYSSEY* packages incorporated the vector data formats, several topological analysis routines and were the earliest analysis packages for vector data. In 1967, George Farmsworth at the US Bureau of Census developed the *Dual Independent Map Encoding* (*DIME*) data format. The internal structure of *DIME* was essentially similar to the *CGIS* and *CALFORM* data structures except that it was specifically designed for arc shapes in urban areas e.g., street networks. Later, in 1988 US Bureau of Census extended the *DIME* and produced the *Topographically Integrated Geographic Encoding and Referencing* (*TIGER*) model. *TIGER* format can store attribute (e.g., street names, addresses).

Scott Moorehouse who worked on the *ODYSSEY* project joined the Environmental Systems Research Institute (ESRI), founded by Jack and Laura Dangermond in 1969. ESRI developed the following vector data formats, which were very much based on the arc-node data structure:

- Arc Export: Arc Export is a transfer format, either ASCII or compressed into binary.
- *ArcInfo* Coverage: An ARC/INFO "coverage" is a set of internal binary files used to represent point, line and polygon features.
- Shapefile: A shapefile is a set of binary and ASCII files and stores non-topological geometry and attribute information for the spatial features in a data set. This format was introduced with the desktop GIS ArcView in 1992. Shapefiles can store point, line, and area features.

In 1977, USGS developed the *Digital Line Graph* (*DLG*) format for digital representation of cartographic maps. *DLG* stills remains a standard used by the USGS and other US government agencies for a large number of digital maps. A DLG file contains the usual arc-node structure for node, line, and area-identifier elements and also stores attribute codes. The following extract from the USGS website[13] describes the *DLG* format.

The DLG data are organized in graphic, optional, standard, or Spatial Data Transfer Standard (SDTS) formats. The table .. describes the difference between the graphic, optional and standard data types in a DLG file.

Format Criteria	***Graphic***	***Optional***	***Standard***
Character Set	*8 - Bit ASCII*	*8 - Bit ASCII*	*8 - Bit ASCII*
Logical Record Length	*20 Bytes*	*80 Bytes*	*144 Bytes*
Block Size	*Multiples of 20*	*Multiples of 80*	*Multiples of 144*
Coordinate System	*Geographic Latitude and Longitude*	*Planimetric*	*Cartesian*
Topological Linkages	*None*	*All Elements*	*Line Elements*

In 1992, USGS developed the SDTS format to store all types of geographical features and the corresponding cartographic information. The SDTS format differs from the graphic and optional formats. The character set consists of a binary data file that contains non-ASCII characters. The characters are formatted according to International Organization for Standardization (ISO) standards 646 and 2022. The binary floating point is determined according to the ANSI/IEEE 754-1985 standard, and binary integers are determined according to the ISO/IEC 8632-3 standard. The logical record length is variable as defined in ISO 8211 and the block size is media dependent. The coordinate system is geographic (latitude/longitude) for the small-scale DLG data and planimetric for the large- and intermediate-scale DLG data; the topological linkages are only in lines (though SDTS supports all objects, DLG to SDTS supports only in lines).

In 1982, Autodesk released *AutoCAD* and with it came the de facto standards for CAD drawings, namely the *Data Interchange File* (*DXF*) and *Drawing File* (*DWG*). Both *DXF* and *DWG* are internal proprietary formats used in the *AutoCAD* software. DXF is one of the most widely used formats for CAD drawings and to a small extent for spatial maps. *DXF* allows multiple layers, 3D objects, extensive set of object types however one the prime drawbacks that has remained till this day which restricted its use a GIS data format (e.g., in contrast to non-topological formats such as shapefiles) is the lack of standards to link attribute tables. In 1985, a CAD software which became quite popular with GI technologists is the Bentley Systems Inc.'s *MicroStation* when it came to be largely owned by Intergraph. The Microstation data format called *DGN* provided storage of element attributes as links to an external database.

MapInfo was founded in 1986. It defined the MapInfo Data Transfer Files formats (MIF/MID). These can store all three main types of geospatial information viz. geographic, attribute, and display.

One less talked about but widely used format behind the scenes developed by the US Defence Mapping Agency (DMA) in 1992, is the *Vector Product* Format (*VPF*). VPF is a binary format based on the arc-node principles and it only allows the storage of geographic and attributes information but no display data. It was generally designed to store very large maps e.g., the Digital Chart of the World (DCW) is published in this format.

In 1995, Ordnance Survey (OS) started the digitisation of its archive of paper maps and it developed the *National Transfer Format* (*NTF*) which later became the British Standard 7567. NTF files are made up of lines with attributes that describe the feature type and annotations. However, the datasets suffered digitisation errors and were not multi-scale. In 2001, OS converted its entire digital map archive into a multi-scale, hierarchical object-oriented, multi-thematic polygons-based *MasterMap* format.

Amongst the raster data formats there are mainly two types that are used in GIS, namely the non-geocoded digital image formats and geo-referenced data formats. Decent digital image formats and geo-referenced data formats appeared round about the same time in the mid-late 1970s. The early years of the digital image formats are not well documented hence only a speculation can be made about their releases. It is possible that graphics format existed around late 1960s – early 1970s but they were most likely internal (to the companies) formats and not widely used. Although the earliest 1960s GIS (*CGIS* and *SYMAP*-breeds) produced map outputs, the data input/output were based on punch cards and involved a crude raster data entry procedure.

With the launch of satellites in 1960 – 70s georeferenced satellite image data had started to appear, these generally remained confined within remote sensing community (which was yet to become a part of GI technologies formally). The most common formats for satellite images are the *Band Interleaved by Line* (*BIL*), *Band Interleaved by Pixel* (*BIP*), and *Band Sequential* (*BSQ*). The primary difference among the different satellite image formats is the manner in which the brightness values for different spectral bands are stored in the file. In the *BIL* format, each line in the file stores the brightness value of a different spectral band, in the *BSQ* format different spectral bands are stored in one continuous block and in the *BIP* format, and spectral bands are stored in consecutive pixels/cells. At present, other popular formats used in remote sensing community are mostly software specific formats such as the ERDAS *IMAGINE* (since 1978), *GRASS* (since 1985) and *IDRISI* raster formats (since 1987).

It appears ESRI was the first to develop a system for the GI raster data format and raster analysis systems in the late 1970s, called the GRID. The existing GRID data, for example of a terrain or landuse map, are binary files and stored (excerpts from *ArcInfo* Help Manual) "*using a tiled-raster data structure in which the basic unit of data storage is a rectangular block of cells. Blocks are stored on disk in compressed form in a variable-length file structure referred to as a tile. Each block is stored as one variable-length record. Both categorical data, such as land use and soil type, and continuous data, such as elevation, population density or scanned gray-scale images, can be represented as grids. A grid is accompanied by registration information that positions the grid in the cartesian coordinate space. Grids also carry additional data description information, such as the map projection associated with the grid and summary-value statistics.*" ESRI also developed a transfer format called *Arc ASCII grid* (*ASC*), to store the binary grids as ASCII grid. The ASC files had the basic georeferencing information such as geographic location of the lower-left corner of the raster, cell size, number of rows, number of columns and a special value assigned to null data cells.

The 1980s observed many developments in the digital image formats e.g., Microsoft included the *Bitmap* (*BMP*), *Picture* (*PCX*) formats with Windows operating system in 1983, Apple developed the *PICT* format, Aldus and Microsoft developed the *Tagged Image File Format* (TIFF) in 1986, and in 1987 CompuServe and Joint-Photographic Experts Group respectively developed the *Graphics Interchange Format* (*GIF*) and *JPEG* formats. GIS used these images as backgrounds to transparent vector layers. These digital images were generally the output of scanned paper maps and aerial photographs. ESRI worked out a way referencing these digital images in GIS using the "world file" and image georeferencing or registration. A world file essentially stores the parameters of affine transformation, required to do the image-world transformation of images. The term GeoJPEG is used to denote the JPEG which has an associated world file.

By 1990s GIS had gained popularity and as a consequence several sophisticated spatial raster data formats arrived in the scene. In the early 1990s the USGS and DMA developed the *Digital Elevation Model* (*DEM*) and *Digital Terrain Elevation Dataset* (*DTED*) respectively by rasterising earlier vector height datasets, stereogrammetry and radar interferometry. Both *DEM* and *DTED* datasets were ordinary raster with elevation values sampled at various resolutions. In 1999, USGS introduced a huge, single, seamless file known as *National Elevation Data* (*NED*). The *TIFF* format was also updated in 1994 when the *GeoTIFF* format was introduced, which allowed the embedding of geographic data as tags within the *TIFF* file.

In the late 1990s GI technologists realised that the terabytes of satellite digital image data and large digital elevation datasets are unmanageable and too big to transfer over networks. In response Earth Resources Mapping and LizardTech separately released the compressed image formats called *ERMapper compressed wavelet* (*ECW*) and *MrSID* respectively. Both compressed raster binary formats are based on the concept of wavelet compression. The latest addition to the compressed image formats is the *JPEG2000*, which also uses wavelet compression.

So much so for the developments of the tools that shaped the GI technologies but the community of GI technologists have also undergone evolutions from new-age cartographers to an interdisciplinary research group with expertise in several topics e.g., computer science, psychology, mathematics, geography. The following sections examine various attempts to define a theoretical or computational foundation for GI technology.

1.2 Advent of the new concepts of GeoComputation, GIScience – A sign of GI Technology maturity or mid-life crisis?

There has been a growing desire amongst GI researchers and technologists to come up with a unifying idea/statement about GI research. Most recently, Rita Colwell, the ex-director of National Science Foundation eloquently described the tentacles of GI research in mapping fear of crime to mapping health epidemics (Colwell 2004). Colwell admires the idea of "geographic portal" that "defines, encloses, and makes sense of space". Colwell's was the latest in a series of essays to appreciate the vastness of GI research. It roughly started off in the late 1990s, when there was a feeling amongst senior researchers to define the scope and aims of computational practises in geography. Longley et al. (1998) and Openshaw et al. (1999) compiled several works that could fit under the GeoComputation umbrella. Another version of this debate is the discussions over the definition of GIScience, which still attracts some refinements (Mark 2003, Goodchild 2004). Was this perhaps the first sign of maturity amongst GI technologists, a kind of self-reflection leading towards clearer aims and achievements? A more amusing view could be that the discipline was experiencing what typically happens in our life around the age of 30-40 known as the Mid-life crisis, trying to figure out the meaning and purpose of everything. But it was a serious issue because the borders were being marked between researchers who preferred a more theorist view (i.e., preferring ideas development over software) of GIScience to be included in the GeoComputation definition against the researchers who wanted to include a more computational techniques oriented viewpoint. However some of the definite positive things to come out from the publication of the books on GeoComputation with a wide agenda were the increased interdisciplinary collaboration especially from computer science researchers, formation of the GeoComputation conference series and its more scientific counterpart GIScience Conference series, and publication of GI related magazines.

At this point, it's difficult to resist the urge to compile the various definitions of the terms GeoComputation and GIScience. According to the GeoComputation.org,

> GeoComputation is "*...not intended as a synonym for GIS or spatial information theory*". Further, the aim of GeoComputation is "*...to move the research agenda back to geographical analysis and modelling, with or without GIS in tow. Its concern is to enrich geography with a toolbox of methods to model and analyse a range of highly complex, often non-deterministic problems. It is about not compromising the geography, nor enforcing the use of unhelpful or simplistic representations. It is a conscious effort to explore the middle ground from the doubly-informed perspective of geography and computer science. A true enabling technology for the quantitative geographer, a rich source of computational and representational challenges for the computer scientist*". Finally, GeoComputation aims to fill the "*gap in knowledge between the abstract functioning of these tools (which is usually well understood in the computer science community) and their successful deployment to the complex applications and datasets that are commonplace in geography*".

An excellent statement but doesn't that definition seems to include everything what there is to know so how is it different from GIScience? There have been several definitions put forward for the term GIScience. The University Consortium for Geographic Information Science (UCGIS) states that GIScience involves understanding the geographical processes and spatial relationships using the theory, data, methods and technology. According to Goodchild (1999a) GIScience is the science behind the GIS, which resonates well with the thoughts of Mark (2003). Some simpler definitions include by Clarke (1997) where GIScience is "*the discipline that uses geographic information systems as tools to understand the world*".

Other than the differences in the opinions about the applications, both GeoComputation and GIScience definitions seem rather similar. If GeoComputation is only supposed to mean the *act* of developing computational solutions (i.e., implementation more important theory) to problems in GI research then it's logical to state that it can't exist without the science. But even with such narrow definition GIScience needs GeoComputation to produce the deliverables. While names that clearly designate the exercises of GI technologies and theories are essential in order to distinguish it from larger geography discipline (which includes social sciences) it seems it is difficult to avoid the basic fact that GI technologies are strongly interdisciplinary. Any stricter definition would restrict the scope, appear confusing and alienate researchers who have a different opinion. A broader definition would seem too general and hence pointless. Nonetheless, a clear distinction is essential and urgent as we discuss in the last chapter how the current status of GI technologies as a small appendage to geography departments is perhaps causing a stunted growth.

Overall it appears that (a) researchers are attempting to define the scope and nature of science behind the GI technology and (b) the common realisations are that there is a need for –inter-disciplinary research; advances in theory (including computational theory) to create better tools; innovative and better practise in the use of existing tools; adopting or adapting techniques and best practises from non-geographic disciplines; apply or adapt techniques from geographic technology to other disciplines (e.g., knowledge discovery/data mining). These are signs of a maturing discipline.

1.3 Role of GI technologies in modern society – A necessary evil?

The success and rapid evolution of GIS since the 1960s is clearly a sign of its immense relevance and popularity. It would not be an exaggeration to state that modern society is made of private citizens and corporate governments as this appears to be the most successful model for running countries and achieving the aspirations of its residents. We all want choice with correct and quick access to information when we are shopping, looking for a school to the children, looking for a house, navigation in the motorways, and the list goes on. Governments want to be able to keep track of our movements, create long-term socio-economic models, build robust transport systems and all at low costs and maximum profits. Guess what? GI technologies are poised to be at the heart of this new-age living from the post-WWII socialist or capitalist extremes. It must have been an immense relief to the GI technologist to see the paper by Gewin (2004) in the reputed journal "Nature" on the potentials of GI technologies. Academics have been shouting this message for four decades. The US Department of Labor now recognises GI technology as one of the top three emerging disciplines.

On the ground, the public itself have started to benefit from GI technologies in mass production e.g., GPS receivers, Location Based Services (LBS), e-governments websites, map interfaces to "store locators", public navigation systems, better delivery services, better transport accessibility plans, and so on. Governments are the biggest consumers of GI technologies and literally run the entire GI technologies industry. Almost all national and regional city or state governments have a dedicated group of people dealing with GIS. Application of GI technologies in the governments range from the mappable inventory of entire natural resources to daily data mining of census statistics.

While most of us like the bells and whistles of GI technologies, there are some who are also scared of the idea that governments and companies are able to trawl through our spatial movements and personal data effortlessly. It is the fear of losing one's privacy and the sense of individuality. Some people are also genuinely scared of ending up in an unfortunate end of "Mr. Buttle" in the movie "Brazil" by Terry Gilliams, due to a silly technological error.[14] No technology is faultless and abuse of information is not rarity. The trust in governments and politicians is generally low around the world. At the same time, arguments put forward by the governments in favour of these intrusive GI technologies make sense. For example, the use of GPS tracking to monitor the cars would help reduce car accidents, identify stolen vehicles, and vehicles without insurance. Similarly, CCTV helps to monitor suspicious behaviour, criminals, and road traffic. Data mining of census statistics helps in understanding migration patterns, development of plans to address regional differences in economy, and better targeting of commercial products (thus reducing spam). It is unclear what type of approach can satisfy the desires of both the public and the administration systems. In Chapter 13, Alastair Beresford highlights the various dangers of GI technologies.

1.4 Layout of the book

The aim of the book is to demonstrate the implementation of ideas, presented in early books on GI technologies and GeoComputation, such as edited volumes by Longley et al. (1998) and Openshaw et al. (1999). It also presents a flavour of more recent developments. The chapters are broadly arranged by similar topics. There are five main themes in the book, namely (i) geospatial software systems designed for simulation and modelling; (ii) distributed geo-computing systems including geospatial databases, semantic web, grid computing; (iii) Geovisualisation systems including augmented reality, multimodal interfaces, mobile mapping; (iv) privacy issues; and (v) challenges facing the GI technologists in developing countries. The remainder of this section presents a brief summary of each chapter contributed to the book.

Until the early 1990s artificial intelligence (AI) and fuzzy logic were two of the most popular topics amongst GIS researchers. The lack of compelling applications and perhaps consequent indifference by major GIS vendors had taken the limelight off these topics. This situation is now changing. Hence, the book starts with two chapters that demonstrate two important practical applications of AI and fuzzy logic. Perfect data are rarely obtainable in the real world since most data contain errors and missing values. Traditional computing methods employed in GIS have difficulty in dealing with these kinds of imperfect data. Soft Computing (SC) methodologies utilise and process imperfect data that characterise real-life ambiguous situations. In Chapter 2 Yingjie Yang and others provide a brief introduction to two popular methods used in SC: artificial neural networks and fuzzy sets. Based on an analysis of the kinds of uncertainty associated with GIS operations, they explain the feasibility of artificial neural networks and fuzzy sets. SC differs from traditional computing methods in that it employs uncertainty models to handle the computation involved. In these models, data are permitted to be imperfect, with errors, noise and missing values. SC can therefore provide a powerful tool for GIS to overcome the imperfect data problem. The chapter concludes with a demonstration of the application of neural networks to two case studies: (i) evaluating aircraft noise disturbance, and (ii) predicting harbour sedimentology characteristics.

Autonomous systems that operate on harsh, natural terrain must deal with the added complexity of using geospatial information in their control logic. For example, field mobile robots must safely traverse uneven, rough, and rugged terrain; spacecrafts must reason about terrain safety constraints while preparing to land on planetary surfaces. In general, these systems must have the ability to assess the terrain and determine the risk associated with traversal, whether roving on the surface or descending to the surface. Few researchers have addressed the inherent problems of real-time assessment of terrain for machine control in a natural environment. In Chapter 3, Ayanna Howard and Ed Tunstell discuss techniques for incorporating geospatial information for control of autonomous systems. The chapter starts off with a discussion on the inherent problems that natural terrain environments pose for mobile autonomous systems. Specific examples are then given for robot navigational guidance and control on natural terrain and terrain characterization for spacecraft autonomous landing.

Lately, the ideas of cellular automata (CA) and agent-based modelling (ABM) systems have attracted a lot of attention amongst quantitative geographers, particularly those inspired from the work done in biological systems modelling and physics. In Chapter 4, Mike Batty introduces a new class of urban model which simulates the location of spatial development through time. These models have emerged partly in response to the need for dynamic simulations of urban growth in contrast to the conventional class of urban model which simulates cities as though they are in equilibrium at a cross-section in time. The chapter outlines the development of these models, beginning with their method which is largely based around cellular automata principles in which land use change is considered a product of change in local neighbourhoods. The chapter examines the extent to which these models can be calibrated to existing data and we then sketch the problems of adapting these to GIS technologies in terms of data input and visualization of outputs. The chapter also demonstrates applications and proposes ways of using these techniques in urban policy making, again showing how GIS can be used in these models in relation to the wider planning process where visualisation and the dissemination of digital data and predictions are central.

Chapters 5 through 8 involve one of the most popular topics in GI technologies, namely geospatial databases and data standards. Chapters 5 and 6 respectively introduce the principles of distributed geospatial databases. In Chapter 5, Phil Yang and Vincent Tao propose the idea of a Distributed Geospatial Information Service (DGIService) as a new paradigm for performing geospatial computing and developing geospatial information applications. The chapter quotes several pointers to further research and is a rich source of primary readings on this topic. They suggest that DGIService harnesses the power of large-scale networked computing resources, reuses existing GIS services, and offers on-demand access and integration of widely distributed spatial data and information. The chapter discusses the construction of DGIService requiring the development and design of a new architecture and system framework. Research on DGIService has focused on the study of interoperability for sharing computing resources and on the development of driving technologies (e.g., agent-based, grid-based GeoComputing, etc.). The chapter presents the basic issues of DGIService from a geospatial computing perspective, and provides professionals and practitioners with an introduction to DGIService concept and technology. In Chapter 6, Liping Di discusses a particular example of DGIService, namely Grid Computing and presents a detailed explanation of the idea of Grid Computing and its deliverables. Grid Computing is a rapid developing technology, originally motivated and supported by science and engineering requiring high-end computing, for sharing geographically distributed high-end computing resources. The vision of Grid Computing is to enable resource sharing and coordinated problem solving in dynamic, multi-institutional virtual organizations. This chapter describes the history and some technical details of Grid technology. Then it discusses problems of directly using Grid technology in the geospatial disciplines, i.e., geospatial Grids, and the features of geospatial Grids. The chapter briefly summarizes some of the current geospatial Grid projects around the world and provides details of a NASA-funded project that applies Grid technology to the Earth observation environment through integration of the Globus toolkit with the NASA Web GIS Software Suite (NWGISS).

In Chapter 7, Yaser Bishr presents yet another currently popular topic of the Geospatial Semantic Web (GSW). In simple terms, a semantic web is an extension of the current web in which information is given well-defined meaning, better enabling computers and people to work in cooperation. The GSW is an application of the Semantic Web which intends to enable better integration between distributed, heterogeneous and web-enabled Geospatial Enterprise systems. The chapter describes the problem of semantics and provides a wider view of the semantic web that goes beyond the Web Ontology Language.

Data management of large volumes of data has become a major requirement in GIS. Therefore Database Management System (DBMS) occupies a central place in the new generation GIS architecture, in which spatial data is maintained in an integrated DBMS architecture and can be accessed by different types of front-ends. In Chapter 8, Sisi Zlatanova and Jantien Stoter, explain the role of DBMS in this new architecture such as what spatial models (including geometrical primitives and topological structure), and spatial functionalities should be offered at DBMS-level to support the new architecture and what is the state-of-the-art concerning these aspects. A separate section addresses the maintenance of spatial data with geometrical primitives and topological structure. In this discussion special attention is on 3D data, since there is a growing need for 3D data in GIS community. The chapter concludes with a discussion on how DBMS functionality and GIS front-end functionality could be balanced in the new generation GIS architecture.

One of the classical challenges in Geovisualisation community has been the development of compact and interactive visualisation interfaces which could respond to touch and sound. These are broadly called the multi-modal interfaces and are particularly useful for disabled users and users with limited scope of interactivity (e.g., a pilot in a plane). In Chapter 9, Reg Golledge and others provide an extensive survey of the multimodal interfaces for representing and accessing geospatial data.

Existing navigation systems, e.g., in-car navigation systems, are tied to the means of transportation. Recent technological developments such as Smart Phones, wearable computing devices, or Personal Digital Assistants (PDA) with integrated Global Positioning System (GPS) have made it feasible for a person to carry their own personal navigation and information system with them. A personal navigation system may further take advantage of specific knowledge about its user, e.g., the PDA may use the agenda to set up a route in which different tasks are spatially and temporally scheduled. In unfamiliar surroundings the PDA could also request more detailed information from the service. In familiar environments the personal navigation system may avoid areas with negative associations or point out familiar landmarks along an otherwise unfamiliar route. In short, the device will become adapted to its user. However, most of these functionalities represent as yet unresolved research questions. In Chapter 10, Sabine Timpf dissects the concept of navigation, classifying the process of navigation into various types with examples. The chapter concludes with examples of some work in progress within the research community and point out avenues of further research.

Geovisualisation technologies have observed immense advances over the years mainly due to corresponding leaps in the computing technology in particular the computer graphics and downscaling of computing hardware. In Chapter 11, Daniel Holweg and Ursula Kretschmer propose the opportunities and needs for Augmented Reality visualization of geospatial data. They propose that by using this visualization technique the real and virtual environment can be brought together to an integrated view. In the past, Augmented Reality visualization techniques have usually been used in indoor environments, whereby several research projects in the last few years have been focused on the transfer of this technique into outdoor environments. The research on accurate tracking techniques and three-dimensional city models which can be used as a reference for video tracking or as a data source for the visualization supports this development. Augmented Reality visualization of geospatial data is connected to the development of three-dimensional GIS and Location based Services. The chapter introduces the visualization method and presents the requirements and the potential of the technique, based on the experiences of the GEIST project. The GEIST project involves Digital Storytelling in urban environments using an outdoor Augmented Reality information system for education purposes of students in history.

In Chapter 12, Satya Prakash and others deal with a sensitive and urgent issue of the challenges to the development of GI technologies in developing countries. It is a hard reality that "*future varies across space*" e.g., while some of us are craving to get the latest 3G handsets and holographic hard disk drives, there are still people in remote parts of South America who are amazed to see LEDs (Fairley 2004). Although with increasing globalisation the time difference between the dispersal of latest technologies has become considerably shorter, there are still several hurdles to a smooth integration of GI technologies in developing countries. Physical and human development strategies of any area are often a collective decision-making process that affects the residents (a group of people residing in the area) where such strategies are put forth. Hence development strategies have a collective rather than an individual well being at its core. Having rationality is pragmatic in such a situation. To a large extent, rationality can be achieved through utilization of technology. In developing countries, the domain of development and of technology–more specifically geospatial and information & communications technologies–are conceptually complimentary but practically yet to be fused. This chapter attempts to establish the strong need to amalgamate and internalise technology platforms in development processes of the developing world and demonstrate through the description of a project. The chapter also discusses an 'inverted pyramid syndrome' in this respect prevalent in developing countries that keeps thwarting the amalgamation of GeoICT and development unless addressed categorically.

The second last chapter by Alastair Beresford of the book presents a detailed and incisive review on the various hidden dangers of GI technologies. The chapter demonstrates with example, several models of interactions/relationships between persons and their environment that influence their privacy. The chapter comments about the ways in which the privacy can be maintained with access control and anonymisation.

The final chapter presents several scenarios related to the future of GI technologies. The scenarios are based on the data on research funding, current political events, integration of technologies and a bit of speculation.

Figure Credits:

All images have been reproduced with the permission of the sources listed below. To the best of our knowledge and unless stated otherwise, copyright of the image resides with the sources. Please contact us for contact information and details on any specific issues.

Figure 1.1 Yahoo Inc.

Figure 1.3 Defra, UK Government. Background topographic map is Crown Copyright, Ordnance Survey.

Figure 1.4 NCSL International.

Figure 1.5 Cambridge University Press.

Figure 1.6 Permaculture Institute of Australia.

Figure 1.7 Cambridge University Press.

Figure 1.8 Mrs. Margaret Dilke.

Figure 1.9 K G Saur Publishers.

Figure 1.11 Peter Ilfand.

Figure 1.12 Public-Domain hosted by Wikipedia.Org.

Figure 1.13 Peter Ilfand.

Figure 1.14 Peter Ilfand.

Figure 1.15 Peter Ilfand.

Figure 1.16 Dukaru, California.

Figure 1.17 Smithsonian Natural Museum of American History

Figure 1.18 Smithsonian Natural Museum of American History.

Figure 1.20a Massachusetts Institute of Technology.

Figure 1.20b Boston Public Library.

1.20c Deustches Museum, Munich.

1.20d Deustches Museum, Munich.

Figure 1.20e Smithsonian National Air and Space Museum.

Figure 1.20f Public-Domain hosted by Library of Congress.

Figure 1.20g Public-Domain hosted by Aeronautical Systems Centre History Office.

Figure 1.20h Public-Domain hosted by National Archives.

Figure 1.20i National Aeronautical Space Administration.

Figure 1.20j National Aeronautical Space Administration.

Figure 1.22a National Epigraphical Museum, Athens, Greece.

Figure 1.22b Qiming Zhou, Hong Kong Baptist University.

Figure 1.22c Qiming Zhou, Hong Kong Baptist University.

Figure 1.22d Public-Domain hosted by Wikipedia.Org (Original Photo from Printing History Museum, France by George H. Williams).

Figure 1.22e Public-Domain.

Figure 1.22f Public-Domain.

Figure 1.22g Mitre Inc.

Figure 1.22h University of Amsterdam.

Figure 1.22i Qiming Zhou,Hong Kong Baptist University.

Figure 1.22j 3DGI, University of Aalborg.

Figure 1.22l GeoVista, Pensylvania State University

Figure 1.23a National Epigraphic Museum, Athens, Greece.

Figure 1.23(b) Luis Fernandes.

Figure 1.23(c) Computer History Museum.

Figure 1.23(d) Andrew Dunn hosted by Wikipedia.Org under the Creative Commons Attributrion Sharealike 2.0 License.

Figure 1.23(e) British Public Records Office, Crown Copyright

Figure 1.23f (f) Worcester Polytechnic Institute.

Figure 1.23(g) Hewlett-Packard.

Figure 1.24 Computer History Museum.

Figure 1.26 Controller of the HMSO, Crown Copyright.

Figure 1.27 Toposys.

Chapter 2

Soft Computing in Geographical Information Systems

Yingjie Yang, David Gillingwater and Chris Hinde

2.1 Introduction

With the continuing development of Geographical Information Systems (GIS), the issue of data quality has become a major concern in its further application (Fisher 1995; Goodchild and Gopal 1989; Goodchild 1999b; Heuvelink 1993; Lodwick et al. 1990). Perfect data are rarely obtainable in the real world since most data contain errors and omissions. The traditional computing methods employed in GIS have difficulty in dealing with these kinds of imperfect data, and as such a methodology is required which allows for and that can handle imperfect data. This gives rise to the need for the application of Soft Computing (SC) in contemporary GIS technology.

Soft Computing consists of a consortium of methodologies that works synergistically and provides a flexible information processing capability for handling real-life ambiguous situations (Zadeh 1994). The aim of SC is to model human perceptions of the world with inexact expression. It differs from traditional computing methods in that SC considers the world as imperfect, and hence employs uncertainty models to handle the computation involved. Models typically used in SC include artificial neural networks, fuzzy sets, rough sets, genetic algorithms and the like. In these models, data are permitted to be imperfect, with errors, noise and missing values. Soft Computing may therefore provide a powerful tool for GIS to overcome the imperfect data problem. There have been considerable efforts in recent years to apply SC within GIS (Fonte and Lodwick 2004; Yang and Rosenbaum 2001; Guesgen and Albrecht 2000; Leung 1997; Chang and Burrough 1987). This chapter presents an overview of Soft Computing and its use in representing uncertainty in spatial analyses. The chapter also demonstrates the principles and methodology of Soft Computing with two case studies of the application of neural networks within GIS.

2.2 Uncertainty and GIS

The first obvious uncertainty associated with GIS is the spatial errors arising from the inclusion and use of imperfect data. There are many sources for spatial errors, such as positional accuracy problems with measurement, human interpretation problems resulting from different data providers and data integration problems as a result of multiple data sources (Chrisman 1991). These errors are not necessarily static, but may also be subject to change with time (Burrough and McDonnell 1998). In addition, different operations of GIS may produce new errors related to the organisation of computer storage and data structure. Taking these errors together, the existence of spatial errors can seriously erode the quality of GIS output, especially when many outputs emanate from derived results rather than original measurement. However, most users of GIS functions assume perfect data quality in their operation, and as a result little attention has been paid to data quality issues (Burrough 1999). Most efforts concerning data quality still remain at the research level (Fisher 1995, Goodchild and Gopal 1989; Goodchild 1999b; Heuvelink 1993; Lodwick et al. 1990) and are consequently not implemented widely in GIS.

The second important uncertainty is human perception. The most important difference between human representation and machine representation of the world is the so called accuracy or completeness of the respective representations. In our daily communication, for example, it is rare for us to exchange precise or accurate numerical information. In most cases, we use a flexible language that employs loose concepts like 'highly', 'probably', 'very possibly', etc. This kind of human language is not compatible with the current generation of GIS therefore we have to state clearly and in numerical terms what we mean by, for example, "near to a river" – i.e., 10.3 kms. This vagueness or fuzziness does not influence our communication, but it presents a big challenge for a machine to understand and represent spatial data. Like any other computing systems, GIS represents geographical data as exact numbers, and hence its information retrieval is based on the traditional two-value logic i.e., TRUE or FALSE. For instance, although aircraft noise around an airport is in fact a continuous distribution, we have to draw a line on a noise contour map somewhere to say that, on one side of this line, people exposed will be disturbed but, on the other side, they will not. This line can pass between two houses, with the implication that people can be classified as noise sufferers depending solely in which house they reside. This can mean that while one household receives compensation (e.g., double glazed windows) whereas the other does not. This rigidity of the Boolean logic leaves little scope for a middle ground (Burrough and McDonnell 1998).

In recent years, owing to the increasing use of GIS applications in industry and public life, the demands for decision support using spatial data have increased dramatically. More and more decision support systems based on GIS are being developed and decision support functions are becoming embedded in GIS (Leung 1997). In fact, the concept of GIS itself has also arguably changed, with decision support functions becoming an integral feature (Leung 1997; Malczewski 1999).

To this end, GIS is no longer simply a special database for manipulating spatial data, but also a more fully functional information system with data analysis and decision supporting functions as well. For instance, knowledge extraction or data mining functions are now seen to be a necessary part of any contemporary GIS (Hand et al. 2001). Such functions require capability in reasoning with available data, and an ability to mimic human intelligence in making a decision under given constraints. Here, uncertainties like missing values, incomplete or vague information, unknown mechanisms and factors are the main issues influencing the quality of outputs. Unlike geographical data, the spatial distribution of attributes in GIS is much more complicated, and it is controlled not only by their spatial location, but also involves many other known or unknown factors. For instance, the distribution of atmospheric emission pollutants is controlled not only by geographical features but also by several other dynamic meteorological factors such as wind direction, wind speed, and ambient air temperature. Consequently, the uncertainty modelling in these cases is inherently complex and worsened by the Boolean logic. However, uncertainty is clearly an inevitable aspect of any GIS and a modern system must be able to deal with it in order to provide a more reliable and human friendly service. Nonetheless, Soft computing provides us an overview of a range of possible solutions instead of rigid rule based ones.

2.3 Soft computing methodologies

There are many different methodologies in Soft Computing including, for instance, fuzzy logic, neural networks, rough sets and genetic algorithms. The most commonly used methods in application today are fuzzy sets and neural networks.

2.3.1 Fuzzy sets

Before we discuss fuzzy sets, we should clarify first what is meant by a crisp set. By crisp sets, we mean the traditional sets where only two classes exist: TRUE or FALSE. For a subset *S* in a domain $U=\{x_1,x_2,\ldots,x_n\}$, a crisp set has the following mapping function:

$$\mu(x_i) \rightarrow \{0,1\} \tag{2.1}$$

For example, we consider the spatial relationship of some entities with a pollution source *P*. We want to find all entities which are near to *P*. With crisp sets, we can only classify the entities into two classes: 'near' or 'not near'. Therefore, we have to clarify what is "near" in the first place. In GIS, we can give a user defined distance *d*, and query all entities which have a distance less than *d* from *P*. The result is shown in Figure 2.1a. With distance *d*, we can draw a circle around *P*: all those entities outside this hashed circle are not near to *P*, and all entities inside it are near to P. According to crisp sets, only A and B would be included in the output for entities near to P. However, the entity D is adjacent to A and it is really difficult to comprehend intuitively why they are treated so differently. In fact, they have only a very minor difference in their distances to P, and so it would be more reasonable to classify them together. This is a typical problem arising from the use of crisp sets, and so we need a different approach to recognise this kind of conceptual fuzziness,

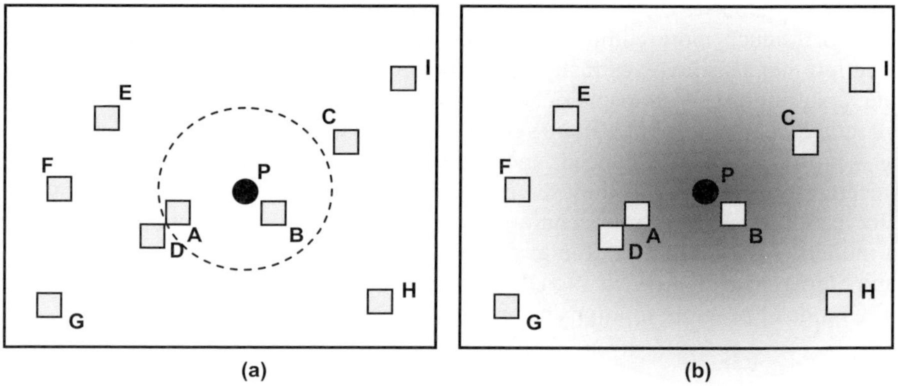

Figure 2.1 Crisp sets and fuzzy sets

Fuzzy set theory was proposed originally by Zadeh (1965). Fuzzy sets are designed to provide a presentation for those values in the middle ground between the two values of traditional sets. They are designed to represent values accurately rather than precisely. For the entities in Figure 2.1b, their distances to P are indicators of their spatial relationship to P. Instead of having two classes 'near' and 'not near' to P, we can allow some entities to be near *to some degree*. For instance, entity G could be classified as 'definitely not near' P whereas B would be 'definitely near' to P, with other entities in between. Compared with traditional sets, the characteristic function can therefore take not only 0 and 1 values, but also a number between 0 and 1.

Formally, a fuzzy subset S in a domain $U=\{x_1,x_2,\ldots,x_n\}$ is defined as a set of ordered pairs:

$$S = \{\langle x, \mu_S(x_i)\rangle : x_i \in U\} \tag{2.2}$$

where $\mu(x_i) \rightarrow [0,1]$ is the membership function of S and $\mu(x_i)$ is the grade of belongingness of x to S (Zadeh 1965). With fuzzy membership, a fuzzy concept can be represented as,

$$S = \sum_{i=1}^{n} \mu_S(x_i) / x_i \tag{2.3}$$

The membership $\mu(x_i)$ provides a powerful tool in representing imprecise boundaries of the kind shown in Figure 2.1b. We can define a membership function as a function of distances of an entity to *P*, reaching 1 when the entity is adjacent to *P* but 0 when it is far away from *P*. Then we get a continuous spatial distribution of membership as shown in Figure 2.1b. It is clear that *A* and *D* have similar memberships in a fuzzy set. Therefore, we can say that *B* is the nearest entity to *P*, *A* and *D* are near to *P* to a high degree, but *G* has the lowest degree near to *P*. Using Eq. 2.3, we can express the concept of 'near to' *P* as,

$$S_{near} = 0.9/B + 0.6/A + 0.55/D + 0.5/C + 0.3/E + 0.2/I + 0.2/F + 0.2/H + 0.1/G \tag{2.4}$$

which gives a more complete description of the relationships between these entities and *P*. A vague description of a spatial phenomenon such as shown in Figure 2.1b is particularly useful to represent classification errors (Chrisman 1991; Goodchild et al. 1992) and imprecise boundaries (Leung 1987). However, traditionally, the representation and analysis of spatial error were generally done using statistical models. Although clearly different from the traditional two valued logic, fuzzy sets are not so sensitive to small spatial errors (Heuvelink and Burrough 1993), and more suitable when representing ambiguous boundaries. For this reason, fuzzy sets have often been confused with probability models (Fisher 1994a). Hence its application has been investigated for most GIS operations involving uncertainties, such as the reasoning process (Guesgen and Albrecht 2000; Chang and Burrough 1987), viewshed operations (Fisher 1992, 1993), information representation (Wang et al. 1990), object modelling (Molenaar and Cheng 2000; Cross and Firat 2000), map similarity (Hagen 2003), area calculation (Fonte and Lodwick 2004), information retrieval (Petry et al. 2003) and data integration (Ahlqvist et al. 2003).

2.3.2 Neural networks

Artificial neural networks (ANNs) are inspired by the mechanisms of the human brain when establishing interrelations between a variety of information sources. This refers to intuitive reasoning rather than the logical reasoning normally executed by machine. One of the most popular training schemes is the back-propagation (BP) network (Rumelhart and McClelland 1986). The back-propagation neural network architecture is a hierarchical design consisting of fully interconnected layers or rows of processing units (Figure 2.2). The interconnections are called weights and provide the means for ANN to save knowledge, the process of "learning". This process modifies the weights by incorporating the errors in the mapped output. Based on the calculation of error gradients, such errors are then back-propagated from the output neurons to all the hidden neurons; subsequently all the weights are adjusted with respect to the errors. The BP process is repeated until the error output has been reduced to a specified minimum value. The weights are then fixed and saved as a record of the knowledge pertaining to this system. Thus for a given input, an output is then associated with the fixed weight system.

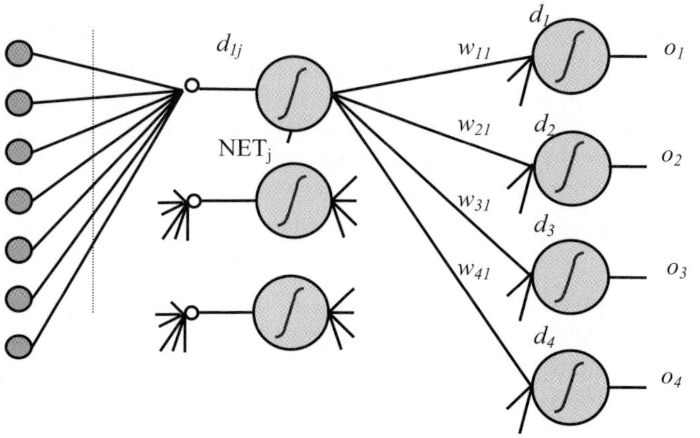

Figure 2.2. Structure of Backward Propagation Neural Networks

The information processing operation facilitated by back-propagation performs an approximation of the bounded mapping function $f{:}\boldsymbol{A} \subset \boldsymbol{R}^n \rightarrow \boldsymbol{R}^m$. This function is from a compact subset $\boldsymbol{A}$ of **n**-dimensional Euclidean space to a bounded subset $f[\boldsymbol{A}]$ of **m**-dimensional Euclidean space, by means of training with examples (x_1, y_1), (x_2, y_2), ..., (x_k, y_k), ... of the mapping, where $y_k=f(x_k)$. It is assumed that the mapping function f is generated by selecting x_k vectors randomly from $\boldsymbol{A}$ in accordance with a fixed probability density function $\rho(\boldsymbol{x})$.

The operational use to which this network is put once training has been performed (on a set of experimental or observed data) makes use of the random selection of input vectors $\boldsymbol{x}$ in accordance with $\rho(\boldsymbol{x})$. ANN then models the mapping by utilising simple neurons based on either a linear or a non-linear activation function. Because of the large number of neuron connections, model behaviour is characterised by co-operation between neurons. Thus an ill definition introduced by a few neurons does not influence the outcomes from its associated mapping.

ANN has a robust quality with respect to uncertain or deficient information, even though such information influences many aspects of a complex system, for example the propagation of uncertainty. ANN can apply additional neurons and weights as required to take full account of such influences, and thus possesses an in-built capability for including any relation once it has been trained using a reference data set.

ANN may well have a large number of nodes, yet the activation function at each node is very simple. The complex knowledge contained within the training data is saved in the form of connections between the various nodes. The connections and activation functions determine the behaviour of the neural network. Thus, no matter how complicated the mechanisms, ANN has the capability of mapping it without having to incorporate a prior supposition or simplification.

The existence of large numbers of nodes needed to represent knowledge provides the robust structure for uncertain or incomplete inputs. The limited connection weights have the advantage of dramatically reducing the requirement for computer memory. Kolmogorov's 'Mapping Neural Network Existence Theorem' (Hecht-Nielsen 1987, 1990) has demonstrated that ANN is capable of implementing a mapping function to any desired degree of accuracy, and can thus enhance the decision support ability of GIS.

Neural networks are good at mapping a 'black box' between inputs and outputs. In environmental GIS applications, the spatial attributes only come from a limited number of monitoring stations over limited time intervals. Hence, it is necessary to obtain data for other spatial locations and time spans, and the only thing we can do here is to derive these data from monitoring stations near by or by monitoring over time. However, because of the complexity of the real world, there are no suitably precise mathematical models to simulate these data exactly. Therefore, neural networks provide a powerful tool for this type of situation, and it can also benefit a variety of decision making processes. Neural networks have mainly been applied to spatial interpolation (Rigol et al. 2001), spatial attributes mapping (Yang and Rosenbaum 1999; 2001; 2003, Yang et al. 2001c) and error simulation (Brunsdon and Openshaw 1994).

In addition to fuzzy sets and neural networks, there are other models which can be deployed in SC, including rough sets (Pawlak 1982), grey systems (Deng 1982) and genetic algorithms (Goldberg 1989). Fuzzy set and neural network models also have many extended versions, such as interval valued fuzzy sets (Sambuc 1975), intuitionistic fuzzy sets (Atanassov 1999), type 2 fuzzy sets (Mendel and John 2002), Hopfield neural networks (Hopfield 1984), adaptive resonance theory neural networks (Carpenter and Grossberg 1995), radial basis function neural networks (Park and Sandberg 1991) and fuzzy neural networks (Kasabov 1996). In recent years, these methods have also attracted attention from the GIS community, and some have already been investigated, such as rough sets (Ahlqvist et al. 2000), genetic algorithms (Wilson et al. 2003) and rough fuzzy sets (Ahlqvist et al. 2003).

2.4 Applications of neural networks

Although there has been a huge amount of interest in the use of fuzzy sets in the literature, here we will demonstrate the application of SC to GIS using neural networks. The first case illustrates a framework for modelling an airport environment, and the second case is an application to a harbour environment. Both cases are illustrative of the problem that their spatial attributes come only from a limited number of observations over limited time intervals. In addition, because of their real world complexities, there are no readily available mathematical models to simulate these data exactly. Therefore, neural networks provide a powerful tool to help model these types of situation; in addition, they provide insights and support for decision makers who have the task of remedying the environmental consequences of, respectively, aircraft noise around airports and sediment accumulation in harbours.

A neural network excels at learning from data and does not require the prior specification of a mathematical model. This feature makes it an ideal candidate in environmental analysis where a large amount of monitoring data exists but where the interactive mechanisms are too complicated or little understood to specify an accurate mathematical model. As a spatial analysis tool, GIS is ideal in dealing with environmental simulation and analysis. Hence, it is a logical step to integrate both into one system. The general structure of this integration is shown in Figure 2.3. Here neural networks provide a powerful function to GIS, and is embedded within the functionality of a GIS system therefore there is no difference to the users of GIS from the point of view of operation as long as a trained neural network is already installed. The system only calls a neural network when no other method can meet the requirement or where the user selects the involvement of a neural network.

Figure 2.4 provides more detail on the role of neural networks. The data required as inputs to the neural networks are fed from the GIS operation, following which neural networks map out an intuitive solution, generating outputs as attributes and images as well as corresponding analytical values (Yang and Rosenbaum 1999, 2001, 2002; Yang et al. 2001c). These results can then be fed into the map algebra function together with other maps to produce synthesised results or provide direct results to the final decision making process. In the overall process, neural networks play an advanced function to solve those otherwise difficult mapping problems.

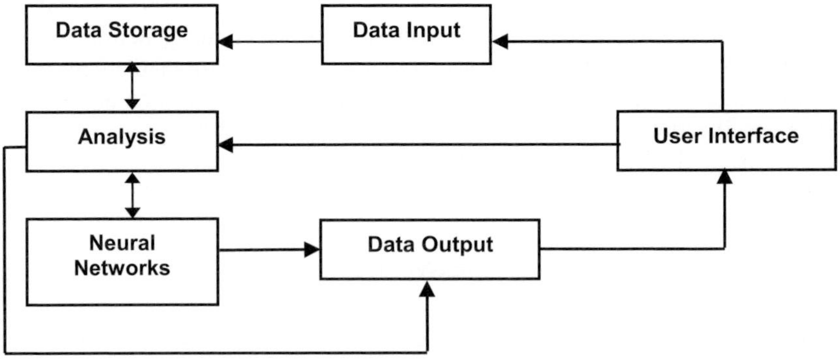

Figure 2.3. A framework for linking neural networks with GIS

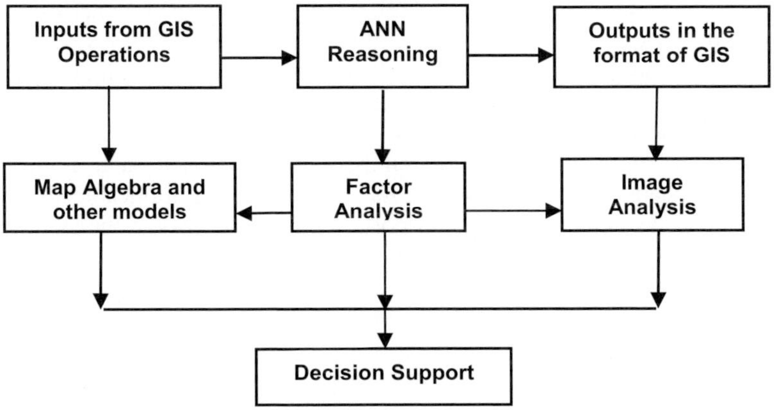

Figure 2.4. The role of neural networks in GIS analysis and decision making

2.4.1 Airport Environmental Impacts

As transportation hubs, airports comprise many transport modes, including air and surface transport operations. Changes in the magnitude of these operations directly and indirectly lead to corresponding social and environmental changes, not least of which are employment opportunities and adverse environmental impacts (Upham et al. 2003). Therefore, it is important to know the likely scale and significance of environmental impacts if any operational changes are planned or proposed (Upham et al. 2004). The basic methodologies available follow one of two classes: (i) the laboratory model – based on laboratory based experiments and standard in-situ tests under given conditions; or (ii) the replication/simulation model – based only on in-situ test data. However, it is not feasible to monitor each impacted location around an airport and the interactions between the key factors are too complicated to enable reliable mathematical models to be developed. The laboratory models are in fact the dominant models in use. For instance, aircraft noise calculations around airports are dominated by some calibrated models based on standard condition tests and aircraft engine manufacturers' data, such as the US FAA integrated noise model (INM) (Connor 1980). These standard models are very useful in simulation analysis at a general level, but they do not incorporate specific local conditions, and thus the reliability of their results are subject to standardised assumptions regarding geographical and weather conditions. From this point of view, a location-specific model established with data from that site is likely to yield more realistic results although its generality is likely to be poor when compared with the laboratory model.

As part of our work on a decision support system for sustainable airport development (Yang et al. 2001b, 2001d), we applied neural networks as a mapping tool to establish the impact of aircraft noise in a GIS environment. Aircraft noise around an airport is influenced by many factors, such as its geographical location and the location of aircraft in 3-dimensional space, the power thrust of an aircraft's engines at any given moment, the number of engines on an aircraft, its weight and speed, as well as wind speed and direction, ambient air temperature and geographical features around the point at which aircraft noise is being measured. This list is not complete, and there are many more factors that affect real airports. These differ also from airport to airport, hence the relationships are very complex. Neural networks approach such complicated functions by means of learning from data, hence it is an applicable tool when training data are available. At those airports where awareness of the significance of environmental impacts like aircraft noise is increasing, more and more environmental data are monitored but collected only from a very limited number of geographical locations. These data can provide a basis for the application of neural networks.

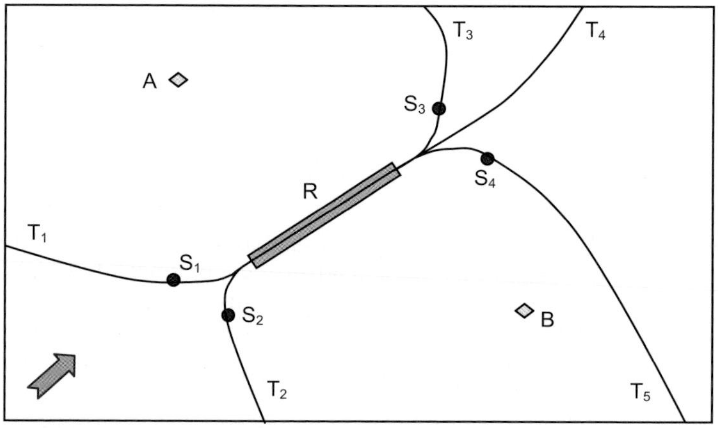

Figure 2.5. Spatial relationships between airport runway, trajectories and monitoring stations

Figure 2.5 illustrates a simple spatial representation of an airport and the relationships between a runway, *R*, aircraft flight trajectories T_1, T_2, T_3, T_4 and T_5, and the monitoring stations S_1, S_2, S_3 and S_4. It is very common for an airport to locate its monitoring stations close to aircraft flight trajectories, like the monitoring stations shown by S_1, S_2, S_3 and S_4. Consider two locations, *A* and *B*, where monitoring data are not collected. In an airport environmental GIS, its users may be interested in examining the environmental effects of its operations on any location around the airport, for instance *A* and *B*. Therefore, one of the important tasks for airport environmental GIS is to derive environmental data for *A* and *B* from the available data collected at S_1, S_2, S_3 and S_4. Traditional interpolation methods in GIS are not very helpful in this case because of the spatial distribution of the very limited number of monitoring stations. We cannot identify a useful monitoring station in proximity to either *A* or *B*. Therefore, in this case a neural network can be used to represent the environment and the uncertainty.

Considering the distribution of aircraft noise, all locations in proximity to the same airport experience similar weather conditions and the same noise sources; their spatial difference is the main factor in determining their difference in the distribution of aircraft noise impacts. Therefore, we establish a set of neural networks using distance to consider the macro spatial relationships between aircraft noise and its distribution. A simplified diagram for inputs and outputs is shown in Figure 2.6.

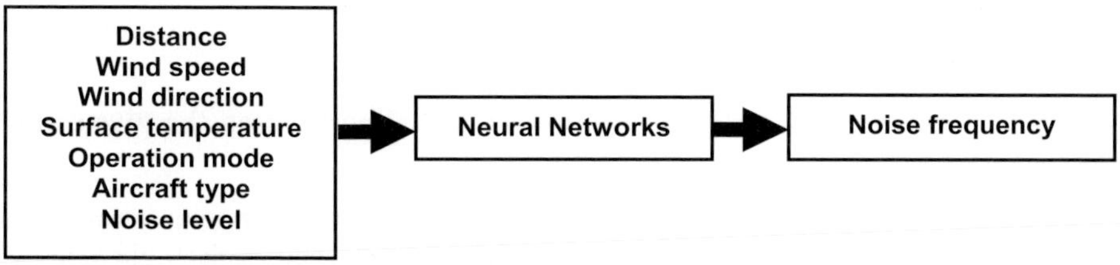

Figure 2.6. Neural network inputs and outputs

Small spatial details can be excluded from the model since, following trial tests, it was found that the noise data from aircrafts (e.g., Boeing 747-400) are not influenced by such features. It is in fact very difficult to establish a satisfactory neural network with different aircraft types included, hence it is considered separately from the input factors. The output of the model is the frequency of aircraft noise rather than noise level. In our tests, it was obvious that an accurate noise level approximation is very difficult to ascertain because of the complexity of the airport environment. For example, according to the monitoring data from a major UK airport, for the same aircraft type and engine type, the data collected at the same monitoring station for the same aircraft operation showed large fluctuations. Therefore, it is not possible for a neural network to provide a reliable prediction for the level of aircraft noise based on such data. The flights regarded as really noisy are those flights with a noise level higher than some threshold. Thus, the same aircraft may not necessarily be as noisy for each of its operations, since its noise output will depend on many other complex interactions within the system. Although, it is practically difficult to monitor all factors which influence aircraft noise levels, but it is relatively apposite to test the probability of an aircraft event being noisy for each kind of aircraft during its operation at the same airport. By probability of an aircraft event being noisy, we mean the probability for the aircraft noise level to go beyond some given threshold for a specific geographical location at the airport. Given the noise monitoring level $\{L_1(x), L_2(x), L_3(x), \ldots, L_n(x)\}$ when the distance between an aircraft flight trajectory and the location under scrutiny is within $[d_1, d_2]$ for the same aircraft in the same operation mode under similar weather conditions, the 'noisy probability' $p(x,t)$ is calculated as

$$p(x,t) = \frac{|\{L_i(x) : L_i(x) > t\}|}{n} \qquad (2.5)$$

Here, x is the specified location and n is the cardinality of the data monitored for that location with distance in $[d_1,d_2]$. Note that $[d_1,d_2]$ is a 'grey number' (Liu et al. 2000) and its degree of greyness determines the resolution of the noisy probability. L_t is the threshold for the 'noisy' noise level. However, a lower degree of greyness has less data available, hence it has to be balanced between the high resolution and reliable probability. With noisy probability, the noisy frequency $F(x,t)$ is easy to derive:

$$F(x,t) = p(x,t) \times N_f \qquad (2.6)$$

where N_f is the number of flights operated at the airport during the given time interval.

For simplicity, we consider that each flight takes a similar time duration in its taking-off operation or on approaching the airport. Hence each flight can be counted only once in the frequency calculation. For the situation where we have more than one trajectory, the probability is calculated separately for each trajectory. In the final analysis, a general probability is calculated according to their weight in terms of number of operations. The operation of neural networks in deriving the noisy frequency distribution in a GIS environment is shown in Figure 2.7.

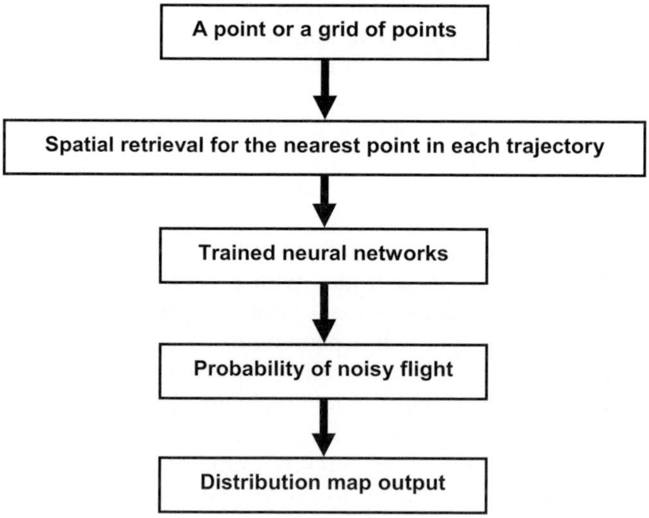

Figure 2.7. Neural networks for noise distribution in GIS environment

The training of a neural network is often a difficult task, especially for problems like noise where the data have such large fluctuations. Partly in response to such problems, we have undertaken extensive work in improving the quality of training using different methods and analysing the different impacts of input factors (Yang et al. 2001a, 2003a, 2003b). For a trained neural network with data from a single aircraft type, output for some virtual flight trajectories is shown in Figure 2.8. The network was trained using monitoring data from a large UK international airport for a Boeing 757 aircraft. The figure to the right highlights residential areas subject to noise disturbance by this aircraft.

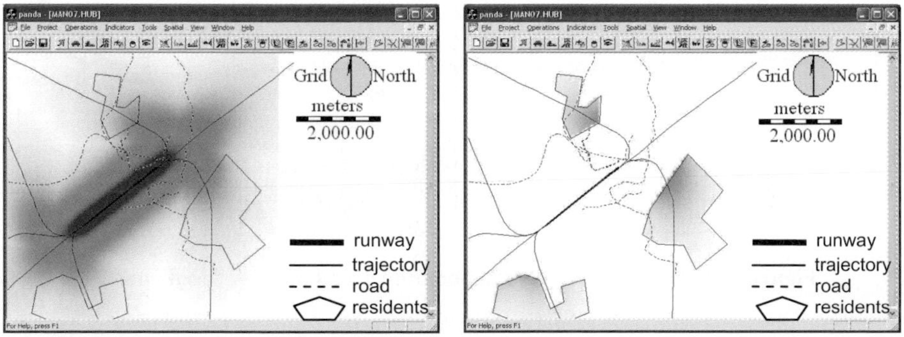

Figure 2.8. Noisy frequency distribution derived from neural networks

2.4.2 Geoenvironmental Change within Harbours

Neural networks have also been applied to the prediction of sedimentological characteristics within the harbour area (Yang and Rosenbaum 1999, 2001, 2002; Yang et al. 2001c). Here sediment accumulation is the result of a number of environmental processes which are difficult to identify, yet their effects can be measured (Stevens 1999). Such harbours, where the tidal influence is low, can create effective traps for sediment deposition, yet turbulence caused by river currents or shipping manoeuvres can cause renewed erosion, possibly re-introducing older, formerly buried and polluted, sediment into the harbour environment. Such behaviour is likely to be influenced by the seabed profile, water depth, and distance from possible sources of pollution, the local biochemical environment, shipping lanes, dredging, and climate, as well as a number of anthropogenic factors. Neural networks have been employed to examine the sedimentological aspects of this system, comparing natural and anthropogenic sediment sources, transport pathways and geochemical changes.

The harbour investigation – in this case, for Gothenburg in Sweden – resulted in 139 sample stations, which were adopted to train and test the neural networks. As an example, the influences of water depth and distance from the river mouth, bank and shipping lanes could be investigated on the distribution of sediment grain size on the harbour floor. In this case, the resulting network would have four input nodes and, say, three output nodes (one for each sediment grade: sand, silt and clay). Figure 2.9 shows an example of the output image from a neural network driven GIS and its training samples.

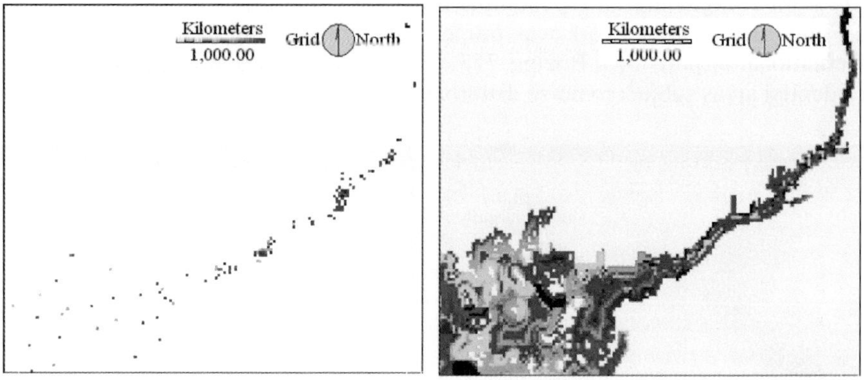

Figure 2.9. Harbour sand distribution from neural network based GIS.

2.5 Developing trends in intelligent geographical information systems

GIS has become, as Sui and Goodchild (2001) point out, a medium between different systems with people as its primary users. As a medium for integrating diverse components, GIS has to provide a reliable service to all users, not only those with a sound knowledge of its operation. In this sense, a GIS has to be sufficiently intelligent to make its operation sensible with the minimum intervention by its users. At the same time, a GIS needs intelligence to be able to incorporate human perceptions as well as computing by words and speech, and its output should also be indicative of its limitations. As Malczewski (1999) has pointed out, the ultimate aim of GIS is to support spatial decision making - an attractive visualisation is useless unless it conveys information deemed necessary by its users. All these require a capability in dealing with unknown situations or uncertain data, hence in our view the application of SC to GIS is essential to the future of GIS itself.

As demonstrated in our case studies, the use of traditional computational methods are not particularly useful or appropriate when significant uncertainties are involved in the operation of GIS. For example, it is often difficult to make realistic predictions of noise frequency around an actual airport based on the application of a generic model populated by aircraft engine manufacturers' test data. It is also very difficult to estimate mathematical models applicable to the sedimentological characteristics of a harbour. Under circumstances like these, SC can be used in treating uncertainties caused by imperfect data.

Considering the recent rapid developments in GIS, it is difficult to offer anything other than a conjecture regarding the future for GIS at this moment; that said, incorporation of the following functions in relation to SC would appear to be important when examining developing trends:

- User friendly data input, such as input data in the form of human speech and handwriting;
- Human style communication interface and outputs, such as a rough map presenting indicative conclusions as well as human language output;
- Learning capability based on imperfect data, such as the ability to incorporate individual preferences arising from an individual user's operation;
- Non-determined spatial and temporal inference mechanisms; and
- Intelligent communication with agents in different fields in a grid environment.

The integration between GIS, Internet and satellite-based (GPS) technology is also a potentially significant developing trend in the context of SC. Such integration is capable, for example, of linking GIS with public users through the Internet, and large amounts of data can be garnered from GPS. Public users need not have any prior knowledge about GIS technology; they do, however, need a more human style communication interface in order to exploit its potential. In addition, GPS generated data requires a robust capability that can cope with the demands of unknown data which has yet to be determined let alone collected. Based on these types of potentiality, it is clear that SC methodologies are poised to become an integral feature of future mainstream GIS applications.

Chapter 3

Using Geospatial Information for Autonomous Systems Control

Ayanna Howard and Edward Tunstel

3.1 Introduction

The modern techniques for data capture in a Geographic Information Systems (GIS) typically involve collection of information from remote sensing systems and digitized geographical maps to determine spatial relationships between features. This captured information allows GIS to perform complex analysis ranging from land site selection to topological modelling. For planetary exploration missions, the functioning capabilities that a GIS provides includes land site selection, localization of feature points, and determining spatial relationships between terrain features. In order to make use of geographical information, data is extracted from a wide range of sources — orbiting satellites, landers, and/or mobile platforms on or near the planet surface. An additional complexity associated with the use of this data is derived from the need to correlate the different sources of imagery, retrieved at varying resolutions, into a unified representation for use in performing complex tasks.

A contemporary example in news is that of the NASA's Mars Global Surveyor, a spacecraft orbiter tasked with the mapping of the surface of Mars. Data from the spacecraft's laser altimeter were instrumental in providing the first 3-D views of Mars' north polar ice cap. In 1997, cameras on the Mars Pathfinder Lander were used to associate terrain features for navigation of a mobile robot on the surface of Mars. The Mars Exploration Rovers (MER) launched in 2003 used onboard visual data to perform geographically related operations. Image data acquired by the MER landers during descent towards the surface was combined with Mars Global Surveyor camera imagery data to make a stereo anaglyph of the landing site. This correlation of surface imagery with orbital imagery exemplifies the complexity of the tasks required of GIS for space exploration missions.

Autonomous space systems that encounter harsh, natural terrain must deal with an added complexity of using geospatial information in their control logic. Spacecrafts must reason about terrain safety constraints while preparing to land on planetary surfaces. Field mobile robots must safely traverse uneven, rough, and rugged terrain. These systems must have the ability to assess the terrain and determine the risk associated with surface interactions, whether descending to the surface or roving on the surface. Furthermore, autonomous space systems must be able to navigate in geographical environments in order to visit geographical locations of relevance to a given mission. For space systems destined for other planets, such as Mars, the problems are magnified by a lack of facilities such as a Global Positioning System (GPS) or the inability to use a compass due to weak magnetic fields. Few researchers have addressed the inherent problems associated with self-assessment of terrain by a machine for its autonomous control in a natural environment. Imprecision in sensor measurements and uncertainty in data interpretation are often not accounted for in the terrain assessment. Comprehensive systems for real-time terrain assessment and classification for robust and safe operation have been inadequately addressed in previous research.

In this chapter we discuss techniques for incorporating geospatial information for onboard control and decision-making by autonomous space systems. The first section discusses the inherent problems that natural terrain environments pose for mobile autonomous systems. Several means for dealing with the challenges by clever use of terrain and geospatial measurements made by the autonomous system are described. Specific examples are then given for real-time decision-making based on terrain characterization for spacecraft autonomous landing and robot navigational guidance and control on natural terrain surfaces. The examples reveal how multi-scale geospatial information from multiple sources can be used for choosing the safest locations to land in rugged geographical regions, for robot localization and safe terrain traversal, and for long-distance path planning based on terrain maps. Limitations and next steps are highlighted in the concluding section.

3.2 Problem Statement

Robotic spacecrafts and autonomous mobile robotic vehicles (rovers), instrumented with a variety of sensors, are employed as surrogate explorers on remote planetary surfaces such as the desolate and rocky terrain of Mars. Spacecraft landers are equipped with all the necessary scientific instruments and communication devices to perform a mission at the site where they land. Their utility is centered around their ability to safely land at targeted geographical locations of interest. The utility of autonomous rovers is a function of their ability to move about and explore geographical regions to which they are delivered by a lander. In either case, these robotic spacecraft must conduct missions intelligently without frequent contact with Earth-based mission operators.

Safe landing of a spacecraft on a planetary surface is of critical importance for the success of exploration missions. The selection of an appropriate landing site for a spacecraft touchdown is therefore of fundamental significance. The current practice for site selection is only performed off-line in which mission scientists visually examine hundreds of pictures of potential sites obtained from previously acquired orbiter imagery. These data constitute geospatial information such as rock sizes and distribution, slope composition, and location and type of terrain features. Based on examination of such geospatial information, appropriate sites are then selected by considering both engineering and science goal criteria. It is assumed that once a site is selected, the terrain quality and related geospatial conditions for safe landing at the specified landing location will permit a physical touchdown of the spacecraft. Typically, there is no re-evaluation during spacecraft descent to determine whether terrain conditions have changed and whether the site is still suitable for safe landing. Additional obstacles require that the system is designed to operate in real-time and in a highly robust manner, even with the possible large data set that may be available during the decision-making process. In addition, due to the extremely fast descent speeds and relative proximity to the surface, only a limited number of corrective manoeuvres can be made by the decision-making system before touchdown.

Autonomous rovers designed for planetary surface exploration must be capable of navigation in the presence of varying surface obstacle distributions (rocks, boulders, craters, etc.), surface characteristics, and hazards. Mobility and navigation hazards include extreme slopes, sand/dust-covered pits, ditches, cliffs and otherwise unstable surfaces. Rovers must avoid surface hazards and negotiate obstacles to be of practical use for carrying out the goals of scientific exploration in natural environments. To navigate successfully on challenging terrain, rovers must be able to detect mobility hazards and assess terrain traversability. Additional problems to be addressed for navigation include maintaining knowledge of rover position and attitude on the terrain, and mapping the local environment and its prominent features or landmarks. Hence, terrain sensing and perception capabilities are essential for missions requiring autonomous land reconnaissance or surveying for the purpose of scientific exploration.

In addition to maintenance of accurate position estimates, robust guidance, navigation, and control, autonomous systems require accurate geo-referenced heading measurement or very good heading estimation. Errors in heading estimates directly affect the accuracy of rover position estimates. Earth-based robots can rely on geomagnetic reference devices, such as compasses, for reliable heading information since they provide good absolute heading measurements with respect to true north of Earth's magnetic field. Geomagnetic information is not as useful on Mars, however, due to the planet's weak magnetic field.

Geospatial issues or aspects of wheel-terrain interactions and their effects on position estimation via wheel odometry must also be addressed. Wheeled mobility systems are subject to undesirable wheel-terrain interactions that cause wheels to slip on rocks and soil. Loss of traction due to excessive wheel slippage can lead to wheel sinkage and ultimately vehicle entrapment. Frequent loss of traction due to wheel slip during traverses from one place to another will also detract significantly from the ability to maintain good rover position estimates. Undesirable wheel-terrain interactions also can cause wheels to sink in soft soils. As the load-bearing strength of terrain/soil varies under rover wheels (such as when transitioning between hard-packed soils and soft sand), so does the amount of wheel sinkage. This has the effect of varying the effective radius of the rover wheels. The accuracy of kinematic models used to compute rover position updates depends on accurate knowledge of the effective wheel radius, which is used to compute the equivalent linear distance travelled by a wheel after some measured rotational displacement on the terrain.

These inherent problems with operating in natural terrain environments combine to make the design and application of autonomous robotic systems for space missions a quite challenging undertaking.

3.2.1 Information Sources

Some of the problems mentioned above can be overcome by using a variety of terrain and geospatial measurements. Autonomous space systems typically operate in environments that are initially unknown and/or unstructured. Thus, any additional information that increases the knowledge of the environment and terrain characteristics in which the system operates enables the system to operate more efficiently and robustly. Terrain information provides additional resources to the autonomous system in dealing with a previously unknown environment or one that is uncertain with respect to sensory data. For space exploration missions, this data may be extracted from various sources with different resolutions and perspectives. Geospatial data and information may correspond to landing sites and regions explored by autonomous mobile systems. Some of the information is available prior to a mission from orbiting spacecraft while other information is acquired *in situ* by surface systems during the mission. While some data is acquired and used by onboard sensing and data processing systems, most of the data is transmitted to Earth, processed, stored, and analyzed by a Ground Data System (GDS).

A GDS can be thought of as the equivalent of a GIS for planetary missions. In general, it is a comprehensive collection of computer systems, scientific data and information and, in particular, software tools that aid decision-making and management of geospatial attributes (among other mission operations functions). More specifically, the GDS encompasses the NASA Deep Space Network (DSN) (a worldwide spacecraft tracking facility managed and operated by Jet Propulsion Laboratory (JPL)), Ground Communications Facilities (which provide data and voice communications between JPL and the three DSN tracking sites spaced about equally around the Earth for continuous tracking of deep-space vehicles), Deep Space Mission System (the system of computers, software, networks, and procedures that processes data from the DSN at JPL), and project data processing systems[15].

For Mars missions, the GDS communicates with and receives data from deployed orbiting assets such as Mars Global Surveyor, Mars Odyssey, and surface systems such as landers and rovers. Orbiters and surface assets at Mars are equipped with imagers, spectrometers, radiometers, and other scientific instruments and communications devices. This variety of equipment provides the GDS with the essential information needed to operate the mission and analyze data.

3.3 Real-World Examples

This section discusses state-of-the-art techniques for incorporating geospatial information into autonomous space systems. In the first example, we reveal how multi-scale geospatial information from multiple sources can be used for choosing the safest terrain locations to land in rugged geographical regions. The second example discusses techniques for robot localization and safe terrain traversal whereas long-distance path planning based on map information is presented in the last example.

3.3.1 Safe Spacecraft Landing

Spacecraft landing sites are considered safe if they have minimum slope, are free of hazards, and have acceptable roughness constraints (Golombek et al. 1997). It is assumed that once a site is selected, the geographical terrain conditions found at the specified landing location will permit a physical touchdown. During execution of a landing sequence, there is no re-evaluation during descent to determine whether the selected site is still suitable for landing.

Typically, engineering criteria established for ensuring success of the mission are constructed by analyzing terrain characteristics that affect the ability of the spacecraft to land safely on a planetary surface within a geo-referenced ellipse of specified uncertainty. Cameras and instruments on the NASA Mars Global Surveyor and Mars Odyssey orbiters have routinely captured recent images of Mars' surface from orbit. Specifically, they have captured images of the landing sites where spacecraft are expected to land. Each spacecraft is navigated to land within a specified landing ellipse of size commensurate with the precision capability of the spacecraft navigation system and with very high probability. For example, the landing ellipses for each MER spacecraft had a major and minor axis of roughly 100 km and 20 km, respectively, and the spacecraft were predicted to land within the ellipses with 99% probability. Within the landing ellipse, the roughness of the terrain and the size/concentration of rocks must be minimal. The angle of approach must be consistent such that there are no large hills, high cliffs, or deep craters that may impede the safety of the vehicle from landing at a desired scientific site of interest. In addition, the surface slope must be within acceptable limits to deter instability of the landed spacecraft, which occurs at certain angles.

To extract all terrain characteristics associated with satisfaction of engineering constraints, a suite of heterogeneous sensors is utilized. As such, data retrieved from multiple sensors must be fused in real-time to ensure safe spacecraft landing during mission operations. During descent, sensor data is used to analyze the approaching terrain for hazards, and sites within the landing ellipse that are deemed safe for spacecraft landing are adaptively selected. This enables any trajectory adjustments to occur in the thruster command sequence in order to minimize the risk to the spacecraft at touchdown.

3.3.1.1 Computing a geologic hazard map

The Entry-Descent-Landing (EDL) operations of a spacecraft occur over a very short period of time, typically on the order of 1-5 minutes. Therefore, the computational speed of any algorithm used for terrain analysis is of utmost importance. The framework employed for selection of safe landing sites combines terrain hazard information extracted from different sensor sources into a global scene description (Howard and Seraji 2004). To obtain hazard information, geological terrain characteristics are identified from sensor data and combined to determine the risk associated with landing on an area of the surface within the landing ellipse with the given terrain characteristics. In order to effectively combine heterogeneous data from the different sensor sources, the hazard maps are first aligned using a combination of rotation, translation, and scaling. Once transformed, individual hazard maps are combined into a fused hazard map representation of the geologic terrain characteristics.

To enable real-time implementation, the terrain hazard maps are constructed using an efficient reasoning methodology called fuzzy-logic. Fuzzy sets and logic (Zadeh 1965) provide flexible tools for modelling the relationship between input and output data of a system, and is distinguished by its robustness with respect to noise and imprecision in the data. Fuzzy sets and conditional statements allow the system to manage heuristic rule-based knowledge, imprecise information from sensors, and the uncertainties in the knowledge about the environment. Fuzzy logic is ideally suited for this application because it naturally copes with ambiguities and imprecision that exists in construction of the terrain maps due to motions and vibrations of the spacecraft. Fuzzy logic rule evaluation involves only simple arithmetic calculations that can be performed very rapidly. Therefore, the computational time required to create a geological terrain map is quite manageable for a real-time decision system, making it feasible for landing operations.

The data extracted from the heterogeneous sensor suite on-board the spacecraft is associated with the height of surface features embedded within the viewable terrain regions. Sensors provide range data that can be converted into an elevation map for extraction of terrain characteristics such as terrain slope and roughness (Figure 3.1). The derived elevation data is used to extract slope and roughness characteristics of the terrain using a least-squares plane-fitting algorithm (Johnson et al. 2001). The slope of the plane which best fits the elevation points is used as the terrain slope value and the roughness is then computed as the residual of the fit. The hazard map is represented using a grid of cells in which values are represented by fuzzy sets with linguistic labels {*SAFE, RISKY, VERY-RISKY, UNSAFE*}. Each cell is associated with a region physically located on the terrain surface. A typical grid resolution is approximately 3m x 3m for a grid map of 100 x 100 at an elevation of 1.4 km above the surface. Once the slope and roughness values are calculated, these terrain characteristics are converted into a linguistic representation using fuzzy sets and fed into a fuzzy logic rule base (Howard and Seraji 2001) to compute numerical values for the terrain map. The roughness parameter is represented by fuzzy sets with linguistic labels *{SMOOTH, ROUGH, ROCKY}* whereas the terrain slope parameter is converted into the

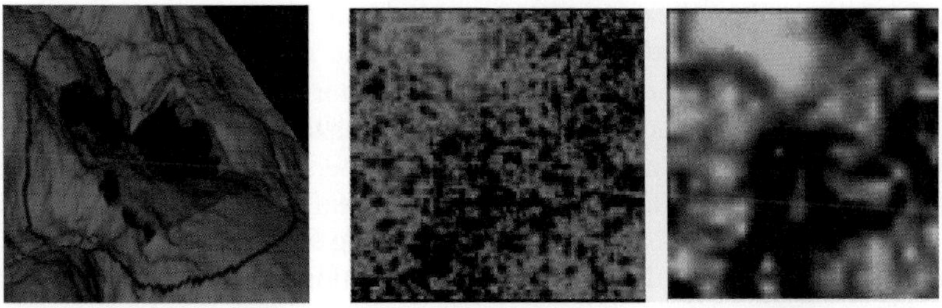

Figure 3.1. Computing the terrain hazard map, where black cells in the slope image represent a STEEP value and black cells in the roughness image represent a ROCKY value.

linguistic representations *{FLAT, SLOPED, STEEP}*. The membership functions of these fuzzy sets are input to a set of fuzzy logic rules used to classify the terrain (Table 3.1). The output from the rule base represents the relative level of safety associated with the viewable area.

Table 3.1. Fuzzy rule base for hazard map construction. Empty fields in the fuzzy rule base indicate the specified input parameter has no effect on the rule outcome.

Slope	Roughness	Hazard Map Value
FLAT	SMOOTH	SAFE
FLAT	ROUGH	RISKY
SLOPED	SMOOTH	RISKY
SLOPED	ROUGH	VRISKY
STEEP		UNSAFE
	ROCKY	UNSAFE

3.3.1.2 Selecting Safe Landing Sites

Individual hazard maps are created independently for each contributing sensor on the spacecraft (e.g., LIDAR, camera(s), radar, etc.). The hazard map information derived from multiple heterogeneous sensors is fused in such a way that each sensor is allowed to influence the final terrain representation. The final map is thus computed using the following equation, which effectively blends individual hazard maps together:

$$H_{i,j} = \frac{\sum_{n=0}^{S}\left(\beta_{i,j}^{n} \sum p_{i,j}^{n} A_{i,j}^{n}\right)}{\sum_{n=0}^{S}\left(\beta_{i,j}^{n} \sum A_{i,j}^{n}\right)} \tag{3.1}$$

where i, j is the index of each cell in an individual terrain hazard map, $H_{i,j}$ is the fused map value computed for each cell, S is the number of relevant on-board spacecraft sensors, $\beta_{i,j}$ represents the certainty factor associated with each cell, $p_{i,j}$ is the peak value associated with fuzzifying the hazard map values ($h_{i,j}$) for each sensor and $A_{i,j}$ is the area under the hazard membership function associated with the hazard value. Once calculated, the fused terrain map is used to select a safe landing site in the landing ellipse. The safe landing site is chosen as the safest site located near the originally designated landing location (Figure 3.2).

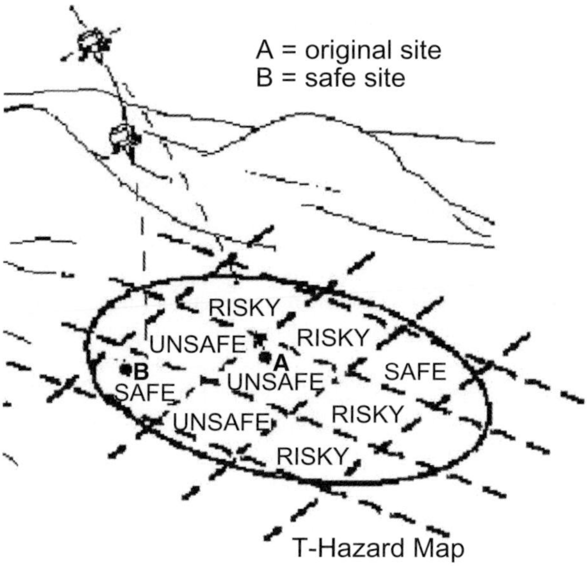

Figure 3.2. Selecting a safe landing site for spacecraft touchdown

3.3.2 Robotic Surface Navigation

Mobile surface systems are delivered to the surface by landers. However, the use of geospatial information as discussed above does not end with safe landing within a geo-referenced landing ellipse. Surface navigation and operations are referenced to a coordinate system whose origin is fixed at the final touchdown location of the lander. Mobile robotic missions on planetary surfaces involve localisation of the robot with respect to inertial reference frames, modelling of rugged terrain traversability, path planning for long distance traversal using geospatial features and associated maps, onboard perception for navigation, as well as sensing and reasoning for safe mobility. In this section, we discuss these aspects in turn, highlighting the various uses of multi-scale geospatial information for autonomous navigation and control.

Autonomous Mars rovers are equipped with sensors that facilitate safe and reliable natural terrain mobility and navigation. The navigation sensor suite used by the two autonomous rovers that conducted the NASA MER mission in 2004 is a representative example (Goldberg et al. 2002). The navigation sensors on the MER vehicles (see book cover page) include front and rear passive stereo vision for local hazard detection and avoidance, as well as stereo cameras mounted on a mast at a height of 1.3 meters above ground. The rovers also use an Inertial Measurement Unit (IMU) consisting of a 3-axis accelerometer and a 3-axis gyroscope. Its accelerometers are used for measuring the local gravity vector (i.e., which way is down?). They provide rover tilt/attitude measurements and are useful for establishing rover heading (yaw). The gyroscope measures angular rates of change of the vehicle. Cameras on the mast are also used to image the sun in order to determine absolute heading relative to true north (also given local time of day and other essential information). These sensors, along with encoder-based wheel odometry, contribute to a fused estimate of position and heading. The rovers are self-contained systems powered by a solar panel or batteries.

3.3.2.1 Robot Geo-Localisation

Because of various mobility challenges (briefly described above) there is a need from time to time to perform and use geospatial measurements/perceptions to establish and improve rover localization to facilitate accurate navigation. For Earth-based applications, common sensing infrastructure for position estimation and localization is the satellite-based GPS, which provides accurate outdoor position knowledge in terrain where sufficient reception can be achieved. Of course, there is no existing analogue available (yet) for global positioning at Mars, and this is a major limiting factor of localisation accuracy achievement for rover systems. It should be noted that potential viable solutions are on the horizon. Namely, studies and plans are underway to establish a telecommunications network infrastructure at Mars called the Mars Network (Cesarone et al. 1999). One of the primary functions of the proposed network of satellites would be to provide enhanced navigation support for surface rovers.

Following safe landing, one of the foremost activities involving the use of geospatial data is landmark and radiometric localisation of the spacecraft/rover in geospatial coordinates. This activity does not yield information that is required throughout a mission for onboard autonomous decision making per se. However, it is a necessary first step to establish the initial local geospatial reference system(s) needed by the autonomy system. Four types of data are generally useful in this regard. They include orbital elevation data and images of the overall landing-site terrain, descent images taken by the landing spacecraft, surface-level rover camera views of the horizon, and spacecraft navigation data. The latter may include in-flight data acquired as the spacecraft entered the atmosphere and by orbiting satellites at Mars during their flights over the rover landing site, as well as data associated with communications sessions between Earth and the spacecraft while stationary on the planet surface.

Use of available and newly acquired data permits continual refinement of the spacecraft location on the surface. As mentioned earlier, the landing ellipse size is commensurate with the precision landing capability of the spacecraft navigation system. As such, the initial gross estimate of the rover geolocation is known to within the accuracy represented by the landing ellipse size. At the next level, orbital images of the landing site that have been assembled into mosaics are registered to digital elevation data such that latitude and longitude coordinates of visible surface features in the images can be estimated. The accuracy of these coordinates can be ±100 meters; this is the horizontal accuracy of the Digital Image Models generated from data acquired by the Mars Orbiter Laser Altimeter (onboard the Mars Global Surveyor orbiter) (Li et al. 2002). Descent images captured beginning at an altitude less than a mile, for example, above the landing site are also compared with and contrasted against orbital images. An approximate location of the actual touchdown within orbital images can be estimated by correlating surface features (e.g., hills, craters, etc) with features visible in the descent images. This enables estimates of better precision than the landing ellipse alone; the location can be narrowed down to areas of much smaller extent.

Once the lander opens and exposes the landscape to the rover cameras, higher resolution rover images at the surface level can be used to further refine its location. The spacecraft/rover establishes a local inertial reference frame fixed to its initial location and geo-referenced using the local gravity vector and the direction to true north (after finding the sun at a known time of day). Scientists and engineers may then use relative measurements to surface features as landmarks in rover images that are on the horizon (e.g., hills or local peaks). Using image-based range, azimuth, and elevation measurements landmarks can be correlated with and registered to features in orbital and descent images. This enables the actual geographical rover location, in global Mars inertial coordinates, to be further approximated by triangulation. In this case, the most important constraint on triangulated solutions is knowledge of the rover's azimuth (as measured by its IMU and sun finding functions). Within several days after landing, radio signal Doppler tracking during communication with the rover can provide localization in the same global coordinate system to an accuracy of about 30 m (Li et al. 2003). Radiometric Doppler techniques compare radio frequencies transmitted by the rover with the frequency of later receipt at Earth. The same is also done with Mars orbiters in the communications path. Using all available options, including Doppler shift measurements and round-trip radio signal ranging, can lead to localization accuracies within 1–10 m. All the various sources of gross, coarse, and fine localization information are combined to refine the location of the spacecraft/rover on the surface. Example orbital and rover images used in a NASA rover field trial conducted in Gray Mountain, Arizona from JPL are shown below in Figures 3.3 and 3.4.

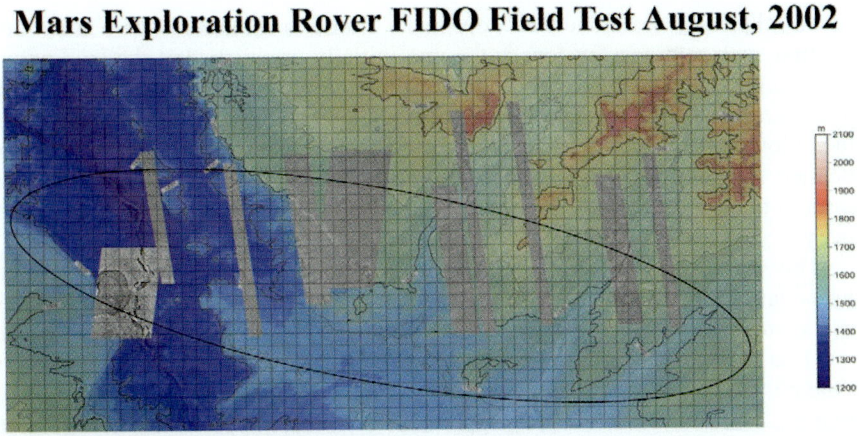

Figure 3.3. Orbital image for use in planning a rover field trial mission

Figure 3.4. Panorama mosaics of rover images useful for localization (lower image corresponds to rectangular annotation in upper image)

3.3.2.2 Localisation during traverses

During traverses from place to place on the planet surface, autonomous rovers continue to make use of geospatial information to maintain a sense of location with respect to local coordinate frames of reference. Throughout a mission, new local coordinate frames are established on the surface after considerable distances have been traversed. This facilitates dealing with the considerable errors in rover position that accumulate with distance traversed since the rover's position is reset to the origin of the new local coordinate system. In turn, this facilitates the accuracy of camera and science instrument pointing; the better the rover position is known, the better the accuracy of pointing at desired locations and science targets in the vicinity. Each new location is referred to as a site (typically numbered sequentially) and the associated coordinate frame is referred to as a site frame. The first such reference system is the one initially established at the landing site. Within any newly established site the localization problem becomes one of maintaining knowledge of rover position within the site coordinate frame. This is handled autonomously by the rover onboard systems in general. However, when unintended terrain interactions (such as wheel slip or sinkage) induce significant errors in position, re-localization by mission operators on Earth may become necessary.

The most basic position estimation done onboard the rover is odometry (dead-reckoning) based on wheel encoders. Although rover heading is typically well known when using accurate IMUs and sun finding techniques, this approach is not very accurate in rough terrain, or over considerable distances, due to the complexities of wheel-terrain interactions. Additional and more sophisticated methods must be used to supplement this basic approach to ensure that a rover can reliably navigate to desired locations. There are a number of candidate techniques that are effective in this regard. Among them are visual servoing algorithms (Das et al. 1999) and wheel traction modulation techniques based on visual perception of terrain textures (Tunstel and Howard 2003). Such methods serve to circumvent or minimize accumulation of position errors induced by undesired wheel-terrain interactions. Better alternatives are methods that not only help to cope with position errors but also help to maintain knowledge of true position. Methods that provide this functionality can be referred to as visual odometry methods. They have been developed and used on planetary rovers (Olson et al. 2000; Huntsberger et al. 2002) to overcome the inaccuracies and limitations of wheel odometry in natural terrain environments. Visual odometry works by determining changes in rover position and orientation between consecutive stereo image pairs taken while traversing and tracking multiple features detected in the images. The algorithm (Olson et al. 2000) can be summarized as follows. Given a pair of stereo image pairs taken at two different positions, automatically locate and track features between the four images. Use feature motion in the image to estimate the overall six degree-of-freedom motion of the rover. Finally, perform self-consistency checks on the detected features before computing a final rover motion estimate.

Both wheel odometry and visual odometry are done in real-time onboard the rover. When necessary, non-real-time localization is done by mission operators on Earth. They make use of triangulation, site vector addition, and least squares estimation methods on stereo range data from rover imagery acquired at spatial locations before and after a traverse. In this case, an overlap is required between images and stereo range data acquired at the separate locations. Site frame coordinates and ranges of multiple (manually selected) geospatial features are recorded from rover stereo image data at the pre-traverse location. Rover-relative coordinates and ranges are recorded from rover stereo image data at the post-traverse location for a subset of the same features that are visible from the new location. These data and necessary transformations permit localization of the rover in the site frame. This is in some sense a larger-scale and non-real-time version of the onboard visual odometry. Re-localized position estimates determined by mission operators are used to update the rover's onboard knowledge of its position prior to its next traverse.

3.3.2.3 Long-distance path planning using terrain maps

To enable long-range traverse of mobile robots on hazardous terrain, global path planning methodologies must be utilized to ensure goal achievement. The process of long-distance planning of paths for robots operating on high-risk access terrain is a difficult task. In order to allow successful completion of robotic exploratory missions, intrinsic properties of the terrain must be directly integrated into the path planning logic. Global terrain features that are risky for mobile robot traversal must be identified and paths must be planned to circumnavigate unsafe regions. By analyzing terrain characteristics before traversal, global paths can be constructed that will minimize risk to the robot and ensure robot survival, and ultimately mission success. As a by-product of the initial post-landing rover localization activity described earlier, the relative geospatial location of such global terrain features is typically known. This information can be used to plan long distance rover traverses throughout the mission.

The approach adopted is based on integrating two main concepts: traversability map building and global terrain-based path planning. The methodology uses a safe path constraint based on terrain features to traverse from two designated start and goal locations. Global terrain attributes are first analyzed and used to segment the terrain into regions of different traversability using the concept of the traversability map, a multi-valued map representation that denotes the ease-of-traversal of different regions of the terrain. Information from the traversability map is then utilized to compute a traversal cost function to ensure robot survivability. A global path planner incorporates the cost function into a search methodology to find an optimally safe path through the terrain.

The traversability map-building process involves identifying major terrain features (such as hills, lakes, valleys) as observed in images obtained during prior aerial imaging or land surveying. This terrain information is passed onto the traversability map-building algorithm, which assigns traversability indices based on terrain characteristics. A grid map based on the traversability map is then constructed in which each grid cell is assigned the *minimum* traversability value of any region encompassed by that cell. This ensures that a cell is marked untraversable if any part of the cell is untraversable.

Once grid traversability indices are computed, the derived terrain map characteristics can be directly incorporated into the global path planning strategy to enable long-range traverse of mobile robots. This is accomplished by using a search algorithm to find a path from the robot's start position to a designated goal position, while ensuring robot survivability (Figure 3.5).

(a) (b)

Figure 3.5. Long distance path-planning for a mobile robot (a) long-distance path (b) example robot traversal

3.3.2.4 Map Representation

The particular form of map representation employed by the robot navigation system is an important issue as it directly affects real-time map-based navigation. Traditional maps for autonomous vehicle navigation have attempted to capture explicit metric information about the environment at the expense of real-time motion performance. Explicit world maps require significant onboard memory and processing power for data storage and effective maintenance. Since field mobile robots typically carry limited computational resources onboard for dealing with the numerous sources of uncertainty encountered outdoors, a concise approximate map of large-scale terrain features is favourable.

To represent a regional-scale traversability map to the robot navigation system, a fuzzy spatial map representation can be employed that does not does not require precise terrain measurements (Tunstel 1995). For this representation, a 2-D grid of a regional area is partitioned using fuzzy sets along the X (West-East) and Y (South-North) axes such that any location (grid cell) designates a fuzzy coordinate. The number of fuzzy sets used for an equally-spaced partition may be selected according to the resolution of orbital imagery and/or another area scale larger than the robot sensor range. Given an image covering the area where the robot is located, mission operators may use a tessellation of the image as a basis for the regional map. Regional-scale terrain features are then registered in the map using fuzzy coordinates, and assigned a level of risk (or safety) associated with traversability at each coordinate (grid cell).

Regional traversability at a fuzzy coordinate can be classified using four fuzzy sets with the linguistic labels {POOR, LOW, MODERATE, HIGH}, for example. This third dimension of the map is multi-valued in that the highest degree of traversability at a fuzzy coordinate corresponds to the center of the grid cell and gradually tapers off with the associated level of uncertainty. Such a representation accounts for uncertainty in the traversability assigned to terrain features due to limited resolution of orbital and descent images and associated human interpretation of them. The representation also accounts for uncertainties in location and size of terrain features at the regional-scale while providing sufficient information for map-based navigation. An example is shown in Figure 3.6, where a tessellated terrain image (covering an area approximately 6 km x 6 km) is shown. The area is partitioned using fuzzy labels designating varying degrees (**N**ear, **M**id, and **F**ar) of **N**orth, **S**outh, **E**ast and **W**est. The computer-generated surface in the Figure represents the traversability distribution of the area. Regional traversability at any location is represented by the height of the computer-generated surface and by the grey-scale shading such that high index values and lightest shades correspond to highest traversability.

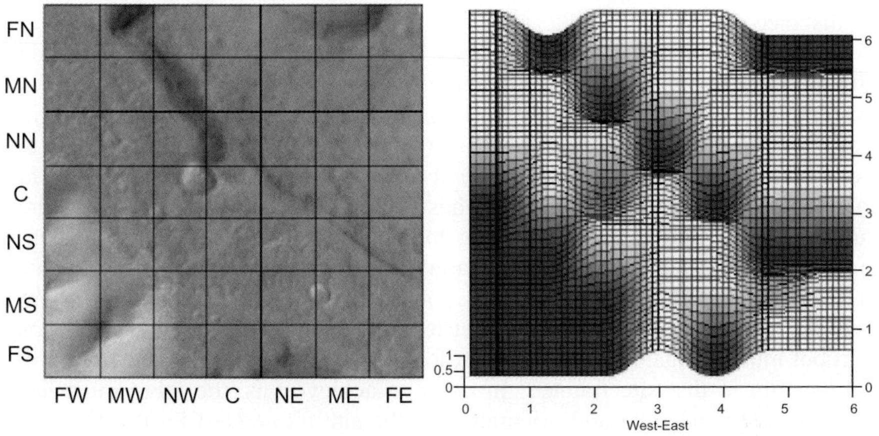

Figure 3.6. Regional-scale terrain and traversability map

Map information can be accessed by the onboard software via a simple representation using linguistic statements in the form of fuzzy if-then rules. For instance, using the linguistic terminology from the example above we can write the following fuzzy rule: IF *X* is *NW* and *Y* is *C* THEN *regional traversability* is *LOW*, where the fuzzy coordinate (*NW*, *C*) corresponds to the area containing the crater near the center of the map. A finite set of rules of this form represent the approximate distribution of terrain features throughout the global area of interest. This allows for concise representation of global traversability that is easy to retrieve and maintain. Without much difficulty the concept generalizes to the case where multiple attributes existing at a fuzzy coordinate can be specified. In addition to traversability, uncertain information that may be available about terrain elevation, slope, roughness, and surface type can be similarly represented in the fuzzy spatial map (Tunstel 1995). The spatial traversability map allows informed guidance/planning of long distance map-based navigation by consideration of traversability beyond the robot sensor horizon. It provides the navigation system with approximate spatial references to large untraversable features (including hazards of extended length), free space regions, and potentially useful landmarks, as well as their approximate spatial relationships.

3.3.2.5 Robot Local Navigation

An essential autonomous motion control objective for robotic planetary surface exploration is for the robot to navigate from its current location to deliver science instruments to a different location in the immediate vicinity for investigations, observations, measurements and analyses. This occurs at a more local scale than that for long distance path planning as discussed above. As such, the use of geospatial information is not as prevalent at this local scale. It is limited to onboard terrain modelling in the very near field to facilitate local path selection while executing a long distance path plan.

Within a site frame, a destination may be intermediate (a waypoint) or final (a goal). That is, a path to a local navigation goal may be specified as a series of waypoints that connect the current location to the goal location along a specific trajectory (in a more local sense than for long distance path planning). There are several approaches to navigating between any two locations; they generally fall into categories of feature- or landmark-based navigation (as with visual servoing, visual odometry, and localization using topological terrain features), and metric (Cartesian) navigation based on non-spatial proprioceptive measurements such as wheel odometry. For metric navigation, robot positions and the positions of waypoints and goals are specified using metric Cartesian coordinates in a site frame. The robot must navigate from its current location in a site frame to a different desired location in that site frame. In order to achieve this, the robot must have knowledge of its position and orientation in the site frame (self-localization) and general spatial cognizance (in the Cartesian sense). Therefore, in order to achieve effective local navigation and motion control, an autonomous robot must first benefit from the use of geospatial information on more regional and global scales to properly situate it in its environment.

Establishing the relative spatial offset of the next waypoint/goal is relatively straightforward computationally, as is the motion control for orienting the robot toward the waypoint/goal and traversing to it. The more difficult problem is traversing to the next waypoint/goal while avoiding mobility hazards. Safe traversal to desired waypoints/goals requires that mobility hazards be detected and spatially registered in the robot's local world map in order for the onboard system to perform the necessary spatial reasoning to determine safe motions that avoid hazards. Autonomous navigation and motion control within an area encompassed by the robot sensor range/horizon is facilitated by onboard traversability assessment using computer vision. Stereo or mono images captured by the robot vision system may be used to perceive the local terrain while image data processing extracts terrain features representing robot mobility hazards. Extracted features are represented in a digital spatial representation of the local terrain that is stored and updated in the robot's computer memory. This local world map is spatially referenced with respect to the robot's coordinate system (not the local site frame) and therefore travels with the robot as it traverses terrain; it is periodically updated as the robot acquires images of newly encountered terrain.

With a persistent representation of local surrounding terrain, the robot uses onboard look-ahead sensing and hazard avoidance algorithms to navigate safely within its sensor horizon. As mentioned above, the navigation objective is to traverse from a current location to a different desired location such as the vicinity of a visual feature, topographical landmark, or geospatial coordinate within a site. Moreover, the navigation objective must be achieved while avoiding mobility hazards. The first aspect of this problem to be considered is the spatial representation of local terrain traversability in the robot's onboard memory. The second aspect of the problem is the use of the onboard representation to reason about traversability and then to infer or compute motion controls that achieve the navigation objective. These aspects of local surface navigation in rough terrain have been addressed using a variety of approaches that are well documented in the literature (Goldberg et al. 2002; Huntsberger et al. 2002; Howard and Seraji 2001; Lacroix et al. 2002).

3.4 Conclusions and Future Work

Geospatial information and information processing technology is essential for autonomous system operation in remote terrestrial environments. On planetary bodies other than the Earth, the problem difficulty is compounded by a lack of remote and in situ sensing technology infrastructure (relative to that available for Earth-based uses) that provides full/continuous coverage to feed a GIS. This chapter provide an overview of the various challenges and problems as well as techniques currently employed for dealing with the challenges and solving the difficult problems. The state of the art is advancing at a steady pace as NASA and other international space agencies increase coverage of planetary bodies in terms of remote sensing and in situ missions. The proposed MarsNet project is an example of what is to come in the not too distance future in the way of advances that will enable technological infrastructure akin to what we are privileged to have on Earth.

The challenge of autonomous operation in unknown planetary environments is partially addressed through the inclusion of geospatial measurements in a system's control logic. Geospatial information is used for maintenance of accurate position estimates, robust guidance, navigation, and control. The geospatial information retrieved though is still not equivalent to that of Earth-based Geographic Information Systems. As such, elements of future work involve improving the geospatial information made available to autonomous systems.

One concept for accomplishing this goal includes the MarsNet project[16] (Cesarone et al. 1999). In terms of geospatial applications, MarsNet represents a concept for developing an in-situ navigation capability to enable more precise targeting and location information for Mars through the establishment of a microsatellite constellation called Microsats. These Microsats, along with relay satellites, MARSats, would enable autonomous systems such as landers, rovers, balloons, and planes, to communicate with each other to improve their relative navigation capabilities. In addition, the MARSats, would have similar functional capabilities as Earth-based geostationary communications satellites.

This is one proposal, among others, that represents efforts to advance the state-of-the-art technology required for improving the geospatial information coverage of remote planets, such as Mars.

Acknowledgement

The research described in this paper was performed at the Jet Propulsion Laboratory, California Institute of Technology, under contract with the National Aeronautics and Space Administration.

Chapter 4

Agent-Based Technologies and GIS: Simulating Crowding, Panic, and Disaster Management

Michael Batty

4.1 Emergencies and Disasters in Small Spaces

We are only just beginning to get to grips with developing spatial analysis and GIS at the small scale. Most analysis to date has been at much larger scales, at the scales of cities and regions, in large scale environments, and often at the global or continental scale. At the very fine scale of complexes of buildings in cities, for example, this has been the domain of property analysis and urban design and only recently has good digital data become available for analysis at these scales. However it is not only the existence of data that is forcing us to look at these scales but the fact that important contemporary events occur at the small scale where crowds gather. It is in this domain that we are beginning to develop scientific analysis of related problems and developing structured methods of design and problem solving for ensuring that such events reach minimum standards of design and performance.

Increasingly we are concerned with the design of small spaces where people come together so that safety is ensured and the quality of interaction in such spaces is optimized. Moreover many problems of movement and migration involve action and interactions in small spaces and the kind of crowding and congestion that can occur at these focal points makes such spaces and the people within them extremely vulnerable to natural and man-made disasters. The kinds of spaces that we are interested in are office complexes and shopping malls but also entertainment centers such as theatres, rock concert venues, and football stadiums which are places where standards need to be enforced for safety reasons. Places where crowds gather more spontaneously like street festivals and concerts-in-the-park are also subject to the same problems and it is in this context that large scale disasters can be generated from the vulnerable situations which are always associated with such gatherings.

In this chapter, we will begin by noting the sorts of spaces in which such emergencies and disasters can occur and then discuss one of the most significant of such events of recent years – the Hajj or Pilgrimage to Mecca which has been subject to some severe emergencies and deaths from fire and panic in the last decade. We will then stand back a little and illustrate how we can begin to design models of movement which show us how such events function. We will take an agent-based view and show how geometry, randomness and behaviour come together in simple models of random walks which present rather good bench-line models to simulate the movement in such spaces.

We then add intentions to this behaviour and illustrate how these can be used to generate models of crowding. These can be used to show how panic and fear develop and how safety can be reinforced through various issues associated with evacuating and stabilizing the spaces that are involved. We then turn to our major example, The Notting Hill Carnival which is a street festival held each year in west central London and which has all the problems of crowding and safety posed here. We will show how the model is built around swarms and then we illustrate how it can be used to illustrate the effects of new policies which change the physical structure of the environment in which these events take place. This shows us how we can begin to use such models in policy analysis and it illustrates the way spatial technologies are being stretched to deal with new problems.

4.2 Typical Examples of Large Scale Disasters in Small Spaces

The sorts of spaces which we are concerned with here tend to be smaller than the city block and often at the level of individual buildings and parts of buildings. Usually such spaces are complex in that they involve many corridors and links which relate to movement although the kinds of disasters and emergencies that can take place in such spaces are best visualized in terms of the simplest, often in terms of congestion and crowding which occurs when too many people in larger space compress themselves into a smaller space such as into an elevator or a corridor where the prime purpose of the movement is egress. These are the kinds of movement that one can see in complex structures such as high office blocks, rail stations and airports where there may be several levels and where the movement patterns tend to criss-cross one another due to the complexity of the layouts involved. The same may also occur in hospitals in accident and emergency rooms, and although one might consider stadiums and theatres associated with entertainment to be somewhat simpler in structure than malls or office centres, the key problems in such venues involve the fact that large numbers of people must locate together as closely as possible in order to partake of the entertainment. If the crowding gets too great and people pack into levels in which fear and panic are generated, then disasters can occur as barriers to constrain the crowds are broken and as the crowd seeks to stabilize itself in ways that lead to some being crushed and trampled.

In Figure 4.1, we show some typical examples of spaces where crowds gather – the new Kyoto Station Complex (a), the Jamarat Bridge which is a key stage in the annual Pilgrimage or Hajj (b), a rock concert (c), and a football stadium (d).

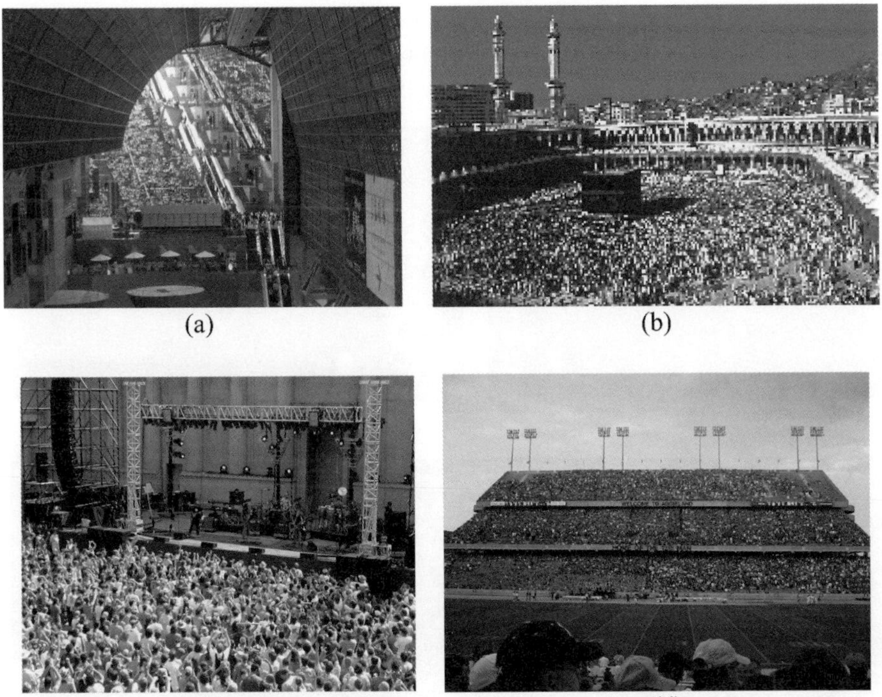

(a) (b) (c) (d)

Figure 4.1. Typical Small Spaces Containing Large Crowds
(a) Kyoto Station Complex (b) The Central Mosque Mecca
(c) Rock Concert in Berkeley, CA, and (d) Texas A&M Football Stadium

Helbing and his colleagues have listed a sample of key events over the last 100 years where major disasters have occurred in that crowds have become too great for the spaces available. These show how important it is to consider ways in which the individuals that form such crowds need to be understood in terms of the way they interact with each other and the geometry of the local situation[17]. We show some of these in Figure 4.2. Such events are dramatic enough and need little comment: they come from many sources – packing too many people into a space is the obvious problem and this relates to a breach of safety standards but natural disasters such as 'Acts of God', terrorist attacks such as 9/11, and human error in driving and navigation are all key issues. The important question is that whatever the cause, the consequences can be quite similar and these usually involve being unable to evacuate a large population fast enough with sufficient space to avoid the disaster. In this chapter, our focus will be much more on disasters that occur through overcrowding rather than terrorist threats or natural disasters and to this end, we will very briefly outline one of the most routine but one of the most problematic of all large crowds – the Pilgrimage to Mecca – which over the last decade has experienced significant and traumatic crowding disasters despite extremely detailed safety measures being put in place.

Figure 4.2. Disasters in Small Spaces
(a) 9/11 (b) Hillsboro' Sheffield 1988 (c) Jamarat Bridge

Let us now look in a bit more detail at the Hajj. This is a pilgrimage that involves upwards of some 2 million Muslims who make their way to Arabia from all over the world to celebrate their faith during a 5 day period which culminates in the city of Mecca. This is an annual event which occurs during the last month of the Islamic year, called "Dhul-Hijjah". The pilgrimage begins at the holy shrines in Mecca itself, The pilgrims then journey some 5 miles to Mina where they camp and then walk another 10 miles to Mount Arafat where they pray. They return to Mina and at the Jamarat Bridge, 'stone the pillar' with stones picked up at Arafat. This, in a way, is the focus of the entire event and it is also the most dangerous.

The pilgrimage finishes at Mecca and sometimes pilgrims then visit the Holy Mosque in Medina. In the last 10 years, there have been a series of disasters which all involve crowding of people in small spaces combined with extreme hysteria involving the various events that comprise the rituals, especially the stoning of the pillar that is said to represent the devil at the Jamarat Bridge. In 2004, 244 people where killed when for 30 minutes the crowd stampeded at the Jamarat Bridge. In fact in each previous year, people are killed in the same locations, 50 in 2002, 35 in 2001, and 107 in 1998. In fact in 1990, 1425 were killed in the pedestrian tunnel in this location. This is all independent of the other disasters that plague the Hajj. In 1997, it was reported that 2000 died in fire that swept through the tent city in Mina. Clearly safety precautions are all important in such situations and it is to this end, that the models we will review in this chapter are devised.

4.3 Walking and Crowding: Behaviour, Randomness and Geometry

We will begin by looking at some simple models of how people move in crowds. These models are based on identifying each member of the crowd as an object – an agent – and simulating its motion with respect to all other agents and the environment in which it moves. There are two critical issues that need to be noted at the onset. This style of model is called 'agent-based' in that it depends on simulating the behaviours of an object or agents individually from the bottom up. Second this style of modelling is based on averages. In short we model how a large number of agents respond although the particular behaviours of each agent is a combination of responses which are based on how average behaviour or routine behaviour is modified by the presence of other agents and by the particular environment in which the agent finds itself. It is this that generates particular situations which can generate quite diverse behaviours given that there are many possibilities of interaction when a large number of agents interact.

The concept of the agent is most useful when it is mobile, in terms of dynamics and processes. Behaviour is not simply a product of intentions. It is as much a product of uncertainty, hence randomness and physical constraints, of geometry. The concept of an agent is also quite diverse in that there are at least four types that can be identified in terms of simulation. There are objects or agents that exist in the virtual world - software objects that move on networks – which are referred to as bots. Objects in the physical world are things like particles and there is much agent-based modelling in this domain, some of which we will draw on for the physics of crowds. There are objects in the natural world which classically are plants and other animals which can be thought of as having behaviour and motion while in our own world, objects are people, perhaps institutions and agencies and it is these that we will deal with here. Agents are thus mainly people, literally individuals, but sometimes other objects, such as physical features like streets and barriers and plots of land, can be treated as agents which is often a matter of convenience in terms of the software used. Finally agents as people can have different kinds of behaviour from the routine to the strategic and it is our contention that agent-based models are much better at simulating the routine rather than the strategic.

Our first models of how people move in geometric spaces begin with randomness. This is at the basis of much movement in most physical systems, and by then adding geometry, we can produce models of walking which in some respects represent our baselines or null hypotheses. Our first model is a one-dimensional random walk which is simply a random deviation from a straight-line which represents the forward direction. This direction is the general intentionality of the walk which is positioned in one or two-dimensional space. If it were in time, then this would be a random series such as those which mirror economic price indices. We show this model in Figure 4.3a but as such, this lacks any memory as the implication is that the agent returns to the straight line direction each time, with the next deviation being with respect to this baseline. If there is a one step memory with the deviation taken from the position just reached, then this gives a series which is much more like the kind of randomness that one sees in stock markets. This is shown in Figure 4.3b. If we then put this into a two dimensional situation – a plane, then we see that the walk with memory criss-crosses the space, In Figure 4.3c, we have kept the walk within the two dimensional square but in the forward direction, and in Figure 4.3d, as we reduce the size of the step, this gives a much more realistic motion, akin if you like to an ant wandering randomly across a landscape.

In Figure 4.3d, it is easy to see how this two-dimensional random walk 'fill' space and can be regarded as a fractal. In Figure 4.3c, the one dimensional line also fills the space and this is a classic example of a shape with a fractal dimension of 1 and a Euclidean dimension of 1. As a digression, it is this kind of model that is at the basis of many physical processes which when constrained in terms of their geometry, and simple rules about connections and adjacency, generate fractals of quite elaborate organization such the dendritic structures that one sees in cities as transport networks and patterns of urban density.

We must now make such walks more purposive by embedding some utility into their structure. We will assume that the walkers are moving to some specific destination and we will encode this into the spatial environment on which the walkers are moving. Let us introduce a source of walkers and move then towards the destination with the walkers climbing a regular gradient surface to the destination. We will add various degrees of randomness to these walks and then constrain the geometry. What we will do is add a light source into a corridor and agents will sense this source using it like a directional beacon and heading towards

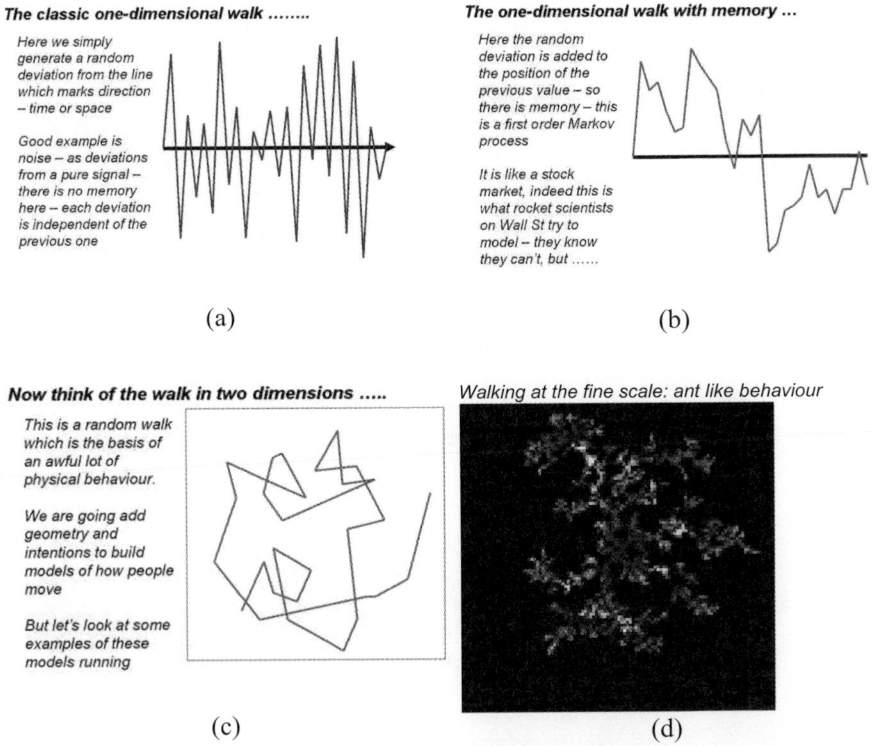

Figure 4.3. Random Walking

it. In essence, this is the easiest way of adding intention and purpose to such walks in that we encode the purpose within the environment and let agents react to it. This light source might be like the Jamarat Bridge for example and we could produce a reasonably effective simulation of the structure of this movement to 'stone the pillar' by simply letting walkers move randomly to this source. In fact we know that we need to add flocking and panic reactions to this movement but we will postpone this until our later discussion of the Notting Hill model. In Figure 4.4, we show how walkers are attracted to a light source moving randomly towards it. We then also introduce various obstacles into the picture and show how walkers circumvent these obstacles, milling round the narrowing corridor until they find a way through.

This kind of experiment is the essence of simulating the effects of geometry on crowds. It is the kind of work which Helbing[18] and his colleagues have developed and which is at the basis of the dynamic models developed by Keith Stills[19]. In Figure 4.5, we show the problem that occurs when two crowds come into contact. This model can be used to simulate the build up of pressure through random motion which then generates a break-through of the watchers into the parade, an event that often leads to disasters of the kind experienced in festivals, rock concerts and football matches as well as ritual situations like the Hajj.

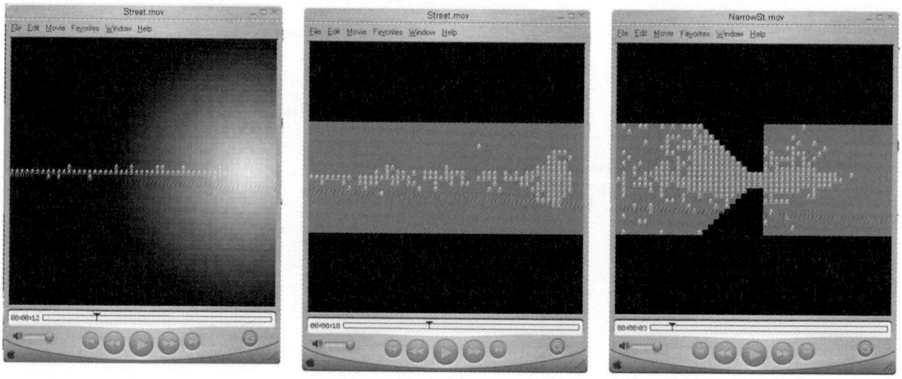

Figure 4.4. Encoding the Purpose into the Environment
A Light Source (a) Movement to the Light Source (b), Introducing Obstacles (c)

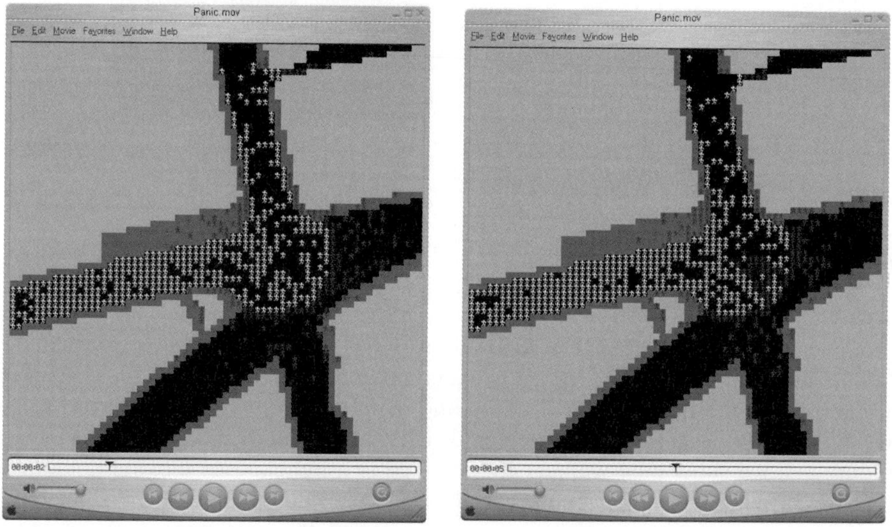

Figure 4.5. How the Watchers of the Parade (a) Build Up Pressure and Break Through into the Parade (b)

4.4 Models of Crowding: Typical Examples

Before we launch into our own model of a street festival and carnival where public safety is a major problem, let us review very briefly a number of different examples. All these models are agent-based or particle-based in that individuals are simulated with respect to the way they move and interact with other individuals and their environment. In essence, these models have five key features, namely

- intention or purpose of walking, either encoded into the environment or encoded into a protocol or behavioural profile which the walkers attempt to implement. This is the difference between passive and reactive agents although the line between these extremes is blurred.

- geometry to which walkers react in term of obstacle avoidance: walkers avoid obstacles by seeing ahead and making changes in their direction or by literally bumping into an obstacle and taking counter measures to step around it.

- randomness which is a feature of any direction of walking based on the notion that walkers cannot actually walk in a straight line but also based on the idea that there are local features of interest in the environment which any walker reacts to and that this needs to be encoded randomly. Randomness also enables new directions to be explored and chosen.

- diffusion for dispersing congestion when crowds build up and people become uncomfortable

- flocking for copying what others do with respect to where the crowd is going. This is useful in situations where crowds have a single goal such as getting to a venue or in situations where most of the crowd does not know which way to go. Flocking does not necessarily lead to the right direction being chosen but it is an essential feature of crowd dynamics – the 'madness of crowds' as Isaac Newton once called it.

We will show four examples. The first is the simplest and involves simulating the movement in some 45 rooms in a well-known British art gallery – the original Tate – in which visitors enter at one point only and then spread out to look at paintings in the various rooms. The attraction of the rooms is coded into the pixels which represent the environment in which the walkers move and the object of the project was to assess the impact of how rooms might be closed or opened and how different types of exhibits could attract different volumes of visitors. In this model, the key issue that was explored was the notion that what we are interested is the steady state distribution of walkers which is the average number of visitors in each room over a long period of time. Each walker is in fact an independent agent by the time it begins to walk around the gallery because the accumulation of random decisions and the great variation in room geometries leads to 'unique' paths taken. Walkers do not follow the same paths but overall, the model is tuned so that the observed average behaviours in terms of visitation patterns are replicated. We show a typical simulation and paths in Figure 4.6a,b.

Our second example deals with shopping where we simulate movements from points of ingress – car parks, bus stations and on-street parking – in an English town centre (Wolverhampton ~ 250K population) which is largely a shopping and commercial centre about 1 square kilometre in area. The movement is purposive in that a retail surface is encoded into the environment and this directs walkers as though they are hill climbing to the points where the 'retail pitch' is greatest. In fact this is a particularly simple way of representing retail attraction and a more elaborate model of the same town centre has also been built using the Swarm programming language which enables us to encode retail attraction in a much more specific way in terms of what shops are offering and also lets us incorporate specific buying profiles into the behaviour of the walkers/shoppers. In this way, we are able to simulate a matching between what shoppers want – their demand – and what is supplied. The model that we show in Figure 4.6c is simply a trace of movement from the pedestrian gateways – car parks and so on but this does give some sense that in this kind of model, the issue is to match spatial demand and supply in a dynamic environment which simulates how people explore a space and choose to satisfy their wants. The use of this kind of application is largely to assess how retailers can maximize profits and shoppers gain their wants in the most acceptable way (which may be may the lowest cost). It follows the tradition of retail modelling in spatial economics which began with Hotelling in 1929.

Our last two examples deal with more local problems. Disasters associated with crowds often come from 'Acts of God' as in the burning of the tent city in Mina in 1997. Galea's Fire Safety Group at the University of Greenwich (UK) has developed a variety of models which involve not only how fire burns through building and other kinds of restricted spaces such as ships and aircraft but also how pedestrians can be best evacuated from such situations[20]. In these kinds of problem, the emphasis is on how the physical conditions posed by the building and the hazard interact with natural movement patterns inevitably leading to panic. In Figure 4.6d, we show some of Galea s' examples of a simulation of fire that swept through a night club in Gothenburg, Sweden in 1998 which lead to 63 deaths.

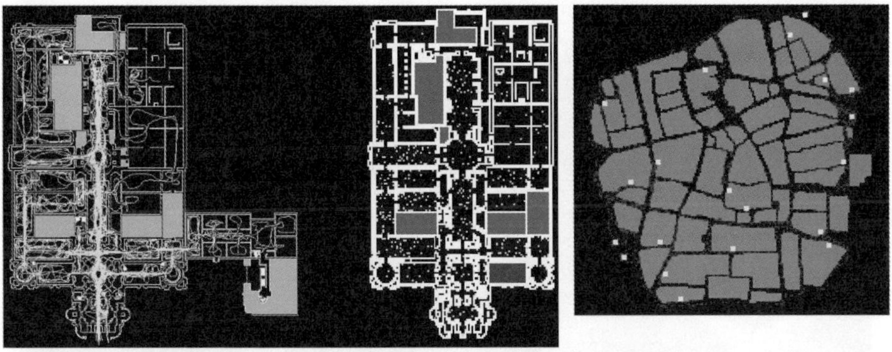

(a) paths in the Tate Britain gallery (b) distribution of agents in the various rooms (c) distribution of shoppers in Wolverhampton centre

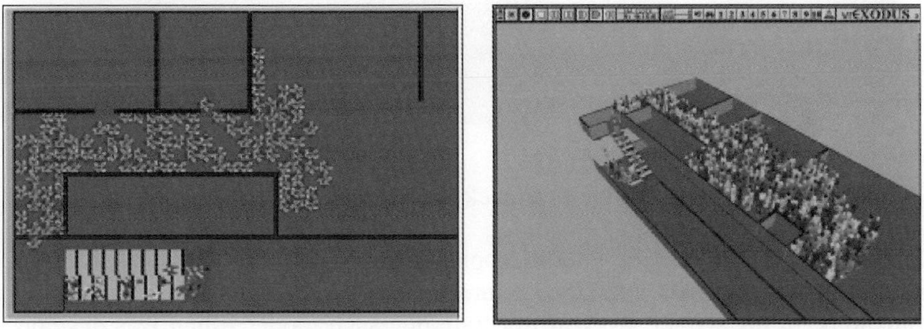

(d) evacuation of disco-dancers from the Gothenburg night club fire in 1998[21]

Figure 4.6. Examples of Agent-based Models in Galleries and Shopping Centres

Our last example simply shows how crowds are formed through positive feedback. One of the main algorithms that we will use in the Notting Hill example we will detail below is one which begins with random walking but in an exploratory fashion with some goal in mind. Once that goal has been reached, walkers who have not yet reached it see signs of how the goal has been successfully reached and act accordingly. A particularly good example of this occurs when a person wishes to cross a field from one corner to the other, cannot see the goal but strikes out in the approximate direction. Those who come next will beat a similar path though the grass and when people are coming from many different directions they will tend to react positively to paths that have already been impressed. In this way, a kind of historical path dependence (forgive the pun) takes place as paths become impressed on the ground. In general, such models lead to good routes which are efficient and parsimonious. These kinds of models have been developed by Helbing and his group and a good illustration of the reality that can be simulated is pictured in Figure 4.7 which shows how paths are established when people cross a grassy square.

Figure 4.7. How Positive Feedback Reinforces Pedestrian Tracking[22]

These examples illustrate a cornucopia of possible examples which involve models of walking with similar structures but with applications to many different situations. In the sequel, we will illustrate one such model which was developed for problems of crowd safety in a context where there were many signals that the event in question was 'a disaster waiting to happen'.

4.5 The Structure of the Pedestrian Model

We are not able to undertake a review here of all the approaches but even within the agent-based domain, very different simulations pedestrian movement have been adopted at different geographical scales, and it is worth noting these. In confined spaces of no more than tens of square meters, models based on analogies with social force and fluid flow have been used[19] (Helbing 1991; Henderson 1971), mainly to predict panic situations in highly focused events such as football matches and rock concerts (Helbing et al. 2000). For events associated with long narrow spaces where the order of the flow is important, queuing theories have been adopted (Lovas 1994). For buildings and urban spaces such as shopping malls, event simulation based on task scheduling, often using cellular automata, has been applied (Baer 1974; Dijkstra et al. 2002; Burstedde et al. 2001). For larger areas measured in square kilometres, accessibility models which simulate decisions between competing attractions have been developed (Borgers and Timmermans 1986). Recently methods which embody properties of self-organization characterizing how crowds form and disperse have become significant (Vicsek et al. 1995; Helbing et al. 1997). In all these examples, there is an emphasis on the density of flows to measure crowding and vulnerability to accidents.

The model we propose for an event over a wide area of something like 3 square kilometres is based on a mixture of these ideas. We first infer the relative accessibility of different attractions which make up the entire event and we then simulate how visitors walk to the event from locations at which they enter. In events such as these, we are never in a position to observe the flow of pedestrians in an unobstructed manner because the events are always highly controlled. Moreover although we have good data on densities, we rarely have data on the actual paths taken and it is unlikely that we will ever do so, despite advances in laser scanning and closed circuit TV. We have thus designed our model in three stages. First we build accessibility surfaces from information inferred about how walkers reach their entry points (origins) relative to their ultimate destinations at the event. Second, we use these surfaces to direct how walkers reach the event from their entry points and then assess the crowding that occurs. Finally we introduce controls to reduce crowding, changing the street geometry and volume of walkers entering the event, operating this process iteratively until an acceptable solution is reached. These three stages loosely correspond to exploration, simulation, and optimization.

We will present a very brief summary of the model at the level of interactions between agents in cells which represent the space on which the event takes place. This space is represented by square cells $i, j = 1, 2, ..., N$. Walkers $k = 1, 2, ..., K$ are defined by the binary variable w_{it}^k at a time $t = 1, 2, ..., T$. In the exploratory stage, we begin with walkers located at their destinations w_{D1}^k where $w_{D1} = \sum_k w_{D1}^k$. Walkers move from i to j in each time period $[t \rightarrow t+1]$ where $j \in \Omega_i$ and Ω_i is the eight cell neighbourhood around i. In general, some of these cells will not be accessible because of obstacles such as buildings and barriers; we thus define a block mask as $b_j = 1$ if cell j is accessible and $b_j = 0$ otherwise. Movement from i to j in search of an origin O is then determined by the probability

$$p_{ijt+1} = \frac{\tau_{jt}\, b_j}{\sum_{j \in \Omega_i} \tau_{jt}\, b_j} \tag{4.1}$$

where τ_{jt} is the route accessibility to origins. A move from i to j is determined randomly according to the schedule of probabilities in Eq. 4.1. If $w^k_{it} = 1$ and $w^k_{jt+1} = 1$, the accessibility surface to destinations is updated as

$$\eta_{jt+1} = \eta_{jt} + \sum_k \left(d^k_{Djt+1}\right)^{-\beta} \tag{4.2}$$

β is a tuneable parameter. The density of walkers at j is then computed as $w_{jt+1} = \sum_k w^k_{jt+1}$.

The process implied by Eqs. 4.1 and 4.2 continues until a walker discovers an origin O. For w^k_{jt+1}, if $j \in \Omega_O$, the walker switches from exploratory to discovery mode $\overline{w}^k_{jt+1}$ and returns to the destination D with knowledge of the discovery. The probability of returning is

$$q^k_{ijt+1} = \frac{\pi^k_{jt}\, b_j}{\sum_{j \in \Omega_i} \pi^k_{jt}\, b_j} \tag{4.3}$$

where π^k_{jt} is the difference between headings in the direction from i to j and from i to the position defined by w^k_{D1}. This move is also chosen randomly and when $\overline{w}^k_{it}$ moves to $\overline{w}^k_{jt+1}$, the walker marks the move by updating τ_{jt} as

$$\tau_{jt+1} = \tau_{jt} + \sum_k \overline{w}^k_{jt+1} \tag{4.4}$$

This process is akin to the walker laying a pheromone trail when a discovery has been made. When the walker comes within the neighbourhood of the destination $j \in \Omega_D$, the walker switches back to exploration mode and the search begins over again.

It takes some time before agents discover an origin. Before this, the search is a random walk with the route accessibility surface set as a uniform distribution. If a walker crosses the edge of the event space, it is absorbed, regenerates at its destination, and begins its search again. In its early stages, this is a random walk with absorbing barriers with the variance of its lengths proportional to $t^{0.4}$. As the process continues, more and more origins are discovered while during exploration, walkers 'learn' to direct their search at routes to origins already discovered. Those origins closest to destinations are discovered first and a hierarchy of 'shortest routes' is built up, continually reinforced by this positive feedback. This is a variant of a generic algorithm predicting trail formation and collective foraging behaviour amongst animal populations such as ants (Helbing et al. 1997; Camazine et al. 2001). The swarms created are extremely efficient in predicting shortest routes in geometrically constrained systems (Bonabeau et al. 1999). Here we do not let the pheromone trail τ_{jt} decay, while the accessibility surface η_{jt} gives the relative attraction of destinations to different street locations. Figure 4.8 illustrates this for the accessibility to various destinations and shortest routes to the tube (subway) stations in Notting Hill. To impress its efficiency, we first show the simulation without obstacles to movement $b_j = 1, \forall_j$ (Figures 4.8c and 4.8d and then with the real street pattern imposed (Figures 4.8e and 4.8f).

The exploratory stage finishes when the difference in path densities $\sum_j \left| \tau_{jt+1} - \tau_{jt} \right|$ falls below a predetermined threshold. In the second stage, we launch walkers from their entry points, and these walkers move towards the event using the surfaces τ_{jT} and η_{jT} as indicators of accessibility. We normalize these as τ_j and η_j and combine them as $\tau_j^{\alpha} \eta_j^{1-\alpha}$. The basic probability of movement is

$$q_{ij} = \frac{\tau_j^{\alpha} \eta_j^{1-\alpha} b_j}{\sum\limits_{j \in \Omega_i} \tau_j^{\alpha} \eta_j^{1-\alpha} b_j} \tag{4.5}$$

where α is a tuneable parameter. Using Eq. 4.5 for selecting directions of movement, new headings are computed as $\overline{\theta}_{it+1}^{k}$ and then used to update the existing heading from $\hat{\theta}_{it+1}^{k} = \lambda \overline{\theta}_{it+1}^{k} + (1-\lambda)\theta_{it}^{k}$ where λ reflects a lag in response.

There are two effects that complicate this movement. The first is herding or flocking (Vicsek et al. 1995; Reynolds 1987). This directs movement as an average $\theta_{it+1}^{k} = \sum_{k \in j} \sum_{j \in \Omega_i} \hat{\theta}_{jt+1}^{k} w_{jt}^{k} \Big/ \sum_{k \in j} \sum_{j \in \Omega_i} w_{jt}^{k}$, of all movement where in the immediate neighbourhood. However a move by walker w_{it}^{k} to w_{jt+1}^{k} only takes place if the density of walkers in cell j is less than some threshold Ψ based on the accepted standard of 2 persons/m^2 (Still 2001; Fruin 1971). If this is exceeded, the walker evaluates the next best direction and if no movement is possible, remains stationary until the algorithm frees up space on subsequent iterations. These rules are ordered to ensure reasonable walking behaviour. This second stage is terminated when the change in the density of walkers in each cell

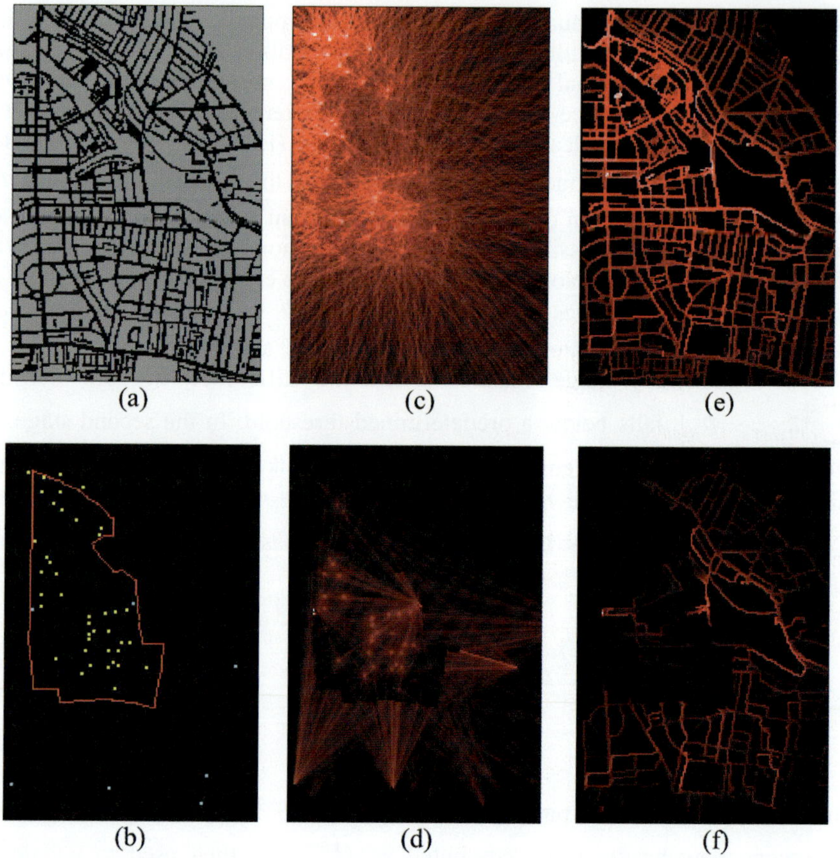

Figure 4.8. Exploration of the Street System and Discovery of Entry Points (Tube Stations). The street geometry in *a*, the parade route (red), sound systems (yellow) and tube stations (blue) in *b*, accessibility $\{\eta_{jt}\}$ from the parade and sound systems without streets in *c*, shortest routes $\{\tau_{jt}\}$ to tubes without streets in *d*, accessibility $\{\eta_{jt}\}$ with streets in e, and shortest routes $\{\tau_{jt}\}$ with streets in *f*. Relative intensities (of accessibility in *c-f*) are shown on a red scale (light=high; dark=low) with $\beta = 0.65$ and $\gamma = 0.01$. Horizontal width of each map is 1.7 kms.

$\sum_k \left| w_{it+1}^k - w_{it}^k \right| / \sum_k w_{it}^k$ converges to within some threshold where it is assumed a steady state has emerged. We can now assess how good the model is at predicting the observed distribution of crowds. We compare the predicted density w_{it} and average neighbourhood density $\widetilde{w}_{it} = \sum_{k\in j} \sum_{j\in\Omega_i} w_{jt}^k / \sum_{j\in\Omega_i} b_j$ in cells where observed densities are available. We then relate these to the number of occupied cells $\sum_i n_i$ (where $n_i = 1$ if $w_{it}^k > 0$, otherwise $n_i = 0$) and the number of available cells $\sum_i b_i = N$ defining system averages as $\rho(t) = \sum_i w_{it} / \sum_i n_i$, $\sigma(t) = \sum_i \widetilde{w}_{it} / \sum_i n_i$ and $\vartheta(t) = \sum_i n_i / \sum_i b_i$. For different threshold values Ψ, if $w_{it} > \Psi$, then $c_{it}(\Psi) = w_{it}$ otherwise $c_{it}(\Psi) = 0$, and the proportion of the population at risk is thus $Z_t(\Psi) = \sum_i c_{it}(\Psi) / M$. Average distance travelled in each time period $[t \to t+1]$ is $U_{t+1} = \sum_{ijk} d_{ijt+1}^k / M$ with the percent actually moving $V_{t+1} = \sum_{ijk} w_{it}^k w_{jt+1}^k / M$.

The third stage of the model is more informal. It consists of examining statistics from the second stage, and gradually making changes to reduce the population at risk by introducing barriers, capacitating entry points, and closing streets. This is achieved by changing the mask b_j to b'_j. As the repercussions of this are not immediately obvious, we make these changes one by one forming $b'_j, b''_j, b'''_j, \ldots$, and re-running the model until an acceptable solution emerges. In estimation, this stage can be also used to assess the efficacy of existing controls. It is not possible to develop a formal optimization procedure as so many additional factors such as resources for policing etc. cannot be embodied in the model. Nevertheless we consider this interactive method of introducing control the best approach so far for assessing alternative routes.

4.6 The Notting Hill Carnival: The Problem of Public Safety

The Notting Hill Carnival has grown from a small West Indian street celebration first held in 1964 to a two-day international event attracting 710,000 visitors in 2001. It consists of a continuous parade along a circular route of nearly 5 kms in which 90 floats and 60 support vehicles move from noon until dusk each day. Within the 3 km^2 parade area, there are 40 static sound systems, and 250 street stalls selling food. The peak crowds occur on the second day between 4 and 5 pm when last year there were some 260,000 visitors in the area. There were 500 accidents, 100 requiring hospital treatment with 30 percent related to wounding, and 430 crimes committed over the two days with 130 arrests. Some 3500 police and stewards were required each day to manage the event. The safety problems posed by the event are considerable[23]. There are many routing conflicts due to crossing movements between the parade and sound systems while access to the carnival area from public transport is uneven with four roads into the area taking over 50% of the traffic. A vehicle exclusion zone rings the area, and thus all visitors walk to the carnival. Crowd densities are high at about 0.25 persons per m^2 of which 0.47 ppm^2 line the carnival route and 0.83 ppm^2 lie inside. We have good data for crowd densities from our own cordon survey, surveys by London Underground of entry/exit volumes at subway stations, and 1022 images of the parade taken by police helicopters in the early afternoon of the second day (Intelligent Space Partnership 2002).

In building the model, we start by finding the shortest routes from the parade and static sound systems to the 38 entry points located on the edge of the traffic exclusion zone (Figures 4.9a and 4.9b). The swarm algorithm predicts the numbers of walkers who 'find' each entry point and we compare this uncontrolled prediction to the cordon survey, thus explaining 64% of the variance. We then use the observed volumes of visitors at entry points to launch these as agents who climb the accessibility surface indicated by Eq. 4.5. We show the crowd density distribution in its steady state in Figure 4.9c, and this identifies significant points of crowding. We predict 72% of the variance of observed density for 120 locations where good observed data is available. At the third stage, we rerun the model with the official street closures and barriers imposed (Figure 4.10a). This stage increases the variance explained to 78%, but not all the points of extreme crowding have been removed. This suggests that even in estimating the model, it can be used in a diagnostic manner to identify vulnerable locations (Figures 4.10b and 4.10c).

More focused analysis requires an examination of average density values for cells and neighbourhoods, pedestrian space occupied, and average distance travelled which we graph in Figure 4.11, where we also show vulnerability to overcrowding $Z(\Psi)$ at two threshold values. These graphs reveal that at each stage, walkers who begin at dense locations, spread out with congestion decreasing rapidly. This then begins to increase, and in the second and third stages, reaches a threshold and settles down. The statistics show considerable temporal volatility around stable trends which illustrate how congestion is continually dispersed. The population at risk stabilizes but most important is the effect of street closures introduced in the third stage (Figure 4.10a) which reduce congestion considerably. The model reveals some potential for more effective management of the existing carnival.

We can now test alternative routes. At the time of writing, after considerable political debate, an interim change in route has been decided where the northern most section of the parade (shown in Figure 4.9a in green) is to be removed. Running the model leads to slightly reduced average crowding but other problems associated with starting and finishing the parade not included in this model, emerge. The process of changing the carnival route is still under review with better data to be collected at this year's event. Many of the problems of using this model interactively with those who manage the event are being improved as we gain more experience of this approach. The model is general

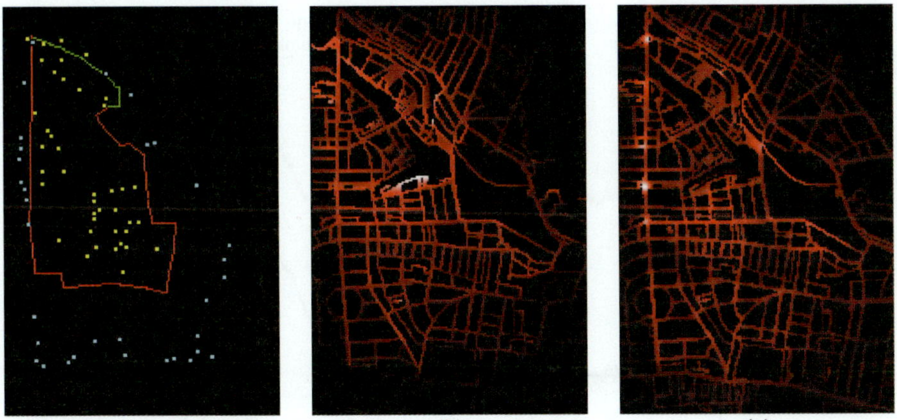

(a) (b) (c)

Figure 4.9. Swarm and Crowd Densities in Stages 1 and 2. The 2001 parade route (red and green) with proposed route in red, sound systems (yellow), and entry points (blue) are shown in *a*. The composite accessibility surface $\{\tau_j^{\beta}\eta_j^{1-\alpha}\}$ from swarms in stage 1 in *b* and traffic density from stage 2 in *c*.

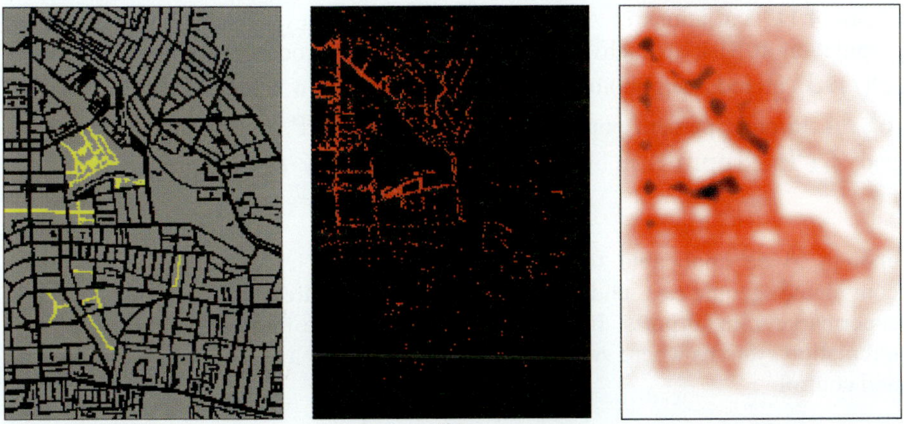

(a) (b) (c)

Figure 4.10. Control of Crowds and Identification of Vulnerable Locations. Areas closed by the police used in stage 3 are shown in yellow in *a*. The location of walkers in the stage 3 is shown in their steady state in *b*. The vulnerability of locations predicted from stage 3 in *c* is shown on a red scale with the darkest – $\Psi \geq 1\,\text{ppm}^2$ – the most vulnerable points. Best parameter values are set at $\beta = 0.65$, $\alpha = 0.5$, and $\lambda = 0.4$.

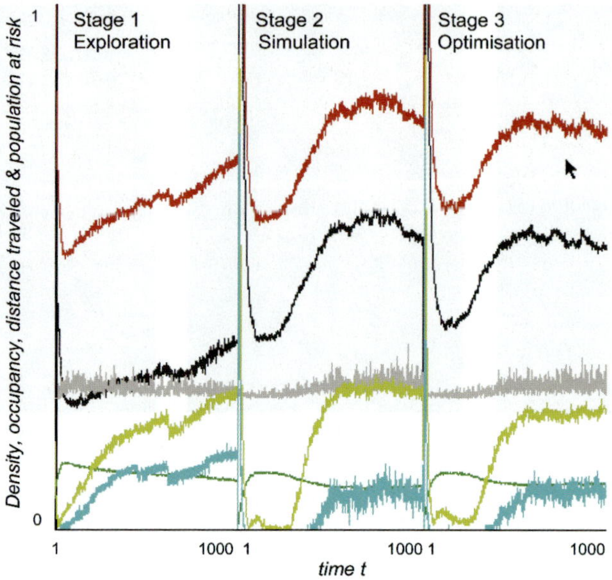

Figure 4.11. Variations in Average Density, Occupancy, Distance Travelled, and Population at Risk at Each Stage of the Modelling Process. Average cell density $\rho(t)/3$ in red, neighbourhood density $\sigma(t)/3$ in black, distance travelled $U_t/4$ in grey, percent population at risk with $Z_t(\Psi \geq 0.5\text{ppm}^2)$ in yellow, and $Z(\Psi \geq 1\text{ppm}^2)$ in blue, and percent occupancy $\vartheta(t)$ in green.

enough to be applicable to crowd situations of a less volatile nature where convoluted geometries and competing attractions are important as in malls, supermarkets, or airports. It may also have relevance to other human or animal circulation systems where swarm intelligence provides a useful way of exploration and control.

4.7 Conclusion

The models presented in this chapter are but a beginning in attempting to simulate crowding at the finest spatial scale. Agent-based approaches are essential at this level because the intricacies of geometry combined with randomness and variations in preferences are such that aggregate models remove too much detail and are not up the task of producing good simulations. In terms of designing to alleviate disasters, there has been much less progress although Helbing's group is working on these problems[22]. Much remains to be done. These kinds of models need to be linked to geodemographics, to the behaviour of various actors and agents in a much more proactive way than is currently embodied in these models. Most of the model strategies are rather physicalist in structure with geometry dominating and the notion of behaviour constituted as social forces. It might be possible to embed much more economic theory and choice theory into such structures but what is clear already is that models of this kind of complexity require extensive visualization if they are to be communicated and used effectively to solve and anticipate problems which require direct and effective disaster management.

Acknowledgements

I wish to thank the various pedestrian modelling groups around the world for the display of images from their web pages and simulations; in particular Keith Stills of Crowd Dynamics Ltd., the Fire Safety Engineering Group at the University of Greenwich, UK, and Dirk Helbing's Institute for Economics and Traffic Group at Dresden University of Technology. All interpretations are solely those of the author. Early versions of this chapter were presented at the Presented at 9th International Seminar on GIS, September 8 – 9, 2004, Seoul Education & Culture Center, Seoul, Korea, and at The Third Great Asian Streets Symposium (GASS), National University of Singapore, December 6 – 7, 2004.

[illegible]

Acknowledgements

[illegible]

[illegible]

Chapter 5

Distributed Geospatial Information Service

Chaowei (Phil) Yang and C. Vincent Tao

5.1 Introduction

Since its inception in the 1960s, Geospatial/Geographic Information System (GIS) has evolved rapidly and has become part of main stream Information Technology (IT). The projected market size is predicted to grow to $30 billion by 2005 (Gewin 2004). One of the main reasons for this fast-paced development is that GIS continually benefits from the advancement of computing technologies (Tomlinson 1998). The birth of computer networks in the 1970s, in conjunction with the emergence of the Internet in the 1990s, has broadened the way people share geospatial data and computing resources.

In the computing and geospatial communities, several important observations/ issues are evident:

1) Use of Computing resources. It is generally the case that computing capacity is very much under-utilized. According to an IBM survey (Conry-Murray 2004), only 6%, 60% and 20% of CPU power of desktops, mainframes and workstations, respectively, are utilized. Collectively this represents an immense, untapped pool of computing power.

2) Compatibility of geospatial data. Geospatial data are highly heterogeneous as they have typically been collected using different techniques, in different formats, and with different resolutions, projections, coordinate systems, etc. (Goodchild et al. 1999; Vckovski 1998). This heterogeneity has limited the degree to which data can be shared and merged seamlessly and transparently.

3) Legacy systems. To date there are many legacy systems that were developed in the absence of system and application standards. The resulting lack of interoperability among these legacy systems is a growing concern in the geospatial user community (Tao 2001).

4) Reliability and scalability. With potentially more and more users concurrently accessing centralized geospatial repositories, reliability and scalability of the centralized repository approach is becoming increasingly difficult.

With the growing use of geospatial data and applications, there is a demand to rapidly develop and deploy new capabilities and new functions based on existing legacy systems[24]. Given the fact that the technology is evolving very quickly, it is not economical or practical to replace existing legacy systems with completely new components in order to build more comprehensive systems.

Interoperability is recognized as a key solution to the integration of multiple legacy systems. Many organizations, (e.g., OGC[25] and FGDC[26]) have been actively working on developing standards and specifications to facilitate sharing of geospatial data and processing capabilities. The International Standards Organization (ISO) TC 211 working group was also created in 1994 to address this issue at an international level and has received strong support from many nations. There are many governments, private and academic initiatives that have become the driving force for interoperability development; for example, The National Map[27], Geography Network[28], GeoConnections[29], Geospatial–One-Stop[30], and Digital Library[31], etc. These efforts have resulted in a series of standards, specifications and successful applications.

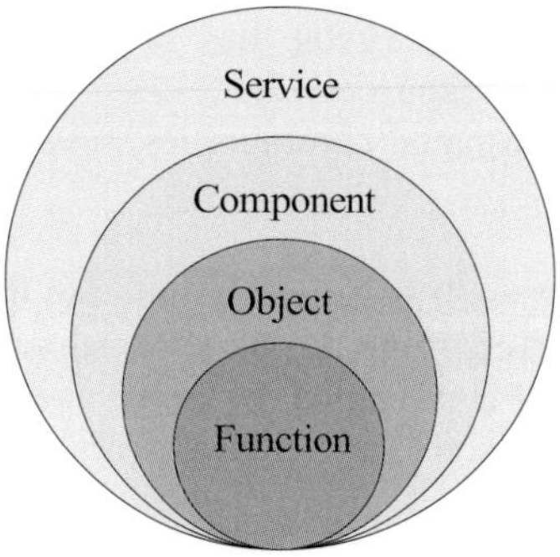

Figure 5.1. The reusability of developed system s has evovled from function , object, component, to service, with an increasing size of reusable unit.

From a software engineering perspective, the cost and time required for software development largely depends on how much previous development can be leveraged / reused for the subsequent development (e.g., Sommerville 2000). Software reusability is a key to assess system development. Reusabiltiy can be classified into four levels, as shown in Figure 5.1[32]: (1) function level, (2) object level, (3) component level, and (4) service level. Early software development was mainly based on software module integration at the function level; for example, early GIS software packages such as Intergraph's MGE[33], etc. Software development in the early 90's evolved from the function level to object-level integration; for example, ESRI ArcView[34], etc. The component level of software development gained popularity in the late 90's. Typical examples are ESRI's ArcGIS[35] and Intergraph's Geomedia[36]. With the growing use of Web Services-based software development approaches, GIServices is becoming the main stream geospatial application in the Internet era. Many GIServices specifications, for example: (1) Web Mapping Service[37] (WMS), (2) Web Feature Service[38] (WFS), (3) Web Coverage Service[39] (WCS), and (4)Web Registry Service[40] (WRS), have been endorsed and deployed.

There were many early research efforts in the field of GIServices. Yang (2000) systematically examined Web-based GIS and GIServices for developing CyberGIS, a WebGIS system. Tao and Yuan (2000) implemented a prototype system, GeoServNet, designed for renting software components and providing a service registry for component providers. Tao (2001) then improved the GeoServNet's capability to offer on-line 2D/3D analysis and visualization services. Luo et al. (2001) and Tsou (2002) have built a series of agent-based GIServicess. Tsou and Buttenfield (1998, 2002) proposed a dynamic metadata GIServices framework to address the problems in dynamic assembly, sorting, and searching of datasets. Yang et al. (2002) proposed a GIServicess environment to organize, manage, share, and utilize distributed geospatial computing resources. Peng and Tsou (2003) systematically examined Internet GIS and GIServices. These and many other relevant research efforts were pivotal investigations for Distributed GIService.

Distributed computing provides an enabling infrastructure for Web Services and Distributed GIService. The combination of GIService architecture with emerging distributed computing technologies, such as Agent[41], Peer-to-Peer[42] (P2P), and Grid[43], has demonstrated advantages by offering more reliable, scalable and efficient capabilities. Currently some research efforts are being devoted to utilization of emerging distributed computing platforms or technologies to build transparent GIServices (Yang et al. 2004a).

In the following sections, we will first introduce the general architecture of Distributed GIService. We then describe emerging distributed computing technologies as well as their use in building Distributed GIService based applications. The fourth section provides an example of Distributed GIService implementation. The last section is devoted to discussion of future research priorities.

5.2 Distributed GIService Architecture

A service is a software application that can be discovered, described, and accessed based on a specific description language and protocol, and does not depend on the context or state of other services or supporting platforms[44]. Service-Oriented Architecture (SOA[45]) is a collection of services with a mechanism to describe, categorize and discover relevant services, provide described services, and integrate an application based on implemented services (Brown et al. 2002). A typical SOA involves three components: service consumer, service broker and service provider (Brown et al. 2002). OGC (2003) adopted this architecture to describe the service trading architecture with 'trader', 'importer', and 'exporter' as illustrated in Figure 5.2. Trader registers offers from the exporter and returns offers upon request to the importer according to the preset criteria. Exporter registers offers with the trader object. Importer obtains offers, if satisfied, from the trader object. In fact, trader plays the role of "matchmaker" in a service-based architecture. This trading interaction process can be modeled in the sequence diagram shown in Figure 5.3.

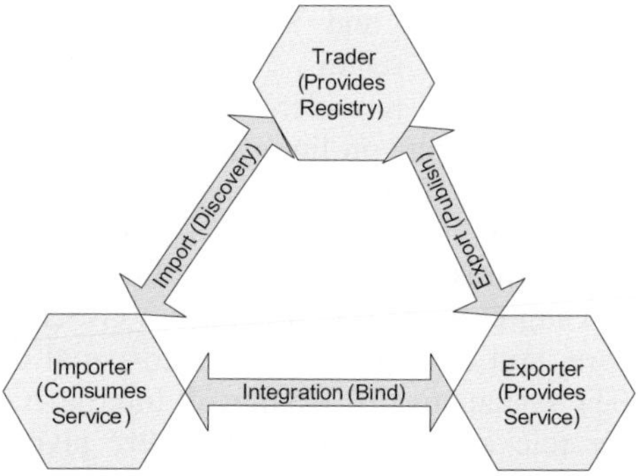

Figure 5.2. Service Trading (revised from OGC 2003)

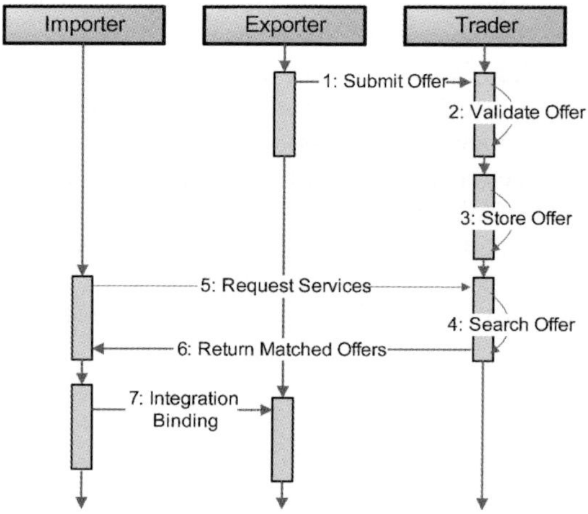

Figure 5.3. Trading Interactions (revised from OGC 2003)

Based on the SOA architecture, OGC (2003) developed a GIService framework as illustrated in Figure 5.4. This framework clearly illustrates the common set of interfaces required for enterprise-wide interoperability. In Figure 5.4, the importer is a multi-source integrated application client. It can be a map viewer, image exploitation client, discovery client, etc; the exporter is a group of service providers. They include Data Services, Portrayal Services and Processing Services providers; the trader is essentially the registry services that provide service type registry, service instance registry, as well as other services. Encoding techniques are unique in the geospatial application. The encoding specifications define the way in which geospatial data is transferred in the GIServices workflow. After the Services providers (Exporter) ***publish*** their services in the Registry services (Trader), the application client (Importer) can ***find*** the requested services from the Registry services. Then the application client will ***bind*** the Service provider and connect to them directly to obtain the services. These OGC services are defined as:

Client Services: the client-side components of client applications that interact with users

Registry Services: provides a common mechanism to classify, register, describe, search, maintain and access information about network resources (data and services)

Processing Services: the foundational application-building-block services that operate on geospatial data and metadata to provide value-added service

Portrayal Services: Portrayal Services provide specialized capabilities supporting visualization of geospatial information. Portrayal Services are components that, given one or more inputs, produce rendered outputs such as cartographically portrayed maps, perspective views of terrain, annotated images, views of dynamically changing features in space and time, etc

Data Services: the foundational service building blocks that serve data, specifically geospatial data. Currently, Data Services include Web Object Service (WOS), WFS, Sensor Collection Service (SCS), Image Archive Service (IAS) and WCS.

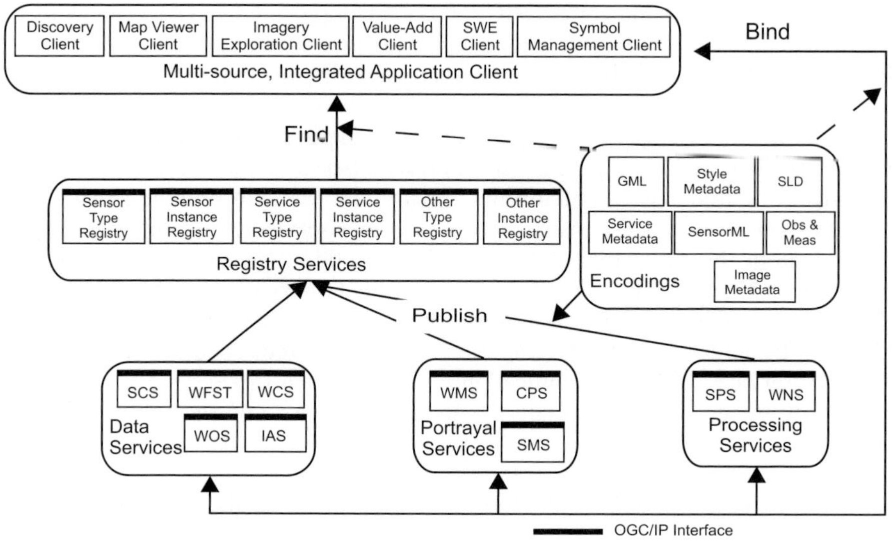

Figure 5.4. OpenGIS Web Service Framework Components (from OGC 2003).

The above logic is clear. However, in the software world, the realization of the above logic requires careful design. From an operational system point of view, Jeng (2001) suggested a middleware-based approach to achieving interoperability. He defined the following requirements for service architecture:

1) The deployment, discovery, and invocation of service-computing applications can withstand everyday use with minimal maintenance effort
2) A set of guidelines and meta-architecture that can be instantiated into service-computing frameworks or applications
3) Services should be described in a universal manner; well-formedness of services needs to be defined
4) Service-computing roles should be definable; examples of roles are: service requesters, service providers, and service administrators.
5) Service computing applications are executed by dynamic composition of services.
6) Service-computing applications shape their collaborations through capability-based lookup at runtime, not static binding. In other words, collaboration among services needs to be defined as framework instead of hard-wired scenarios.
7) A Service computing platform is required to implement the above concepts, relieving service-computing developers from underlying details such as communication mechanisms, programming languages, operating systems, and middleware.

There has been a realization that high performance distributed computing technologies would make Distributed GIService more reliable, scalable, and flexible (Yang et al. 2004b).

5.3 Distributed Computing Technologies

Distributed GIService relies upon the connection of networked computers to ensure high performance services. Issues, such as parallel execution, reliability, flexibility, and fault tolerance, must be considered in building an effective Distributed GIService. Among the emerging distributed computing platforms, P2P, Agent, and Grid are the most popular ones that have drawn much research. These distributed computing technologies are focused on the utilization and sharing of existing networked computing resources for pervasive access. Implementation generally requires the use of middleware. The following subsections will describe the use of these technologies from a geospatial perspective.

5.3.1 P2P Computing

According to a story related by Gelernter[46], "If a million people use a Web site simultaneously, doesn't that mean that we must have a heavy-duty remote server to keep them all happy? No; we could move the site onto a million desktops and use the internet for coordination." P2P computing takes advantage of all networked computers, especially widely distributed PCs (Barkai 2001). The basic strategy is to use all P2P-networked computers to serve as not only clients but also servers.

As illustrated in Figure 5.5, in the traditional client-server architecture, the server provides services to clients, and all services are processed on the server. If the server fails, or the server load is too heavy, response is very slow or requests are blocked. In the P2P computing environment, clients are also configured to provide services instead of only requesting services from the server. When server load is heavy, clients respond to the request, thereby reducing response time. The more clients that are integrated into the P2P computing environment, the better the system performance. This architecture has been used by portals such as Napster for MP3 file sharing and Bit Torrent[47] for file downloading. Although there are some legal issues with the specific applications, the technology itself has demonstrated great potential for sharing large volume information. For example, if NOAA were in distributing a large animation file for a hurricane case, P2P technology would be a good choice.

In order to transfer a client into a server and redirect requests to the transferred client, a middleware layer is used in the P2P computing platform, shown in Figure 5.6 (Barkai 2001). The P2P platform can be constructed on a number of computers (so called a P2P community) in order to support the same

P2P applications. A P2P middleware includes software installed on each PC and some control components installed on a server. The middleware layer schedules the P2P computing, creates processes, manages resources, checks security, and conducts access control. The P2P middleware also provides application interfaces for developing P2P applications.

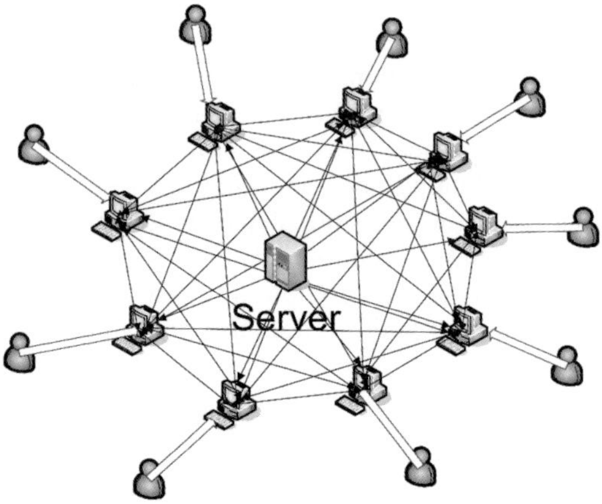

a. Traditional Client-Server model: PC is served by Server and some PC users are not satisfied

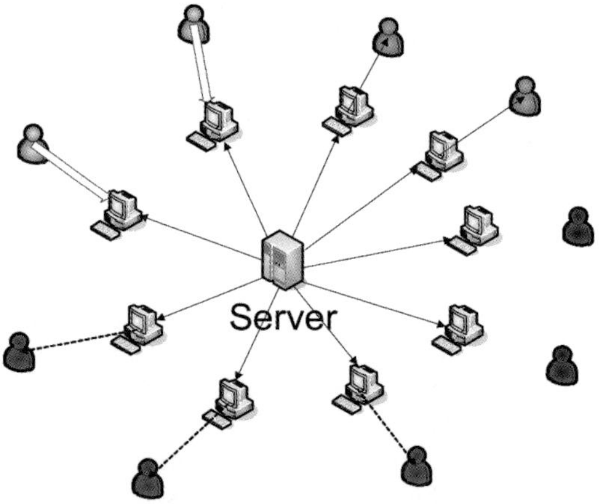

b. P2P Computing Model: All PCs are connected and provide server functions to satisfy more users

Figure 5.5. Comparison between Tradition and P2P Computing Models: In a traditional client-server computing model (a) the Server handles all the client requests. Only a few users are satisfied when many users concurrently access the server. Whenever a server failure occurs, requests are blocked. In a P2P computing model (b), clients also respond to requests as servers. The more clients join, the more requests can be accommodated. When the server is inactive, clients can still respond to other clients' requests.

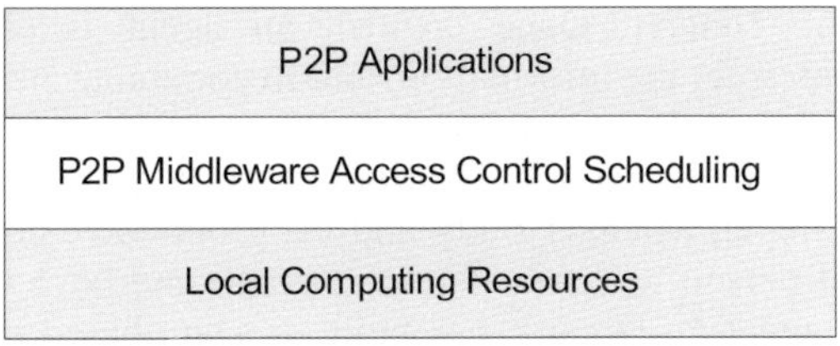

Figure 5.6. P2P Platform Architecture.

Besides file sharing, the middleware could also comprise a series of programs installed on each 'participating' PC to perform processing. When the program operates, it utilizes connected PCs to process client requests when the local user is not using the PC. Once processing is complete, the program returns results to the requesting client and responds to another processing request. When local users begin to use their PCs, the programs will stop and wait until local users' operations are finished. Therefore, P2P computing has the capacity to take advantage of a massive number of networked computers, especially PCs, which are currently only used to 6% capacity.

As an example, Intel has a P2P Program, named Philanthropic, to demonstrate the power of distributed computing by linking millions of PCs around the globe to create one of the most powerful computing resources in the world. Nearly two million PCs joined to provide resources for scientific research. Other companies, such as IBM, are actively involved in software development and commercialization of P2P distributed computing platforms for customers. The information science field can benefit from such a computing platform to process some normally time consuming information analyses and share information (Pitoura et al. 2003).

P2P computing provides an infrastructure for sharing the widely untapped computing power within the Internet. Lowe[48] and Li (2003) discussed adopting P2P to share maps and conduct GIS operations. The P2P advantage of decentralization, pervasive computing, cost efficiency, scalability and reduced bottlenecks provides potential QoS support to DGIService. In the P2P environment, communication and coordination among all peers is a key issue. Agent technology, especially mobile multi-agent system[49, 50] is considered to be a useful alternative to address this issue.

5.3.2 Agent Computing

The Agent concept resulted from the research of artificial intelligence, and refers to a component of software that is capable of acting in order to accomplish tasks on behalf of its user. The objective of an agent is to accomplish a particular task within a specific context. Once created, an agent is able to find needed information, to interpret the information, and to communicate with other agents to accomplish its task in order to provide a specific service.

The introduction of agent technology to GIS produced some impressive research. For example, Michael Batty and his group have conducted research on the simulation of person behavior patterns in GIS and have tested the possibility and usability of agents[51]. Doran[52] has built an agent-based model of ecosystems for sustainable resource management. Luo (1999) researched the possibility of developing an agent-based distributed GIS and developed a prototype at Peking University. Rahimi et al. (2002) designed and developed a complex, administrative infrastructure with a multi-agent system for geospatial information gathering and integration.

Agent technology can be used to build registry search, service discovery and integration, parallel administration, parallel service evaluation, and parallel service register functions. The mobile and multi-agent system approach is also an ideal complement to the P2P platform on the P2P communication and coordination.

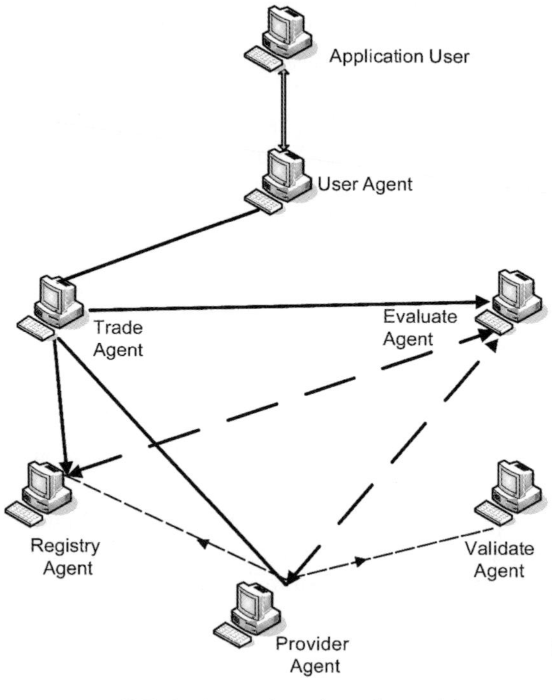

Figure 5.7. An agent-based service architecture.

A potential agent-based system for Distributed GIService is illustrated in Figure 5.7: 1) An application user sends an application request to the user agent, the user agent interprets the application request and communicates with trade agent to discover proper services with registry agent, selects services with evaluation agent, and comunicates with service provider agent to assemble the application (as illustrated by solid lines); 2) Evaluation agent will communicate with registry agent and provider agent to evaluate services (as illustrated by dashed lines). 3) Provider agent will contact registry agent to add service entry after validation by validate agent (as illustrated by thin lines).

5.3.3 Grid Computing

Grid defines a blueprint for a new computing infrastructure (Foster and Kesselman 2004). Grid can help us to develop computing systems to provide and/or ingest computing services; just as we are able to plug any electric device into the electric power grid to share electrical power. Grid envisions the unification of all computing resources to provide universal access. Data storage from widely dispersed servers, for instance, can be managed in the grid and can be accessed in the same way in which our local data are accessed. Grid also makes storage more reliable and more efficient, for example, a failure of one server will redirect data access from that failed server to another active server transparently. unified computing context embodies many research topics. The construction of such a data grid or computing grid[53] includes the development and deployment of a series of middleware, advocated by NSF[54], on each of the computers participating in the data or computing Grid. These middleware supply functions such as service controls, uniform data access, security services and co-scheduling.

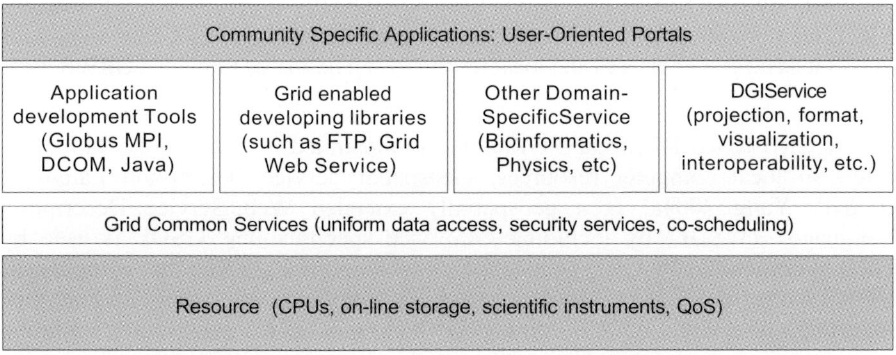

Figure 5.8. Grid Architecture. White boxes are implemented as middleware.

A typical grid platform includes four layers, as illustrated in Figure 5.8[53]: 1) the resource layer includes all the computing resources, such as CPU and online storage. 2) The grid common services layer includes the unification services, such as uniform data access, security services, co-scheduling. 3) The domain-specific layer includes specific systems and developing interfaces, such as those for

Bioinformatics and Physics. 4) User-oriented portal layer includes grid-supported information service applications. Grid computing users access Grid through Grid Portals. The implementation or deployment of a Grid is based on the middleware components developed. For example, Globus toolkit[55] is a set of middleware for implementing grids. NSF Middleware Initiative[56] lists a set of middleware that can be used for deploying Grid. Armstrong and Wang (2004) are developing GeoMiddleware to support geospatial interoperability for geospatial grid. Yang et al. (2004a) are utilizing high performance techniques in GIS (Yang et al. 2004b) to develop Geospatial Resource Broker (GRB), a geospatial middleware, for building geospatial grid.

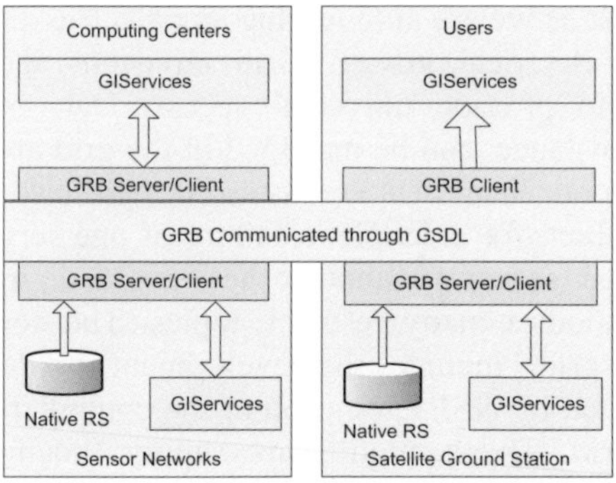

Figure 5.9. A Distributed GIService can be implemented by deploying GeoSpatial Resource Broker (GRB) on any computing devices, such as sensor networks, computing centers, and satellite ground stations to share GIServices. GRBs will communicate through GeoSpatial Service Description Language (GSDL). Users can access Distributed GIService through a GRB Client and human-computer interfacing GIServices, such as GIS interfaces. GIServices and local resources can be shared through a GRB Server/ Client.

A GRB-based Distributed GIService is illustrated in Figure 5.9. Native RS refers to local computer resources; Geospatial Service Description Language (GSDL, Yang 2004a) is a geospatially extended Web Service Description Language[57] (WSDL) by including GIService specifications. GSDL is used by GRB to communicate. GRB is installed on a computer to enable that computer as a GRB Node, which can be either a server for providing GIServices or a client for ingesting GIServices. As a client, a GRB node will access geospatial computing resources, including data and computing. As a server, a GRB node will provide its GIServices, such as geospatial data, analytical, modeling, and visualization services, uniformly to the GRB clients.

An end-user who does not want to provide any service to the Grid could install the GRB client only. For computers where both the GRB client and server are installed, users can access the Distributed GIService in the same way as a computer with only the client installed. GIServices that require intensive computing for a single request are installed on high-performance computers and made available to GRB clients. Users will access the grid through human-computer interfacing GIServices. The grid will provide ubiquitous access to geospatial computing resources.

5.3.4 Supporting Distributed GIService

P2P, Agent, and Grid-based computing provide new ways for utilizing large numbers of networked computers for the purpose of sharing computing resources. They offer powerful mechanisms to support more reliable, scalable, and flexible geospatial data and processing sharing over the network. As discussed, transparency is important in system design so that end users cannot notice any change of their computing environment. This can be implemented by concealing network communication details and by simplifying the middleware deployment process. It is our view that the integration of emerging distributed computing technology will dramatically expand application of GIServices, thereby benefiting the user community. The next section provides an example of a Distributed GIService implementation.

5.4 Implementation of Distributed GIService

Apart from the Services illustrated in Figures 5.3 and 5.4, other computing services that rely upon geospatial characteristics are also required in order to implement an operational Distributed GIService; these must provide: 1) verification to ensure any added services are what they claim to be; 2) evaluation, which tracks service quality, usability, etc,; and 3) balance, to allow for service selection based on performance optimization criteria. Therefore, except for the consumer, provider, and trader (suggested by most SOA applications such as OGC 2003), Verification Service, Evaluation Service, and Load Balancing Service (as part of Trading Service) are also required for enabling Distributed GIService.

Verification Service: Verification service evaluates other services prior to these being added to the service registry. This is necessary to ensure that the submitted service has the claimed functions and that it complies with specific standards.

Evaluation Service: Evaluation service evaluates services based on a service's performance to keep services' historical performance through a credit system.

Balance Service: Balance Service selects services based on specific performance criteria and services' credits in order to ensure that distributed resources are used more efficiently, optimally and uniformly than might otherwise be the case.

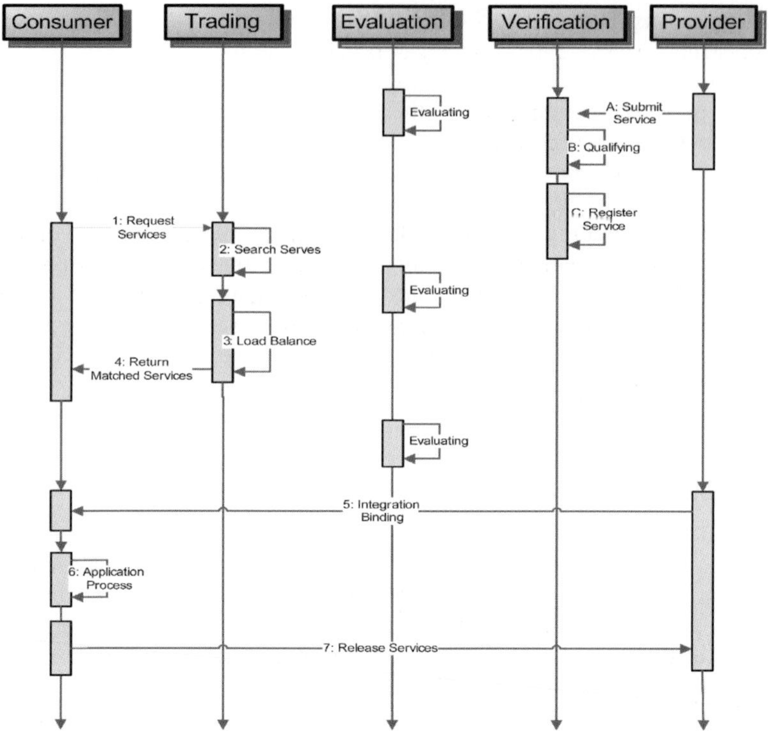

Figure 5.10. Services-evaluating, service-registering, and application-assembling operations within a Distributed GIService

Added these three services, a Distributed GIService has parallel operations as illustrated in Figure 5.10: 1) Service providers submit services to verification service for verification and register as illustrated through A-B-C; 2) The evaluation service will periodically evaluate registered services based on their operational performance; 3) Application assembling within Distributed GIService includes service request, service search, application binding, application operation, and service release as illustrated in steps 1-2-3-4-5-6-7. Although a Distributed GIService, including the services in Figure 5.10, can be implemented using different computing platforms, such as DCOM[58], CORBA[59], and DCE[60], the pervasive access of HTTP services[61] across different computing platforms has resulted in the success of Web Services[62] for implementing Distributed GIService.

Web Services are based on the Web environment and are supported by essentially four relevant technologies:

1) HyperText Transport Protocol (HTTP). HTTP provides a de facto pervasive protocol for conducting web request;

2) Simple Object Access Protocol (SOAP[63]). SOAP provides the protocol for Web Services to communicate;

3) Universal Description, Discovery and Integration[64] (UDDI). UDDI facilitates the universal archive and discovery of services; and

4) Extensible Markup Language[65] (XML)/ WSDL. XML provides a naturally understandable language for describing information and WSDL provides a way for describing the services provided through Web.

These technologies provide SOA-required transmission, messaging, describing, archiving, and binding support to Web Services. Based on GSDL, GRB, and Web Services supported WMS (a GIService), a Grid-based Distributed GIService can be constructed as illustrated in Figure 5.11.

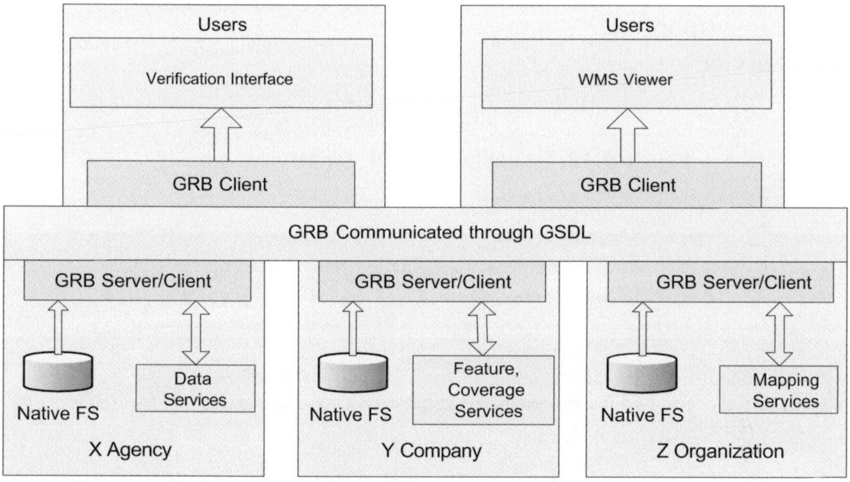

Figure 5.11. A Distributed GIService: Different GIServices, such as data service, feature service (WFS), coverage service (WCS), and mapping service (WMS), will be integrated through GeoSpatial Resource Broker (GRB) using GeoSpatial Service Description Language (GSDL).

While the coordination of Service selection, integration, validation, and evaluation is taken care of by GRB, Human-Computer Interacting GIService provided to Users will be transparent within the Distributed GIService. During the interaction process (e.g., a Zoom-In including overlay), Distributed GIService may integrate services from different GIServices providers. For example, a Zoom-In will require data services, feature service (WFS) from X, coverage service (WCS) from Y, and require mapping (WMS) from Z. The integration processes are operated by the functions provided by the GRB. The service information exchanged are described in GSDL (Sillitti et al. 2002; Yang et al. 2004a), as illustrated in Figure 5.12. Interfaces are illustrated in Figures 5.13 and 5.14. Based on legacy systems, only minor changes are needed to WMS/WFS/WCS-enabled legacy GISs in order to develop applications. Therefore, Distributed GIService architecture will allow a fast deployment of applications across a network.

```
...
<service name="Web Map Service X">
      <documentation> map service server 1</documentation>
        <port name="mapPort" binding="tns:mapbinding">
          <soap:address location="http://X.COM/wms/"/>
        </port>
</service>
<service name="Web Map Service Y">
      <documentation> map service server 2</documentation>
        <port name="mapPort" binding="tns:mapbinding">
          <soap:address location="http://y.com/wms/">
        </port>
</service>
...
```

Figure 5.12. Example of GSDL for services binding

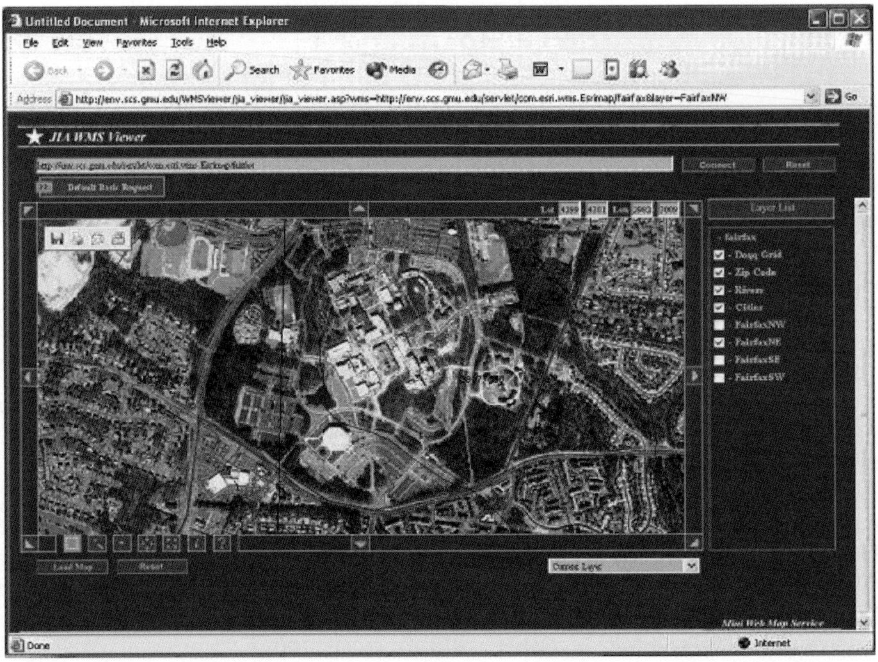

Figure 5.13. A GIService verification interface for testing, qualifying WMS service

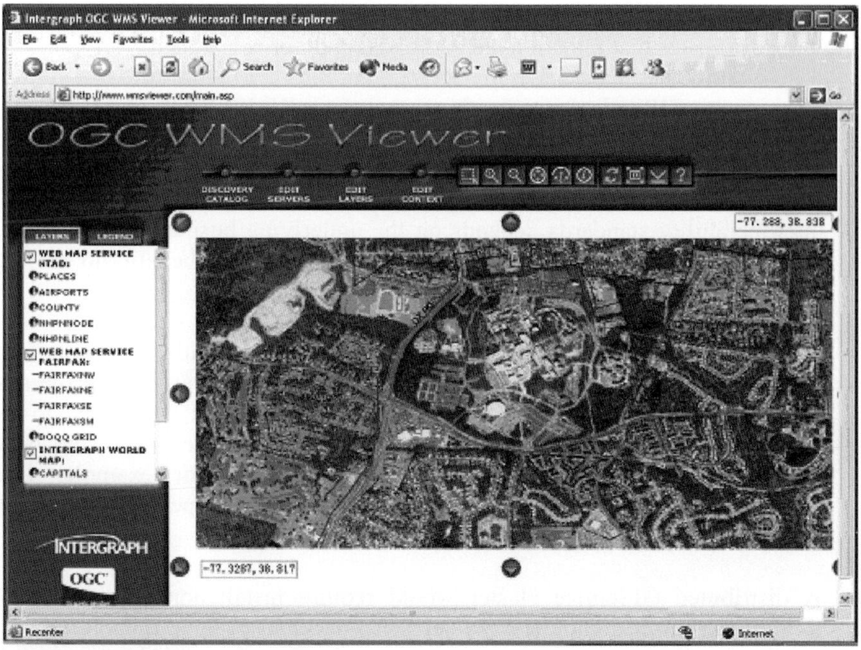

Figure 5.14. The Intergraph WMS viewer is accessing different WMS services and geospatial data services transparently .

In this example, only data service and WMS/WFS/WCS, several GIServices, are used to illustrate the functionality of a Distributed GIService. A fully functional Distributed GIService can be developed using one of two approaches: 1) add functions to expand GIServices; or 2) expand the implementation area from application, department, community, to the Internet, as a route of infra-Grid, intra-Grid, extra-Grid, and inter-Grid suggested by Ernest et al. (2004) for grid evolution.

5.5 Conclusion and Discussion

Technically, Distributed GIService is driven by: 1) GIServices supported by Interoperability and Web Services, and 2) distributed computing technologies, such as Grid, P2P, and Agent. P2P is useful for conducting stateless computing or large volume data sharing. Agent computing can be used for performing parallel system administrating, evaluating, and coordinating. Combination of enabling distributed computing technologies with GIServices will provide more reliable, affordable, stable, and scalable services and solutions. Though distributed GIService is being used largely as part of research experiments or government applications, its commercial potential is enormous and transformational. The following are issues that remain to be addressed and / or researched:

1) Many OGC GIService specifications, including WMS, WFS, and GML, etc. are getting popular in the geospatial industry, with more and more software vendors are adopting these specifications. For example, the OCC WMS standard is now endorsed by the ISO. However, the sustainability of OGC interoperability standards depends on the underlying business model and on the clear demonstration of economic and technological benefits to the end user community.

2) New distributed computing technology (Grid or P2P) provides a new platform to implement Distributed GIService. In order to take advantages of the new computing platform, geospatial middleware (e.g., GRB) needs to be developed. It is important to conduct integration research using the latest advancements in the distributed computing field; for example, Grid Interoperable standardization framework, such as Open Grid Service Architecture[66] (OGSA), needs to be considered.

3) A distributed GIService cluster would require installation of software on participating user computers. As such, it is critical to have reliable security protocols in place for protecting participating users. Also, copyright information from contributing parties is also important in the vastly connected network.

4) Transparent access and integration of GIServices provided by or served for different user communities, as well as a high level of ontology and knowledge among different communities, is required. Systematic research and development on geospatial syntax, semantics, ontology, and schematics is considered as a cornerstone that will eventually achieve strong interoperability among the communities[67].

Clearly, distributed GIService will introduce many new opportunities for both software vendors and end users. Software vendors can focus on their niche and develop more a competitive edge with less concern for system integration. Users can benefit by getting the best performance/price solutions as well as a wide variety of offerings. On the government side, this framework is also attractive as it lowers the cost of development, shortens the deployment time and is open enough for further upstream enhancement. In fact, for information and resource sharing, the framework is in many ways valuable to academic institutions and public organizations (Lambert and Widya 2001; Sui and Goodchild 2001). The extremely broad range of applications and strong technical advantages of Distributed GIService suggest it to be a promising emerging paradigm for geospatial computing.

Chapter 6

Geospatial Grid

Liping Di

6.1 Introduction

Geospatial science is the science and art of acquiring, archiving, manipulating, analyzing, communicating, and utilizing spatially explicit data for understanding both physical, biological, and social systems on the Earth's surface or near the surface. One of the very significant aspects of geospatial science is that it is not unique to any one discipline; instead it is an interdisciplinary science, built on the knowledge of many science disciplines. Because geospatial science can be utilized in virtually all situations and human activities that are spatially related, it is becoming more and more integrated into many aspects of contemporary life.

Three significant features distinguish geospatial research from scientific endeavors in other scientific domains: 1) the research is multi-disciplinary; 2) the research needs a large amount of data and information and may be computationally intensive; and 3) the research can be on regions that span multiple spatial scales and may be micro to macro in nature (e.g., a leaf, a field, or an area of local, regional, continental, or global extent).

Normally, the geospatial knowledge discovery process involves three consecutive data and information flow steps*: 1) Geoquery, 2) Geodata and information assembly, and 3) Geocomputation. Geoquery* is the location (discovery) and acquisition of data from data repositories. *Geodata and information assembly* is assembly of the data and information from potentially distributed and heterogeneous data repositories in a way that satisfies the needs of geocomputation. *Geocomputation* is analysis and simulation of the complex Earth system using the data and information from the geoquery process.

Because of their importance in social and economic activities, large amounts of geospatial data have been collected by various public and private sector organizations mainly using remote sensing methods. Those data must be converted to information and knowledge before they become useful. Because of the multidisciplinary nature of geospatial science, a typical geospatial research and applications project requires access to and manipulation of data and information from multiple sources provided by multiple data and information systems. Currently, datasets from data centers are diverse; the data products may differ in spatial/ temporal extent and resolution, origin, format, name conventions, and map projection. Scientists spend considerable time assembling the data and information into a form ready for analysis in the geo-computation step, even when the analysis is very simple. An estimated 50% ~ 80% of the research time spent by ESS scientists is for data and information discovery and assembly (Di and McDonald 1999). The fundamental problem cause is that system of, and the data, and information available from the different data repositories are not interoperable. This paper discusses the geospatial Grid technology that will significantly reduce the problems associated with archiving, manipulating, analyzing, and utilizing large volumes of geospatial data at distributed locations.

6.2 Grid Technology

6.2.1 What are Grids

Grids are persistent environments that enable software applications to integrate instruments, displays, computational and information resources that are managed by diverse organizations in widespread locations (Foster and Kesselman 1999). They bring together geographically and organizationally dispersed computational resources, such as CPUs, storage systems, communication systems, data and software sources, instruments, and human collaborators to provide advanced distributed high-performance computing to users[68] (Foster et al. 2001).

The recent explosion of commercial and scientific interest in Grid has resulted in much confusion on what a Grid is. Foster[69] proposed a three-point checklist for determining whether a system is a Grid. A Grid system must

1) *coordinate resources that are not subject to centralized control;*
2) *use standard, open, general-purpose protocols and interfaces;*
3) *deliver nontrivial qualities of service.*

6.2.2 The history of Grids

In late 80's and early 90's of the last century, new classes of scientific research were being conducted which required high-performance computational resources not available in a single computer or at a single geographic location. This requirement led, starting in the early 90's, to the construction of the large scale high-performance distributed computing environment. An example of such efforts was the CASA Gigabit testbed[70]. With the availability of fast networks, the construction of *networked virtual supercomputers*, or *metacomputers* (Catlett and Smarr 1992) became possible. Metacomputers are execution environments in which high-speed networks are used to connect supercomputers, databases, scientific instruments, and advanced display devices, perhaps located at geographically distributed sites. One of important experiments for testing the metacomputer concept was the I-WAY experiment (Foster et al. 1996).

While the physical infrastructure to build metacomputing systems is becoming widespread, the heterogeneous and dynamic nature of the metacomputing environment poses new challenges for developers of system software, parallel tools, and applications. The Globus project, started in mid 90s, provides a set of common software tools to address these challenges (Foster et al. 1996). The Globus system is intended to achieve a vertically integrated treatment of applications, middleware, and network. In Globus, a low-level *toolkit* provides basic functions such as communication, authentication, network information, and data access. These functions are used to construct various higher-level metacomputing *services*, such as parallel programming tools and schedulers. The preliminary versions of Globus components were deployed as part of the I-WAY networking experiment (Foster and Kesselman 1997). With the development of the Globus, the metacomputer concept was expanded into *computational grids* that provide dependable, consistent, and pervasive access to high-end computational resources[71,72] (Johnston et al. 1999). Data grids, a concept parallel to computational grids, emerged at the same time to address the needs to manage, access, and analyze geographically-distributed datasets by communities of users (Chervenak et al. 2001).

The Grid research has expanded significantly since beginning of this century. Major applications communities have been emerging, and major infrastructure deployments are underway around the world. A rich technology base has been constructed and industrial involvement has become significant. In order to exchange information and to coordinate the development of Grid technology, the Global Grid Forum (GGF) was established in 2001[73]. GGF represents the merger of Grid technical communities in North America, Europe and Asia/Pacific. It holds meets three times per year, alternating between North America and Europe, and Asia/Pacific. GGF supports mechanisms for formal review, approval and release of best practices guides and standards for Grids.

6.2.3 Globus Toolkit

The Globus toolkit is the de-facto standards for Grids. Most Grid projects use Globus for construction of production Grids. Globus, consisting of a set of services and software libraries, is the key middleware that provides core Grid capabilities. The

Globus toolkit facilitates the creation of usable Grids, enabling high-speed coupling of computers, databases, instruments, and human expertise. With Globus, scientists can run their gigabytes-per-time-step job on multiple high-performance machines at the same time, even though the machines might be located far apart and owned by different organizations. It can be used for large distributed computational jobs, remote instrumentation, remote data transfer, and shared immersive storage spaces[74].

The open source Globus toolkit has been developed by the Globus Alliance[74]. Currently, two major versions of the Globus toolkit are being used in the Grid community, GT 2 and GT 3. The latest version of the Globus toolkit is Globus GT 3.2. GT3 is the first major implementation of Open Grid Service Architecture (OGSA) specifications[68]. GT 2 is the pre-web services version, with the last release to be GT 2.2.2. The Globus toolkit provides the following function components: the Core component for the basic infrastructure needed for building grid services; the Globus Resource Allocation Manager (GRAM) for providing a common user interface to submit a job to dispersed multiple machines; the Monitoring and Discovery Service (MDS) for providing information services through soft state registration, data modeling, and a local registry; the Grid Security Infrastructure (GSI) and Community Authorization Server (CAS) for providing generic security services such as authentication, authorization, and credential delegation for applications that will be run on Grid; GridFTP and Reliable File Transfer (RFT) for providing a standard, reliable, high-speed, efficient, and secure data transfer service; Metadata Catalog Service (MCS) for storing and accessing metadata associated with data managed by a Grid; Replica Location Service (RLS) for maintaining and providing access to mapping information from logical names for data items to target names which may represent physical locations of data items; and other modules such as simple Certificate Authorization (CA) etc.

6.2.4. What Grids Provide

Computing and data Grids are emerging as the infrastructure for 21st century science, engineering and high-performance applications and systems. Grids provide a common way of managing distributed computing, data, instrument, and human resources across administrative boundaries, an underlying security infrastructure facilitating the interaction of many different people from many different institutions, a seamless processing environment, and an infrastructure scalable to a large number of resources. Grid technology is best suited for facilitating large-scale science, engineering, and operational applications that are widely distributed geographically and that are processing- and/or data-intensive.

Grids support various user communities. Grids are invisible to scientists and domain problem solvers who use the Grid-enabled applications and services, although currently implementing and deploying such applications and services in Grids need some efforts. Just as for a regular non-grid application, they need to be able to express a problem or experiment in terms specific to the application domain, specify the input parameters request that the solution be obtained, and manage the resulting graphics, data, etc. However, Grids enable them to solve large-scale scientific problems that could not be solved before because of lack of resources. They will see an application whose development was eased by using Grid functions or Grid-based web services. The ease of development should result in development of more applications or faster availability of applications. To direct users of a Grid system, the Grid enables them to easily execute jobs at one or more remote sites and move data between sites, all with single sign-on security. To the developers of the Grid-enabled applications, Grids supply the core capabilities common to most applications, so that developers do not have to re-implement this core capability with each application. The standards for Grid services also enable application developers to share services built by others and use those services as building blocks to easily develop more complex services targeting specific application areas.

6.3. The Characteristics of Geospatial Data and Information

Contemporary geospatial datasets originate from multiple sources, including remote sensing, simulation models, and in-situ measurements. These products are commonly archived in a variety of legacy public domain repositories. The NASA EOS program alone currently produces more than 2 Tb of geospatial data per day (Di et al. 2002). There is also increasing availability of real time data from in-situ sensor systems. Such real time data are valuable for ground truth, may be input into forecast models, and are useful on their own. Compared with data products in other disciplines, geospatial data products have several unique characteristics:

1. **Time-dependence**. Many data products are typically based upon repeated observations to identify changes over time, both natural and anthropogenic.

2. **Vertical-dependence**. Many geospatial data products span a range of vertical levels. These products may use pressure or other monotonic variables as the vertical coordinate.

3. **Large volume**. Many of these products have data volumes measuring several terabytes, due in part to the time- and vertical-dependence, and to multiple measurement parameters.

4. **Specialized data models**. Remote sensing data products are often captured and stored in the form of a 1-D, 2-D or 3-D swath, as a consequence of satellite orbit

and view angle. Output from numerical weather and climate forecast models and thematic products derived from satellite data takes the form of gridded data with multiple parameters varying in three spatial dimensions and time. Digitized street maps and political boundaries are stored in vector form.

5. **Specialized data formats**. Large numbers of data formats are used for geospatial datasets.

6. **Competing data access protocols**. Many data access protocols exist for geospatial data. Each geospatial discipline may have its private protocols for accessing data. This diversity makes cross-disciplinary data access very difficult.

7. **Preservation of raw data values**. Raw data values are important in scientific investigations. Even when products derived from the raw data are available, scientists sometimes request raw data to validate the derived products, test the refined algorithms, generate new products, or produce improved products.

8. **Multiple parameters**. The same scene may be measured using multiple spectral bands or multiple algorithms. Arithmetic combinations of these bands or parameters are often physically meaningful.

9. **Version control**. Some scientific data products undergo reprocessing, as geophysical algorithms are refined. A means of associating a dataset with its version is desirable.

10. **Real-time data**. Radar and other observation systems may produce real-time measurements at the rate of several products per second, which must be disseminated in a timely manner.

6.4 Geospatial Interoperability Technology

In order to interoperate among the geospatial data and systems, the geospatial community has developed a set of standards at both the abstract and implementation levels through the standard-setting bodies, including ISO TC 211, the Open Geospatial Consortium (OGC), and the U.S. Federal Geographic Data Committee (FGDC). Among those standard bodies, OGC is the one primarily concerned with the establishment of implementation specifications for geospatial interoperability.

The Open Geospatial Consortium, Inc.[75] is a not-for-profit international membership-based organization founded in 1994 to address interoperability among systems that process georeferenced data. Since 1999 OGC has successfully implemented four web-based geospatial interoperability initiatives: Web Mapping Testbed I (WMT-I) in 1999, WMT II in 2000, OGC Web Service Initiative 1.1 (OWS 1.1) in 2001, and OWS 1.2 in 2002. Through those initiatives, OGC has produced a set of web-based geospatial data interoperability specifications. The most important ones related to interoperable data access are the Web Coverage Services[76] (WCS), Web Feature Services[77] (WFS), Web Map Services[78] (WMS), and Web Registries Services[79] (WRS) specifications. OWS 2.0, started in March 2004, addresses issues in geospatial web services and service chaining.

The OGC WCS specification defines the interfaces between web-based clients and servers for accessing on-line multi-dimensional, multi-temporal geospatial coverage data in an interoperable way. Coverage data includes all remote sensing images as well as gridded data such as DEM, land use classification. The OGC WFS specification defines the interfaces between web-based clients and servers for accessing feature-based geospatial data. Feature-based geospatial data includes the vector-based transportation road network, the coastline, political boundaries, utility lines, etc. WCS and WFS together cover all geospatial data. They form the foundation for web-based interoperable data access. The OGC WMS specification defines web interfaces for dynamically assembling maps from multiple sources within a heterogeneous distributed computing environment over the Internet. Maps are the visualization of data. A WMS server normally converts the data in its archive to a visualized form (map) based on the requirements from the client. In many cases, a WMS server may talk to a WCS or WFS server to obtain the needed data for making maps requested by a client. In this sense, a WMS server can be considered as a data visualization service for either WFS or WCS servers. The OGC WRS defines the interfaces between web-based clients and servers for finding data or services required by users from the registries. A data registry consists of data entries available at data archives that might be served by WCS, WFS, or WMS servers. WRS allows data users to find the requested data in vast archives. OGC WRS is designed to be a general interface that can be used not only for data discovery but also for service and resource discoveries.

OGC technology allows data users to specify the requirements for the data they want. An OGC compliant server must preprocess the data on demand, based on the users' requirements and return the data back to the users in the form they specify. At the end, the users will get the data that exactly match their requirements in both contents and structure (e.g., format, projection, spatial and temporal coverage). Because OGC specifications are industry standards, many GIS vendors are implementing those specifications at servers or clients. The proliferating installation of OGC-compliant servers and the increased number of OGC-compliant GIS clients available enable a user to access data from multiple sources that are co-registered and in a form ready for analysis. This will significantly reduce the time needed for users to acquire and preprocess the data before the data can be analyzed.

6.5. The Application of Grid Technology in Geospatial Domain—The Geospatial Grid

The geospatial community is responsible for the collection, management, processing, archive, distribution, and application of geospatial data and information. Because of the importance of geospatial data, many public and private organizations have been engaged in geospatial data and information activities. As a result, geospatial data and the associated computational resources are widely distributed. The multi-disciplinary nature of geospatial research and applications requires the integrated analysis of huge volume of multi-source data from multiple data centers. This requires sharing of both data and computing power among data centers. Most geospatial modeling and applications are both data and computation intensive. Therefore, Grid is an ideal technology for the geospatial community to manage the distributed data and computational resources.

Geospatial Grids are the extensions and domain-specific applications of the fundamental Grid technology in the geospatial discipline. Geospatial Grids include both geospatial data Grids and geospatial computational Grids. Geospatial data Grids emphasize data access and information services on large, distributed geospatial data archives. Geospatial computational Grids are mainly for coordinating computational resources for large-scale geospatial modeling and applications such as climate modeling. A geospatial Grid could be combination of both.

The core Grid technology is developed for general sharing of computational resources and is not especially designed for geospatial data. As discussed in section 3, geospatial data and information are significantly different from those in other disciplines. The geospatial community has developed a set of standards specifically for handling geospatial data and information. The geospatial community has been familiar with those standards and has used them intensively in geospatial systems development and research. Therefore, a geospatial Grid has to be able to deal with the complexity and diversity of geospatial data and the large data volume and to comply with the geospatial standards. In order to make Grid technology applicable to geospatial domain, geospatial domain-specific extensions to the core Grid technology should be constructed.

The geospatial community started to look into the potentials of Grid technology around the year 2000. Currently, dozens of geospatial Grid projects are in progress worldwide. Most of the projects are at the stage of learning Grid technology and evaluating its applicability to the geospatial domain. A few prototype applications are being developed. Some projects are working on extending Grid to meet the specific requirements of geospatial applications. No large-scale multi-organization, operational geospatial Grids exist today. Table 1 provides a brief description of various geospatial Grid projects.

Name	Sponsor	Focus
Information Power Grid (IPG)	ipg.nasa.gov NASA-IPG	Create and apply a production Grid for aerosciences and other NASA missions. Currently, IPG funds various projects to explore the geospatial Grids.
The Integration of OGC and Grid Technologies for Earth science modeling and applications.	laits.gmu.edu NASA-AIST	Integrating OGC Web services technology with Globus to make Grid geospatially enabled. Provide OGC interfaces to Grid-managed geospatial data. Research on virtual geospatial datasets, service modules, service models, and service chaining.
Grid Prototype for Remote Sensing Data	NASA-ESIP Federation	Explore the application of SRB/MCAT to Remote Sensing Data
Earth System Grid (ESG)	earthsystemgrid.org DOE Office of Science	Delivery and analysis of large climate model datasets for the climate research community
Network for Earthquake Eng. Simulation Grid	neesgrid.org NSF	Create and apply a production Grid for earthquake engineering
SCEC	NSF	Integrated geophysics modeling
European Union (EU) DataGrid	eu-datagrid.org European Union	Create & apply an operational grid for applications in high energy physics, environmental science (geospatial data grid), and bioinformatics
Climate*Prediction*.Com	UK e-science	Estimating Climate Uncertainty Harness the power of 10,000's of PCs by providing downloadable model experiments

Table 6.1. Summary of some major geospatial Grid projects

6.6 The CEOS Grid Testbed

The Committee on Earth Observation Satellites[80] (CEOS) is an international organization responsible for coordinating international civilian space-borne missions designed to observe and study the planet Earth. Its membership encompasses the world's government agencies responsible for civilian Earth Observation (EO) satellite programs, along with agencies that receive and process data acquired remotely from space. Inspired by the successful applications of Grid technology in other disciplines, the CEOS Working Group on Information Systems and Services (CEOS WGISS) started a CEOS Grid testbed in September 2002 to evaluate the feasibility and applicability of Grid technology to the EO community. The testbed, currently consisting of seven applications from individual agencies in the U.S., the European Union, and China, aims to address the use of Grid technology for efficient support to diverse users worldwide for easy access and applications of EO data and to EO data providers for improving their efficiency of operation and maximizing the usefulness and benefit of the EO data which they gather[81]. Table 2 summaries the CEOS Grid Testbed applications.

The CEOS testbed is a multi-phase project. Each phase has an objective to accomplish. The phase I objective is to establish a CEOS Grid technology core testbed with at least three major participating agency nodes. The purpose of the Grid technology core is to establish an immediate Grid capability base (including technologies, pilot applications, and knowledgeable people) within the participating CEOS agencies. This phase has been finished. Phase II of the testbed, scheduled to be finished by March 2005, will demonstrate at least three CEOS Grid-enabled applications, each involving at least one CEOS agency and partner site which could include other CEOS agencies. The purpose of the Grid-enabled application demonstrations is to show proof of concept, evaluate benefits, and obtain lessons learned from infusion of Grid technologies from the technology core into real CEOS agency information systems and applications. The phase III objective is to infuse applicable Grid technologies into member agency information systems and into at least one WGISS Test Facility (WTF). The outcome of this Grid technology infusion would be to create a persistent CEOS Grid that would be available to support future CEOS agency initiatives. The ultimate goal of the CEOS Grid testbed is to establish a CEOS Data Grid for support of major international science and EO initiatives that covers most of the CEOS member agencies. The details of the CEOS Grid Testbed can be found CEOS website[82].

Name	Sponsor/implementer	Focus
The NOAA Operational Model Archive and Distribution System (NOMADS)	NOAA/NOAA NCD	A distributed grid framework promoting standards across multiple institutions for access to climate and NWP models for analysis and intercomparison.
NASA EOS Data Pools	NASA/GMU	Data access and services to petabytes of EO data distributed on on-line storages at NASA data centers through geospatial-enabled Grid technology.
NASA GSFC Advanced Data Grid	NASA/NASA GSFC	To demonstrate the execution of scientifically 'meaningful' applications software in a data grid environment that includes two 'collaborating' data systems with SRB/MCAT
Earth Science Data Mining	NASA IPG, NSF/UAH	To test geospatial mining technology in the Grid environment.
USGS Grid Application	USGS/USGS EDC	To explore the feasibility of utilizing GRID technologies for the delivery and reception of data. Primarily this would employ the services of GridFTP and certificate authority.
Interaction with GoD-WM: GRID on Demand for Web Mapping	ESA/ESA ESRIN	To interface CEOS interoperability technologies with GRID environments. Demonstrate the integration of technologies for selected EO applications: the first selected case is for Ozone data handling and exploitation.
EO data sharing and distribution system	China Satellite Remote Sensing Ground Station	To facilitate the sharing of Earth Observation data in China by using Grid technology.

Table 6.2. Summary of applications in CEOS Grid testbed.

6.7 An Example of Geospatial Grids

In the past several years, the Laboratory for Advanced Information Technology and Standards (LAITS), George Mason University, has worked on a project for integration of OGC geospatial Web service technology with Grid technology for geospatial modeling and applications. The objectives of the project are 1) to enable the management of geospatial data by Grids; 2) to provide OGC standard compliant access to Grid-managed geospatial data; and 3) enable geospatial modeling and the production of virtual geospatial products in the Grid environment.

6.7.1 NWGISS

As a member of OGC and a participant in those OGC interoperability initiatives, LAITS at GMU has developed an OGC specifications compliant software package called the NASA Web GIS Software Suite (NWGISS). It is a web-based geospatial data distribution and service system, which implements multiple OGC specifications, for delivering NASA EOS data to broad user communities. Currently, NWGISS consists of a Web Map Server, a Web Coverage Server, a Web Registries/Catalog Server, a Multiple-Protocol Geoinformation Client (MPGC), and a toolbox (Di et al. 2001, 2002). The OWS interface specifications implemented in NWGISS include the Web Map Service (WMS), Web Coverage Service (WCS), and Catalog Interoperability Specifications[83] (CIS). The integration has taken place mainly on modification of NWGISS so that it works with the Globus toolkit. The following paragraphs provide brief descriptions of each component. Detailed information can be found in (Di et al. 2001).

- *The Map Server*: The map server enables GIS clients to access HDF-EOS data as maps. Currently the NWGISS map server complies with OGC WMS version 1.1.0. The OGC specification defines three interfaces, namely GetCapabilities, GetMap, and GetFeatureInfo. All three interfaces have been implemented and all three HDF-EOS data models (Grid, Point, and Swath) are supported.

- *Coverage Server*: the OGC Web Coverage Service (WCS) specification is designed to enable GIS clients to access multi-dimensional, multi-temporal geospatial data. WCS defines three interface protocols: getCapabilities, getCoverage, and describeCoverageType. The NWGISS coverage server has implemented versions 0.5 0.7, 1.0, and 1.0.2.0 of the OGC WCS specification. Three formats for encoding user-requested coverages are available in NWGISS, namely, HDF-EOS, GeoTIFF, and NITFF.

- *Catalog Server*: Both WCS and WMS have the GetCapabilities protocol for clients to find geographic data/maps and services available at servers. This protocol works nicely when a server has a small data archive. If the server has a large amount of data, the capabilities description, which basically is a data catalog, becomes very large. The catalog server allows GIS clients to search and find available geographic data and services in a NWGISS site based on the OGC catalog interoperability specification (CIS). The integration of this OGC catalog server with the MCS and RLS of the Globus Toolkit makes the retrieval of geospatial data and services Grid enabled.

- *Toolbox*: It contains tools for automated data ingestion and catalog creation. Currently, two types of tools are provided: the format conversion tools and XML capabilities creation tools. A third type of tools, the catalog creation tools, will be provided later.

- *The Multi-protocol Geoinformation Client (MPGC)*: It enables the access of multi-dimensional and multi-temporal geospatial data from multiple coverage servers in the form that exactly matches users' requirements, regardless of the original forms of the data in the servers. The easy-to-use and easy-to-install web Java client enables GIS users to easily access the HDF-EOS data as well as other formatted data in a unified way, greatly enhancing the interoperability and public use of EOS data. Coupled with the NWGISS Server packages, MPGC makes HDF-EOS data available to GIS users based on the Open Geospatial Consortium's (OGC) interoperability protocols

6.7.2 The Integration

The integration between OGC technology and Grid technology is taking place mainly between NWGISS and the Globus toolkit. The information model used for the integration is shown in Figure 6.1. The Grid software works as the security and information infrastructures on top of the computer Operating System (OS) and its File System (FS) and Database Management System (DBMS), while NWGISS software works as the application on top of the Grid security and information infrastructures to execute the secure sharing of geospatial data, information, and services. In order to do so, NWGISS servers have been modified at their backend so that they can talk to Grids. We call those NWGISS servers the Grid-enabled NWGISS servers. The Virtual Organization (VO) manages all sharable computers and associated resources. All machines and data users in the VO must be authorized with certifications from VO's Certificate Authorization (CA) center. OWS services will first be authenticated by the Grid security infrastructure as an authorized user and then use the Grid information infrastructure to complete the user requests. If all required resources for filling a user's request are available at the machine where the OWS service request is received, the services will be provided at the local machine without using the Grid infrastructure. Doing so will reduce the response time for filling user requests.

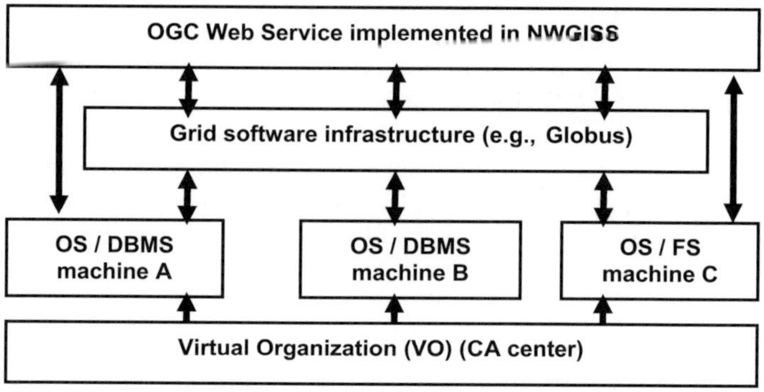

Figure 6.1. Integrating Grid technology with OWS I

In addition to the integration of the NWGISS and Globus for secured data access and processing, we also integrated the geospatial metadata information model with the Globus MCS[84] and RLS[85] information model to facilitate Grid-enabled geospatial data finding. Based on the metadata available at HDF-EOS data files (Heller 1994), and referring to ISO 19115 Geographic Information-Metadata[86] and FGDC Content Standard for Digital Geospatial Metadata: Extensions for Remote Sensing Metadata[87], we proposed a general geospatial metadata information model to describe the HDF-EOS datasets. At the database level, we integrated this geospatial metadata information model with the MCS, making geospatial data retrieval Grid enabled. Users first search the geospatial metadata and MCS/RLS to get the accessible physical file name of data items. Then, users retrieve the data items by the physical file names. After the integration, OGC clients will not notice any difference between Grid and non-Grid managed geospatial data.

6.7.3 The Sequence of Data Access after the Integration

The geospatial data access model in OGC assumes a two-step process; users first search metadata catalogs to find the datasets they want, and then send a data access request with a description of the form of the retrieved data. The data servers responding to the request normally need to process the data in the archive with a set of automated data transformation services in order to fulfill users' request. NWGISS

WCS and WMS servers have been implemented with some of the most commonly used data service functions, including spatial, temporal, and parameter-based subsetting and subsampling, georectification, reprojection, and reformatting. Those functions can be combined to produce data products that exactly match users' requirements for data. In the integration, the two-step model has been retained. Figure 6.2 shows the sequence for real data access.

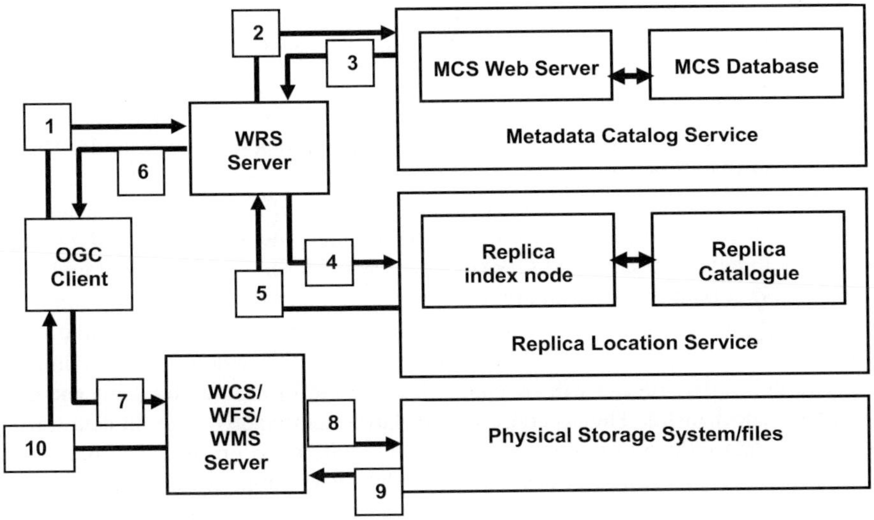

Figure 6.2. The Sequence for Real Data Access after the Integration. Query Data Flow: 1- OGC WRS Query, 2- MCS Query, 3- Logical Filenames, 4- Logical Filenames, 5- Physical Locations, 6- Results of WRS Query, 7- OGC Data Access Request, 8- Access to Physical files, 9- Data from Physical files, 10- Data back to OGC Clients

In our implementation, Grid-enabled NWGISS servers at the machine that receives the user request play the major role in responding to user requests. In the search, a user sends a query to the catalog server to search and find data through a client. User requirements on the data are expressed in the query. The WRS server converts the query to a MCS query and sends the converted query to the Grid MCS server. Then the MCS sever searches the MCS database to obtain matched logical file names (LFN). The LFNs are returned to the WRS server. The server then sends the LFNs to the Grid Replica Location Service to obtain the physical file names (PFN). The matched physical filenames and related metadata, encoded in OGC specified form by the WRS server, are sent back the client.

After the user receives the search results, the user may choose to obtain the data by sending an OGC data access request to one of NWGISS servers (e.g., WCS server) residing at a Grid node (the local node). The WCS server parses the query to find the physical file name contained in the query. Then it uses the name to determine if the requested datasets reside in the local machine. If the datasets exist locally, the data services of NWGISS will process the geospatial data sets locally, and return the results to the client. If the datasets are not available locally but are available on any other Grid nodes in the VO, then the NWGISS server on the local node will send a data request with the physical file name to the Grid nodes which host the data. Some of these Grid nodes may have only data but no services while others may have both. If the remote node contains only the requested data, not the service, the NWGISS server at the local node will initialize Grizd (grid-proxy-init) both locally and at the remote node. Then it invokes the Grid secure copy function (globus-url-copy) at the local node to copy data from the remote node. After the data are copied, the NWGISS server at the local node will perform the data transformation services to complete the user requests. If a remote node has both data and requested services, the NWGISS server on the local node will initialize Grid at both the local and remote node, and then call the Grid job submitting function (globusrun) to submit the user request to Grid services at the remote node. Grid services will invoke the NWGISS data services at the remote node to complete the user request. The results will be copied back to the local node by using the Grid secure copy mechanism. The second case will have significantly reduced data traffic, because the data reduction is done at the machine where the original dataset resides. All of the above processing is transparent to the OGC client which sends the data access request.

6.7.4 Current status of the implementation

Currently the integration of NWGISS software with Globus toolkit 2.2.2 has been successfully implemented. The success has extended the application of Grid technology to the geospatial community and also made OGC technology Grid enabled. Our experience with the project indicates that Grid technology has great potential for applications in the geospatial disciplines. By using Grid software as the foundation for geospatial data infrastructure, we achieve secured sharing of geospatial data and associated services while providing common OGC interfaces to users. This integration takes advantages of both Grid and OGC technologies. It provides the user community a standard, disciplinary-specific access to a huge volume of geospatial data available at space agencies and managed by Grid, while shielding the details of the Grid infrastructure underneath. The NWGISS software mentioned in this paper can be downloaded at our website[88].

The next step in this example project will be the integration of OGC geospatial Web services and the Open Grid Service Infrastructure to enable geospatial modeling and virtual geospatial data services under the secured Grid environment.

6.8 Future Directions in Geospatial Grid Research

Geospatial Grids is a new area in the geo-information science research. Although some work has been conducted so far, research on the traditional areas of geospatial Grids is still needed. The areas of research include the further extension of Grid functions to make them spatially aware, provision of standard geospatial interfaces, data models, metadata, and service catalogs, handling of the complexity and diversity of geospatial data, and the scaling up to production systems which can handle petabytes of geospatial data.

Both Grid and Web service communities have adopted Service-Oriented Architecture (SOA). The Open Grid Service Architecture (OGSA) represents a major step in the evolution of Grid technology[69]. OGSA will converge with Web services through the Web Service Resource Framework[89] (WSRF), which introduces a design pattern to specify how to use Web services to access "stateful" components[89]. Currently, OGC is defining standards for geospatial Web services. One of the major research areas in the development of the geospatial Grid will be to enable geospatial services under OGSA/WSRF, including the implementation of virtual geospatial product concepts, development of geospatial service modules and models, and construction and execution of service-based dynamic geospatial models under OGSA.

Semantic/intelligent geospatial Grids will be able to automatically answer many users' geospatial questions intelligently. Because of the large geospatial data volume, development of such capabilities is important for full utilization of geospatial data. Ontology-driven deductive reasoning for automated service model/workflow construction is one of the most promising approaches for reaching this goal. However, research in this area is just beginning.

In addition to the development of geospatial Grid technology, establishment of large-scale geospatial Grids for supporting operational sharing of geospatial data, information and resources at agency, national, or international levels is also very important. With the current on-going research, establishment of such geospatial Grids has become technically feasible now. However, the implementation of such operational geospatial Grid is not just a technical issue, but also a policy issue, organization issue, resource issue, and operational issue. Careful planning has to be done before such a large-scale operational geospatial Grid system becomes possible.

Chapter 7

Geospatial Semantic Web

Yaser Bishr

7.1 Introduction

The Mars Climate Orbiter was lost on the Red Planet on September 23rd 1999 because the mission's navigation team was unfamiliar with the spacecraft. The team failed to detect a mistake by engineers who delivered navigation information in English rather than metric units. This seemingly simple error led to the loss of the $125 million spacecraft, a loss that has complicated a subsequent Mars landing mission. While the root cause of the failure was the mix-up in units between navigation teams, the real problem was the lack of semantic interoperability between systems supporting the spacecrafts and navigation information teams. This problem is not limited to multi-team scientific projects but can occur in any collaborative enterprise.

During the last couple of years we have seen Geospatial applications become an intrinsic part of the enterprise. The next few years will see spatial capabilities increasingly woven into the fabric of the IT infrastructure to provide a slew of new enterprise applications. The result is a geospatially aware enterprise. As virtual enterprises become more prevalent, information sharing and remote service access across systems becomes a business requirement. The ability to combine systems becomes a challenge as result of hard, software and information heterogeneity.

Open standards have extended classical proprietary GIS into the enterprise, enabling disparate departmental or geographically distributed systems to be viewed or edited from the web, and used with other business systems. Geospatial technologies therefore must be designed to enable the rapid and ever changing business demands of the enterprise. Many experts view the Semantic Web to be a promising technology that will help systems and software engineers to quickly respond to those demands.

7.2 The Semantic Heterogeneity Problem

When we speak of interoperable information systems we mean that systems should be independent and yet can transparently communicate at all levels of the technology stack, specifically the information content layer. Heterogeneity mainly arises as a result of differences between any layer of the technology stack including but not limited to: operating system, network protocol, application interfaces and information content. In this section we will only focus our discussion on heterogeneity as a result of differences between information content in different systems.

Limited semantic-richness and semantic heterogeneity are perhaps the greatest deterrents to application interoperability throughout the IT world. Most systems tend to operate on the assumption that users cognitively "add" this information in the course of exploiting these data in an application, or that the applications themselves will add semantic-richness through hard coding the business logic in the application code. This makes it hard to change and adapt to ever changing needs of the enterprise. The lack of sufficient semantic information can lead to the following problems:

1. A condition of "semantic heterogeneity" between databases (i.e., differences in meaning and significance), which inhibits data sharing.

2. Users must perform tedious tasks that employ semantic-limited data (where the tedious nature of the tasks stem from the lack of semantics)

3. Users employ semantic-limited data inappropriately (in an application for which they are not intended)

4. Users misinterpret the meaning of semantic-limited data while using it in an application

These problems have serious and costly consequences for any enterprise. For example, semantic heterogeneity is widely regarded as the most significant obstacle to the successful sharing and exchange of data. It is difficult to reuse data between agencies or across different applications within the same agency. Data development efforts are often limited in scope, which severely limits the ability to apply available data to effective, multi-discipline decision-making. Worse still, we often undergo laborious data conversion tasks, translating data sources developed for one domain into a new domain, without understanding the limitations of using the data in the target domain. Of equal concern is the tremendous difficulty in integrating data obtained by different organizations. Limited semantics and semantic heterogeneity impede:

- Service personalization – the lack of semantics at the user level limits service customization and configuration, and in particular, the ability to tailor services for a given user and user situation.
- Application extensibility – the lack of semantics at the business process level greatly limits the interoperation and integration of distributed applications and business processes.
- Flexible querying of databases, schema (dictionary) integration, and automatic data translation – the lack of semantics at the database level greatly limits the interoperation and integration of distributed databases under common application frameworks, within an information community.

- Service automation – limited semantics and semantic heterogeneity inhibit service automation by requiring users to supply missing information, overcome interoperable connections, etc.

7.2.1 Data Heterogeneity

The web is dominantly viewed as a library where users search for and access information. Although the web is the newest medium for information flows, it is the fastest growing new medium of all time, becoming the information medium of first resort for its users. According to a study by the University of California, Berkeley, there are now 180,000 terabytes of information on the web. The current challenge is to get the maximum knowledge out the web in the minimum time possible.

7.2.2 Semantic Heterogeneity

Semantic heterogeneity is usually the source of most information sharing problems. Semantic heterogeneity can be in many forms, including but not limited to, differences in naming, scaling, confounding (e.g., "real-time news" does it mean 5 minutes delay, 10 minutes delay or no delay). The classes of conflict that arise from semantic heterogeneity are well documented in literature (Kim and Seo 1991), and beyond the scope of this book.

Semantic heterogeneity can also exist in geometric descriptions of features as a result of merging (integrating) different data sources. For example, suppose that within one domain a line is defined by two points, and in anther domain, a point is defined by the intersection of two lines. Merging these models results in a circular reference to the resultant tree model and therefore it is infinite.

Users from different communities, who share their data, are likely to share interest in a common understanding of the real world. For example, a soil mapping and conservation group is interested in defining detailed classification of TMUs (Terrain Mapping Units), their profiles, horizon, and rate of erosion. This differs from the way land-use group recognizes features. They identify natural vegetation areas, types of land-use at farm and plot levels, soil suitability for agricultural purposes, and irrigation schemes. The land-use group, however, regularly requires soil information, which can be retrieved from the soil conservation group.

7.2.3 Schematic Heterogeneity

Schematic heterogeneity is the difference in the class hierarchies and attribute structure of two independent database schemas. For example, several classification methods exist for ecological land classification (Sims et al. 1996). There are made several methods to classify soil units, e.g., the ASTM and SOTER classification methods (Engelen et al. 1995; ASTM 1985). Moreover, two databases might adopt the same classification method and have different class and attribute structures, which results in the schematic heterogeneity.

It is worth noting here that the line between schematic and semantic heterogeneity remains fuzzy despite the preponderance of literature because of one simple fact: schematic structure and organization often say something about the semantics of the data, and the semantics of the data at minimum suggest and at most dictate its structure and organization. In other words, the two concepts are intertwined.

7.2.4 Service Heterogeneity

Another way to view the web is to think of it as a space that enables users to access services remotely. This view has neither been widely exploited in the literature nor in the marketplace, in the sense that there is a lack of search engines that allow users to search for services rather than web pages. Web services do not merely provide static information but allow one to effect some action such as buying products online. The quest for inference services, which automate tasks, is a current focus of Internet research. The Web Ontology Language for Services (OWL-S)[90] is an attempt in this direction. OWL-S is a fundamental technology that will enables systems to locate, select, employ, compose, and monitor Web-based services automatically. An example of service heterogeneity is in web services that can differ in the service type they provide (e.g., Credit Card Verification vs buying service), or in the tasks they perform to execute the same service. There is also the semantics in the signature of the service that amounts to data heterogeneity.

7.3 The Semantic Web

The role of semantics for interoperability and integration of heterogeneous data, including geospatial information, has been long recognized (Bishr 1997). Associating formal descriptions to provide well-defined meaning to data and other web resources, such as services, captures semantics. Thus information processing (retrieval or integration) can be based on meaning instead of keywords. The Semantic Web aims to alleviate these problems—making Web resources more readily accessible to automated processes by adding information that describes content in a machine-accessible fashion.

A common underpinning for the Semantic Web is envisioned to contain several languages, as in Tim Berners-Lee's "layered Wedding cake" diagram, Figure 7.1, which he first presented at XML 2000. The diagram depicts a Semantic Web Architecture in which languages of increasing power are layered one on top of the other. The basis of a particular way of providing meaning for data is embodied in the model theory for the Resource Description Framework, RDF[91] and Web Ontology Language, OWL[92], the language at the base of the Semantic Web.

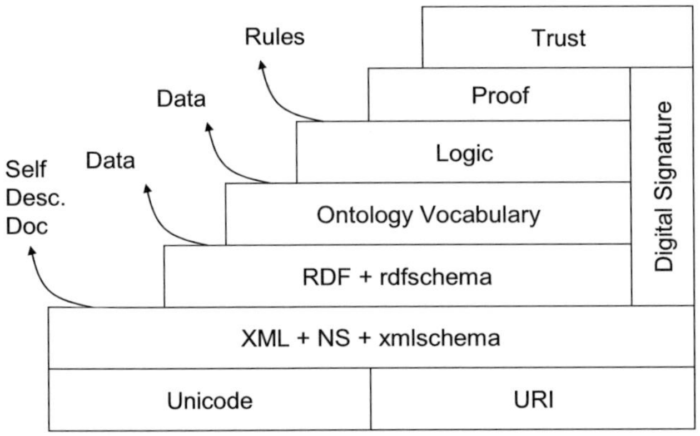

Figure 7.1. The Semantic Web wedding cake (Berners-Lee et al. 2001)

7.3.1 URI and Unicode

The bottom layer of this cake is the alphabet of the semantic web. It is constructed of Universal Resource Identifiers (URI) and Unicode. URIs are a generalized mechanism for specifying a unique address for an item. They provide the basis for linking information on the Internet. Unicode is the 16-bit extension of ASCII text, which has support for non-English characters.

7.3.2 The eXtensible Markup Language (XML)

The next layer up is the eXtensible Markup Language (XML). XML grew out of demands to make HTML more flexible. The technology itself has two aspects. It is an open standard which describes how to encode a tree-based data structure within a plain text file. On a more conceptual plane, XML is a strategy for information management. The rules of "well-formedness" provide enough information that generic XML parsers can be developed to read, write, and manipulate the XML files. The "well-formed" criteria guarantee that the parser can read the XML file, but from the application's point of view, it does not give any confidence that the data in the XML file will be complete or consistent. To solve this problem, the basic form constraint can be extended through the use of Document Type Definitions (DTDs) or Schema. Both of these technologies are ways of specifying a pattern or model, similar to database models, to which the XML document must also conform. At the conceptual level, XML asks that content be separated from presentation.

7.3.3 The Resource Description Framework (RDF)

Recall the example mentioned where a line is defined by two points in one application while a point is defined by the intersection of two lines in another application. XML cannot directly support the merger of these two models. A graph, however, can. Thus we introduce the Resource Description Framework (RDF). RDF defines a standard for describing the relationships between objects and classes in a general but simple way. Within the Semantic Web, for example, a primary use of RDF will be to encode the metadata information such as the title, author, and subject, associated with web pages. An RDF description is composed as a triplet of subject, property, and object. The property can be an attribute, relationship, or a predicate (verb). A simple example of RDF is shown in Figure 7.2.

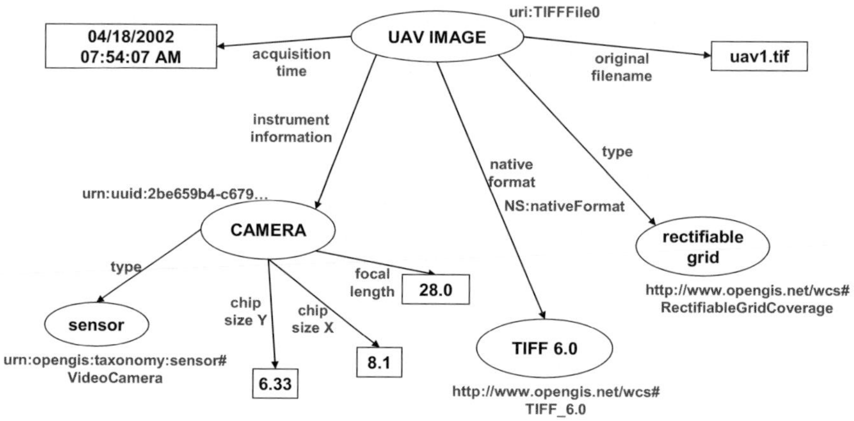

Figure 7.2. Example of RDF model

A key limitation of RDF is poorly defined semantics. RDF has no sense of vocabulary. It does not provide any notion of scope within which a specified vocabulary can be constrained. Any node within a connected RDF graph is reachable by any other node. To support automated reasoning, agreement on a uniform, well-defined, vocabulary is needed.

7.3.4 Ontologies

It is widely accepted in the knowledge engineering community that Ontology consists of taxonomy and semantic interconnections represented as Description Logic axioms. It also includes rules represented as Horn-Logic axioms. An ontology typically contains important concepts in a domain, along with descriptions of the properties of each concept. Ontologies will play a pivotal role in the Semantic Web by providing a source of shared and defined terms.

The recognition of the key role that ontologies are likely to play in the future of the web has led to the extension of web markup languages in order to facilitate content description and the development of web based ontologies. In previous sections we briefly talked about the XML, XML Schema, and RDF. RDF Schema (RDFS) is considered an ontology representation language: it talks about classes and properties (binary relations), range and domain constraints (on properties), and subclass and subproperty (subsumption) relations. RDFS is, however, a very limited language and more expressive power is clearly both necessary and desirable in order to describe resources in sufficient detail. For example, it is useful to be able to state that a property is functional or transitive and it is extremely useful to be able to describe classes in terms of the properties of the individuals that belong to them. Moreover, such descriptions should be amenable to *automated reasoning* if they are to be used effectively by automated processes, e.g., to determine the semantic relationship between syntactically different terms.

Recognition of the limitations of RDFS led to the development of new web ontology languages such as the Ontology Inference Layer, OIL[93], and the DARPA Agent Markup Language, DAML[94]. Both these efforts were later combined into a unified language known as DAML+OIL and submitted to W3C. DAML+OIL was used as the basis of a new W3C web ontology language called the Web Ontology Language, or OWL. As a second language for the Semantic Web, the W3C decided to use RDF/RDFS as the basis for encoding OWL.

To illustrate what ontology is, let us assume that we want to define that a person is a kind of animal, has parents from the same species, and can only have one father:

```
Class Person
Person is subClassOf Animal
Property hasParent
Domain of hasParent is Person
Range of hasParent is Person
Property hasFather
Domain of hasFather is Person
Cardinality of hasFather is 1
```

Note that we use pseudo code that's closely resembles the Web Ontology Language (OWL). Here we have defined sufficient criteria that can be used to identify any person. We have done so without actually talking about a particular person. This is called intensional definition. Ontology is about defining the intension of concepts and their relationships. An extension, for example, would be an actual instance of a person. Knowledge consists of: intensional definition of entities, extensional definition of entities, and rules that govern the behavior of those entities, as shown in Figure 7.3.

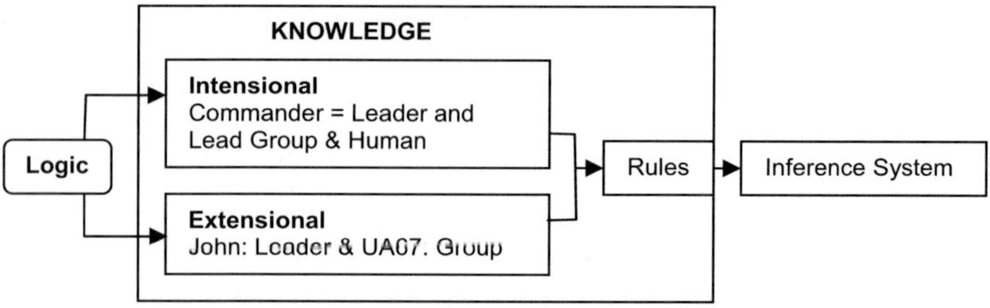

Figure 7.3. Components of Knowledge

One might argue that we still need to define the meaning of "person". Doing this would actually mean that we are going beyond ontology and defining what we call epistemology. Epistemology is the study of theories of knowledge or ways of knowing. What makes something to be truly a person, for example, has been investigated and answered by philosophers and is referred to by ontology. Therefore, ontology assumes some level of epistemological agreement, which is beyond the scope of this chapter.

Now that we have defined the concepts in the ontology we still need to define relationships between them. Suppose we want to state "a Person always walks above ground." The next set of axioms define the intensional relation "above":

```
above(x,y):
Height(Center_of_Mass(x)) > Height(Center_of_Mass(y)) and
Latitude(Center_of_Mass(x)) > Minimum_Latitude(y) and
Latitude(Center_of_Mass(x)) < Maximum_Latitude(y) and
Longitude(Center_of_Mass(x)) > Minimum_Longitude(y) and
Longitude(Center_of_Mass(x)) < Maximum_Longitude(y) and
```

We can then use the term "above" to indicate the intensional relationship between person and ground. The Web Ontology Language (OWL) can be used to encode this type of ontology. OWL has three increasingly expressive sublanguages: OWL Lite, OWL DL, and OWL Full. OWL Lite supports those users primarily needing a classification hierarchy and simple constraints. OWL DL supports those users who want the maximum expressiveness while retaining computation completeness where the axioms are decidable. OWL Full is meant for users who want maximum expressiveness and the syntactic freedom with no computation guarantees.

7.3.5 Logic and Rules

OWL provides a set of formal constructs to define basic semantic and ontological concepts. Ontology needs to describe a formal conceptualization within a particular domain that does three things:

1. Provides a semantic representation of each entity and its relationships to other entities;
2. Provides constraints and rules that permit reasoning within the ontology;
3. Describes behavior associated with stated or inferred facts.

The last two require the ability to create logical statements (rules) that allow the computer to make inferences. To facilitate and enable this capability, the RuleML Initiative developed an XML-based markup language that permits Web-based rule storage, interchange, and retrieval. RuleML[95] encompasses a hierarchy of rules that include reaction rules (event-condition-action rules), integrity constraint rules (consistency maintenance rules), derivation rules (implicational inference rules), and facts (premise less derivation rules). The Semantic Web Rule Language, SWRL[96], is similar to RuleML but uses OWL for its encoding. The W3C is currently using SWRL as the basis to standardize the effort.

7.3.6 Digital Signatures

Digital signatures are based on work in mathematics and cryptography, and provide proof that a certain person wrote (or agrees with) a document or statement.

7.3.7 Proof and Trust

Because the Semantic Web is an open and distributed system, in principle, users and systems alike, cannot determine the reliability of the available information, whether it is inferred or readily available. To deal with the inevitable situation of unreliable and contradictory statements (data and information) on the Semantic Web, there needs to be a mechanism where we can verify that the original source does make a particular statement (proof) and that source is trustworthy (trust). At this point, notions of proof and trust have yet to be formalized, and a theory that integrates them into inference engines of the Semantic Web has yet to be developed.

7.4 The Geospatial Semantic Web

In the previous sections we have introduced the Semantic Web technologies and showed the types of services they provide. In this section we will attempt to consolidate these technologies into a specific larger picture that we call the Geospatial Semantic Web (GSW). We will outline the basic ingredients that required to semantically enable the Geospatial Information Systems (GIS) of the enterprise.

We believe that the maturation of the Semantic Web technologies will lead into a new bread of GIS that is fundamentally different from current systems. GSW will have in its core the basic Semantic Web technologies as depicted in the

Semantic Web wedding cake. In our view, GSW will differ from current GIS in many respects in that it is real-time, reactive, and provides spatial reasoning services. GSW will also enable seamless integration with content management systems so that geospatial content are managed as a seamless part of the larger content assets of the enterprise.

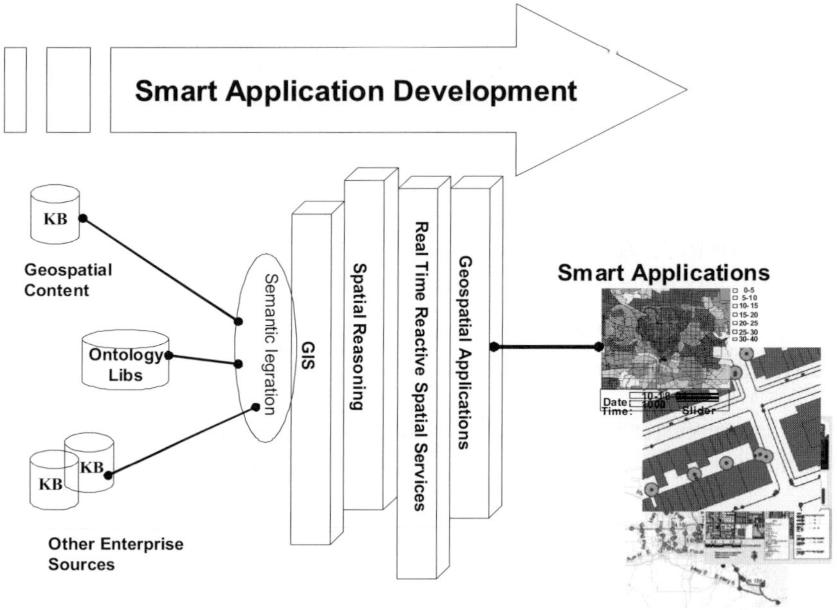

Figure4 Integrating Semantic Web Technology and the Geospatial Technology

Figure 7.4 shows a high level view of the GSW. In its core is a library of geospatial ontology based on well-defined models (see section on Qualitative Reasoning). The classical GIS with its powerful analytical capabilities are layered on top of a set of interactive and runtime tools for semantic integration and annotation. Through the use of Ontology, the tools will enable GIS to seamlessly access other enterprise resources. The tools will also enable enterprise content management systems to seamlessly access geospatial data. These tools may exist in two types:

- The interactive tools will enable users to map (annotate) their database schemas to the corresponding ontologies.
- The runtime tools will enable the integrated view of an enterprise through those mappings and the use of the inference capabilities.

As shown in the figure, the spatial reasoning capabilities add another dimension to the GSW by providing qualitative spatial reasoning. Triggering the GIS analytical and qualitative reasoning capabilities are done through a middleware layer that provides real-time and reactive capabilities. This layer provides well-defined APIs to build specialized GSW applications.

7.4.1 Qualitative Reasoning

The principal goal of spatial reasoning is to make common-sense spatial knowledge explicit, so that given appropriate reasoning techniques, a computer could make predictions, diagnose and explain the spatial behavior of physical systems in qualitative manner without recourse to quantitative methods. Geometry focuses on quantitative aspects of space. Topology, on the other hand, answers qualitative questions about spaces, for example adjacency and directionality. To enable the qualitative reasoning in GSW there must be an ontological model that captures the qualitative aspects of space.

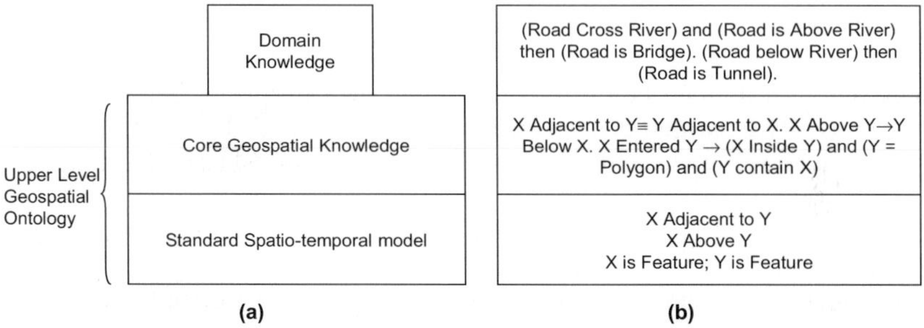

Figure 7.5. Ontological Model of Space and Time

Figure 7.5a depicts a proposed three-layered ontology for GSW. The first two layers constitute what we call the Upper Level Geospatial Ontology:

1. In the first layer a model of basic topologic, geometric types, and the notion of features and feature properties are defined. A simple example of this layer is shown in Figure 7.5b. In the example, *X Adjacent to Y, X Above Y, X is Feature; Y is Feature* are facts that can be encoded in OWL.

2. The default spatial inference and rules between features are defined in the Core Geospatial Knowledge. A simple example of this layer is shown in Figure 7.5b. Here *X Adjacent to Y≡ Y Adjacent to X* can be encoded in OWL. *X Above Y→Y Below X and X Entered Y → (X Inside Y) and (Y = Polygon) and (Y contain X)* are conjunctive rules that can be encoded in SWRL.

3. The third layer is the application layer. These are specialized ontologies for specific application, e.g., roads, rail, environmental engineering, etc. A simple example of this layer is shown in shown in Figure 7.5b. In this example *(Road Cross River) and (Road is Above River) then (Road is Bridge), (Road below River) then (Road is Tunnel)* are conjunctive rules that can be written in SWRL

There exists, in fact, a fourth layer that is not shown in the figure. This is where the domain knowledge is mapped to actual database instances. This step is usually called semantic annotation and is often done using the interactive tools mentioned above.

7.4.2 Spatial Inference Engine

In the previous section we briefly discussed a proposed geospatial ontology framework to represent space and time. The framework forms the knowledge base for GSW. For the geospatial ontology to be useful we need some capabilities that can reason with it. An inference engine tries to derive answers from a knowledge base. It provides a methodology for reasoning about the information in the knowledge base, and for formulating conclusions. The knowledge base will naturally be the ontologies as defined before and the actual data instances. Several inference engines currently exist that can consume ontologies encoded in OWL[97,98]. Generally speaking, an inference engine has three main elements (Mestel and Albus 2002):

1. An interpreter. It executes the chosen agenda items by applying the corresponding base rules.
2. A scheduler. It maintains control over the agenda by estimating the effects of applying inference rules in light of item priorities or other criteria on the agenda.
3. A consistency enforcer. It attempts to maintain a consistent representation of the emerging solution.

7.4.3 Real-time GSW

A real-time GSW is one in which the correctness of the computations not only depends on their logical correctness (i.e., the business logic), but also on the time at which the result is produced. In other words, a late answer is a wrong answer. Real-time applications provide an action or an answer to an external event in a timely and predictable manner (Nissanke 1997). The concept of predictability may vary from field of operation, but for real-time applications it generally means that a task or set of tasks can always be completed within a predetermined amount of time. Depending on the situation, an unpredictable real-time application can result in loss of data, loss of deadlines, or loss of plant production. Examples of real-time applications in GSW include tracking mobile objects, notification service for spatial events (e.g., entering a building, increased water level in a dam, etc.), and automatic vehicle routing.

7.4.4 Reactive GSW

A reactive GSW would respond to external *stimuli* and triggers *events* that may be perceived by its observer (observer here can either be human or software). The stimuli are usually measured using some external sensors (Kulkarni 2002). The events are triggered through a set of actuators. For example in GSW a sensor that monitors the air quality sends information to a GSW application. The actuators may trigger a series of spatial analytical processes to compute the affected area once the value reaches a pre-defined threshold. This kind of applications, integration with sensor technologies with Geospatial technologies cannot be realized without full account for GSW.

7.5 Issues Arising from Implementing GSW

In the previous sections we have seen types of capabilities that can exist in future GSW. The ability to have seamlessly integrated systems and provide real-time reactive geospatial services will open a new window of innovative applications for GIS. As with any new technology, adding semantics will introduce some new issues and concerns that must be dealt with, for example linking ontologies with legacy databases. Subsequent sections will briefly discuss some of those issues.

7.5.1 Ontology evolution

Ontologies are the building blocks for the Semantic Web in general and the GSW in particular as they provide a reusable piece of knowledge about a specific domain. Those pieces of knowledge are often not static, but evolve over time. Domain changes, adaptations to different tasks, or changes in the conceptualization require modifications of the ontology. The evolution of ontologies will hamper their effective reuse. Support to handle those changes is needed. This is especially important in a decentralized and uncontrolled environment like the web, where changes occur without attunement. Therefore, a versioning methodology is needed to handle revisions of ontologies and the impact on existing sources. There are mainly three causes for changes in the ontology:

1. Changes in the domain: For example an ontology that restricts marriage to only opposite sex, may change to relax this restriction.
2. Changes in conceptualization: For example earlier Greek philosophers viewed the earth as flat.
3. Changes in explicit specifications: For example in a new ontology version a knowledge engineer may consider adding some new properties to specific classes.

Multiple versions of the same ontology are bound to exist and must be supported. Not knowing how ontology is being reused means not being able to "force" applications to switch to a new version of the ontology. The management of ontology changes is therefore the key issue in the support for evolving ontologies. The problem of ontology evolution is more apparent when there dependencies between data sources, applications, and ontologies. Changes in ontologies will thus have far-reaching side effects. It is often not practically possible to synchronise the changes to an ontology with modifications to the applications and data sources that use them. Klein and Fensel (2004) defines four types of ontology changes:

- **Full compatible revisions** (upward and backward): the semantics of the ontology is not changed, e.g., syntactic changes or updates of natural language descriptions; this type of change is compatible in both *prospective* use and *retrospective* use.
- **Backward compatible revisions**: the semantics of the ontology are changed in such a way that the interpretation of data via the new ontology is the same as when using the previous version of the ontology, e.g., the addition of an independent class; this type of change is compatible in *prospective* use;
- **Upward compatible revisions**: the semantics of the ontology is changed is such a way that an older version can be used to interpret newer data sources correctly, e.g., the removal of an independent class; this revision is compatible in *retrospective* use.
- **Incompatible revisions**: the semantics of the ontology is changed in such a way that the interpretation of old data sources is invalid, e.g., changing the place in the hierarchy of a class; this type of change is incompatible in both *prospective* use and *retrospective* use.

7.5.2 Semantic Annotation

Sharing data requires sharing concepts and hence sharing ontologies. Semantic annotation is about binding ontologies to actual geospatial resources (data and services). The relationship between ontologies and geospatial resources can be many-to-many. A number of annotation tools for producing semantic markups exist. Ontology and databases are linked to provide powerful utilities to access, interpret and use data. The relationship between Ontology and databases is rather strong. Ontology and DBMS have the following tasks:

1) DBMS deal with persistence of data and with the management of large amount of it.
2) Ontology and Ontology representation manage intensional knowledge.

Advantages of extending DBMS with Ontology are

1) Ontology is accessed by reasoning tools that can bring to the conceptual modeling phase significant advantages, as compared with traditional languages, whose role is limited to modeling. For instance by using consistency rules in DL one can verify at design time whether an entity can have at least one instance, thus clearly saving all the difficulties arising from discovering such a situation when the database in being populated.

2) By expressing the queries to a database in ontology querying language, one gains the ability to classify them and therefore to deal with issues as query processing and optimization

3) Query with intensional expressions rather than only query instances. We evaluate data based on extension, RDMBS does not allow facilities to query the intensional definitions, i.e., ontology.

4) Ability to reason about views. Ontology can be used as a filter with which data in conventional RDBMS can be viewed. The interface to the data was made significantly more appropriate for a data analyst and views that were found to be productive could be saved. In fact they were saved in a taxonomy and could be classified with respect to one another. In a sense this allows the schema to be viewed and queried explicitly something normally not available when using a raw DBMS directly.

7.5.3 Performance and Scalability

Simply introducing languages is not enough. We need an ontology-based computing infrastructure that includes: ontology development and management tools, content creation systems, storage and retrieval systems, ontology reasoning and mediation, and lastly, integration of reasoning with real-world applications. For this vision to become practical, ontology-based technology must be scalable and show acceptable performance. This means that issues associated with striking the balance between expressiveness of description logic must be balanced against tractability of computation.

7.6 Conclusions

Semantic heterogeneity remains to be a major barrier for knowledge sharing, spatial and non spatial. In this chapter we highlighted the different aspects of semantics heterogeneity and introduced the rapidly evolving Semantic Web technology as a solution to the problem. We also provided a high level view of the Semantic Web and its different components.

Semantic Web is not only used to integrate databases; it also provides a set of core capabilities that enable smarter services. Making GIS real-time and reactive and integrate with reasoning capabilities will open the door for a new breed of geospatial applications.

Semantic technology is not mature yet and several problems are still being investigated. We believe that it is going to be few years before we see industrial strength semantic technologies.

Acknowledgement

The sections on Geospatial Ontology Layers and Context where partly developed under Contract: W9132V-04-0020, from the National Geospatial Intelligence Agency (NGA). The Author would like to thank NGA for its support. The results of this work, however, do not reflect the adopted views of NGA.

Chapter 8

The role of DBMS in the new generation GIS architecture

Sisi Zlatanova and Jantien Stoter

8.1 Introduction

Since the early '90, Geographical Information System (GIS) has become a sophisticated system for maintaining and analysing spatial and thematic information on spatial objects. DBMSs are increasingly important in GIS, since DBMSs are traditionally used to handle large volumes of data and to ensure the logical consistency and integrity of data, which also have become major requirements in GIS. Today spatial data is mostly part of a complete work and information process. In many organisations there is a need to implement GIS functionality as part of a central Database Management System (DBMS), at least at the conceptual level, in which spatial data and alphanumerical data are maintained in one integrated environment. Consequently DBMS occupies a central place in the new generation GIS architecture.

An extended description on how GIS architecture has evolved can be found in Vijlbrief and Oosterom van 1992. GISs used to be organised in a dual architecture consisting of separated data management for administrative data in a Relational DBMS and spatial data in a GIS. This was caused by different nature of alphanumerical and spatial data, and the inability of early DBMSs to handle spatial attributes. In the dual architecture (Figure 8.1 left) the two parts are connected with each other via links based on unique id's. The spatial attributes are not stored in the DBMS and therefore they are unable to use the traditional database services such as querying and indexing. In the dual architecture consistency of the data is hard to manage. For example if a parcel is deleted in the spatial part, subjects can no longer have a relationship with this parcel, which is maintained in the non-spatial part.

The solution to the problems of dual architecture was a layered architecture in which all data is maintained in a single RDBMS. Since spatial data types were at that time not supported at DBMS level, knowledge about spatial data types was maintained in middle ware (Figure 8.1 middle). Spatial information was maintained in the DBMS by means of Binary Large Objects (BLOBs). SQL cannot process data stored as

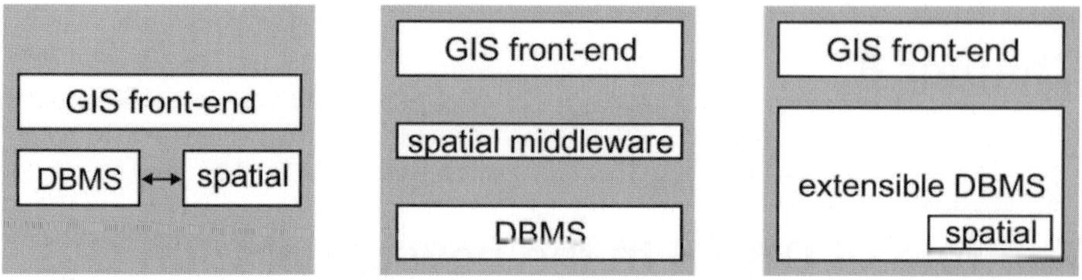

Figure 8.1. Evolving architectures of GIS. Left: dual architecture; middle: layered architecture, right: integrated architecture (courtesy of Vijlbrief and Oosterom van 1992).

BLOBs and therefore the data depends on the host application code, which handles the data in BLOB format. This solution requires data transport from the DBMS to middle ware and consequently queries cannot be optimally implemented.

In recent times DBMSs are evolving towards an integrated architecture in which all data is maintained in one object-relational DBMS (Figure 8.1 right). Presently, most mainstream DBMSs support spatial data types and spatial functions by means of Abstract Data Types (ADTs). This architecture is more beneficial for the integrity and consistency of the data.

This chapter is devoted to the role of DBMS in new generation GIS architecture and focuses on the manner spatial data can be managed, i.e., stored and analysed, in DBMSs. Two important aspects of DBMS functionality are addressed in detail, i.e., spatial models and spatial analysis. Special attention is placed on the third dimension (3D) because of the increased demand for 3D modelling, analysis and presentations in many applications. The discussion in this chapter is restricted only to vector models, i.e., raster models are outside the scope of this chapter.

The chapter is organised as follows: Section 8.2 outlines modelling of spatial features in DBMSs, both using geometrical primitives and topological structure. Section 8.3 makes an overview of possibilities to perform spatial analysis in DBMS. Section 8.4 provides case studies and elaborated discussion on topology versus geometry in DBMS. Section 8.5 is devoted to the third dimension. It starts with examples of 3D GIS applications and elaborates on available techniques and new developments required for full 3D support. The chapter ends with a discussion on the role of DBMSs in the new generation GIS architecture considering both data storage and spatial functionality.

8.2 Spatial models in DBMS

A lot of progress is observed in the management of spatial and non-spatial information for objects in one integrated DBMS environment, called a geo-DBMS. The OpenGeospatial Consortium (OGC) largely contributed to this progress. The OpenGeospatial Consortium adopted the ISO 19107 international standard (ISO 2001) as Topic 1 of the Abstract Specifications: Feature Geometry[99]. These Abstract Specifications provide conceptual schemas for describing the spatial characteristics of spatial objects (geographic features, in OGC terms) and a set of spatial operations consistent with these schemas and with vector geometry and topology up to three dimensions embedded in 3D space. According to the specifications, the spatial object is represented by two structures, i.e., geometrical structure, i.e., simple feature, and topological structure, i.e., complex feature. While the geometrical structure provides direct access to the coordinates of individual objects, the topological structure encapsulates information about their spatial relationships.

The OGC Abstract Specifications have been transformed into Implementation Specifications, of which the most relevant for DBMSs is the OGC Simple Features Specification for SQL (OGC 1999), which provide guidance on how spatial objects have to be maintained in object relational DBMS environments. Sub-section 8.2.3 briefly describes the implementation strategies of mainstream DBMSs. Since the native maintenance of topology by DBMS is still in an ascent stage, special attention is paid to topological models. The problems associated with 3D topological models as well as some prototype implementations are discussed in detail.

8.2.1 Geometrical primitives in DBMS

To date, mainstream DBMSs such as Oracle, (Oracle 2001), IBM DB2 (IBM 2000), Informix (Informix 2000), Ingres (Ingres 1994), Postgres (PostGIS) and MySQL have implemented spatial data types and spatial operators more or less according to the Simple Features Specification for SQL of OGC. The implementation is based on Abstract Data Types (ADTs) that support storage, retrieval, query and update of simple spatial features, i.e., points, lines and polygons, and a set of spatial operations built on top of them.

Currently, no 3D primitive is implemented. However, most DBMSs, including Oracle, Postgres, IBM, Ingres, Informix, support the storage of simple features in 3D space. In general, it is possible to store for example a polygon in 3D. 3D volumetric objects can be stored in a geometrical model as polyhedrons using 3D polygons, i.e., a body with flat faces, in two ways: as a set of polygons or as multipolygon. The multipolygon is one type consisting of several polygons.

Our tests with different representations have shown advantages and disadvantages of both approaches (Stoter and Zlatanova 2003). An advantage of 3D multipolygons compared to list of polygons is that they are identifiable as one object by front-end applications, e.g., GIS and CAD, which can access objects stored in the DBMS. Furthermore, this approach has one-to-one correspondence between a record and an object. The major disadvantage of both implementations is that the DBMS does not recognise the 3D object, e.g., volume cannot be computed. In addition, spatial functions on 0D, 1D and 2D primitives defined in 3D space project the spatial data on a 2D plane. The way out is support of real 3D volumetric data types. A possibility of having a 3D geometrical primitive at DBMS level is shown in Arens et al. 2003. In this work, a 3D primitive as polyhedron is defined as part of the geometrical spatial model of Oracle Spatial, including validation functions and spatial functions in 3D.

8.2.2 Topological structures in DBMS

Topological structures are generally used to represent planar or space partitions without redundancy and to represent (linear) networks. In planar and space partitions, spatial objects are defined on the basis of non-overlapping partitions. A large number of 2D topological structures are already available in the literature, of which some of them have been implemented in commercial and non-commercial systems and populated with data (LaserScan[100]; Oosterom and Lemmen 2001; Oracle Spatial 10g[101]). Many 3D topological structures are also reported but only a few of them are further extended to support spatial operations.

OGC has also recognised the need for topology standardisation. Implementation Specifications for topological structures, i.e., complex features in OGC terms are currently being developed by the OpenGeospatial Consortium in cooperation with ISO. The new interfaces will build on the OGC Simple Features Specification to address feature collections and more complex objects and concepts including curves and surfaces in 2D and 3D, compound geometries, arcs and circle interpolations, conics, polynomial splines, topology and solids. The interfaces will cover creation, querying, modifying, translating, accessing, fusing, and transferring spatial information.

To illustrate current state of the art of topological structure in DBMS, this section continues with the view of OGC and ISO about topological primitives (sub-section 8.2.2.1), a non-commercial implementation of 2D topological structure (sub-section 8.2.2.2) and commercial solution LaserScan Radius Topology[100] and Oracle Spatial 10g (sub-section 2.2.3). Similarly to a 3D geometrical primitive, 3D topological structure has not (yet) been implemented as part of a DBMS. Sub-section 2.2.4 describes non-commercial implementations of a 3D topological structure.

8.2.2.1 ISO and OGC about primitives

'Face' is the topological equivalent of the geometrical primitive 'polygon'. There is a slight difference between the definitions of face and polygon in ISO and the OGC specifications for SQL.

According to the ISO standard a face is defined by edges as the orientation is strictly defined (anti-clockwise). A face can have holes (called inner rings) and the orientation of the edges is in clockwise order. Every edge has a reference to preceding and succeeding edges. The definition of a polygon needs refinements. For example, it is not clear whether the outer boundary of a polygon is allowed to touch itself nor is it clear if inner rings can touch the outer boundaries or other inner rings (Oosterom et al. 2003). However, since only one outer boundary is allowed, a polygon with two outer boundaries defining potentially disconnected areas is certainly invalid.

As was stated before, the OpenGeospatial Consortium adopted the ISO Spatial Schema as Abstract Specifications and transformed these to the implementation level in the OGC Simple Feature Specification for SQL. Since the OGC Specifications for SQL are based on geometry, only the geometrical primitives are defined. A polygon is defined as a simple, planar surface. Rings may touch each other in at most a point and self-intersection of outer and inner rings is not allowed (Oosterom et al. 2003). Inner rings, which divide the polygon in disconnected parts, are also not allowed. Note that OGC does not address orientation of polygons.

8.2.2.2 Non-commercial implementation of 2D topological structure in DBMS

A lot of implementations of non-commercial 2D topological models are currently in use. An excellent example is the topological model developed and implemented by the Netherlands' Kadaster (Oosterom and Lemmen 2001). The model is based on the winged-edge structure (Baumgart 1975) The most important tables are 'boundary', i.e., cadastral boundaries, and 'parcel', i.e., parcel identifiers. The edges in the boundary table contain references to other edges according to the winged edge structure, which are used to form the complete boundary chains, i.e., parcels. The edges also contain a reference to the left and right parcel. A parcel has exactly one reference to one of the surrounding boundaries and one reference to a boundary of each enclave. The structure of the topological references and the relationship between parcels and boundaries is visualised in Figure 8.2.

The apparent disadvantage of non-commercial topological structures organised in the DBMS is the fact that DBMSs are not aware of the geometry of spatial objects. To be able to perform some spatial operation, e.g., compute area, a function that 'realises' the geometry from the topological relationships has to be developed to support the topological model in a GIS environment. Several approaches are possible depending on the underlying topological structure. Two different implementations of such a

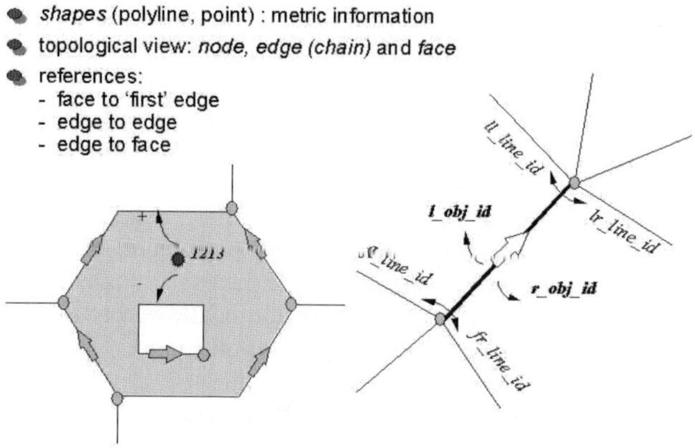

Figure 8.2. Topological structure in the spatial DBMS of the Netherlands' Kadaster (courtesy of Oosterom and Lemmen 2001)

function called 'return_polygon' are presented by Quak et al. 2003. The first solution uses the information on the relationship between edges. The second solution is based on the left-right information of edges.

In the first implementation, the function creates a polygon geometry according Oracle Spatial and ISO rules: the coordinates of the outer ring are listed in anti-clockwise order and the coordinates of the enclaves are listed in clockwise order. To generate the parcel polygon, the function starts with the first boundary, which is referred to in the parcel table. Then the connecting boundary in anti-clockwise order is found. This step is repeated until the polygon is closed. The polygon is constructed by connecting all linestrings of the found boundaries. In case of enclaves, the same procedure is followed but in clockwise order.

The second implementation of the 'return_polygon' function uses only the left-right information stored with every parcel boundary and a (geometrical) comparison to find and join connected boundaries in a ring. Here the boundaries that have the given parcel to the left or to the right are selected. The boundary of the parcel is composed by repeatedly joining boundaries that end at the same endpoint. Enclaves are realised in the same way. The orientation of the rings, i.e., clockwise or counter-clockwise does not follow from the algorithm and must be calculated afterwards.

The implementations differ in the underlying geometrical primitive. In the winged-edge implementation the outer ring of a face can touch itself in the outer boundary at exactly one point and in the left-right implementation this is not possible. This difference can be illustrated by the polygon as shown in Figure 8.3: a polygon that has an island that touches the boundary in exactly one point. The relationship-between-edges algorithm will generate a polygon with one self-touching outer ring which is not valid according to the OGC definition, while the left-right algorithm will return a polygon with a boundary and an island.

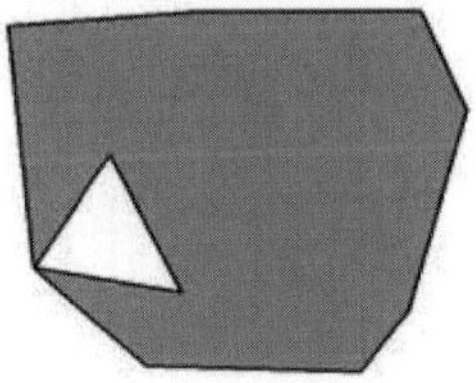

Figure 8.3. A polygon with a hole that touches the boundary

The performance of both implementations also differs and is dependent on the complexity of the data: the more points in a boundary, the worse the performance. This is especially true for the left-right implementation, where the computational cost increases with the number of points. Also the more boundaries there are in a polygon, the worse the performance. This is especially true for the relationship-between-edges implementation since the more boundaries there are; the more select statements need to be performed.

8.2.2.3 Commercial implementation of 2D topological structure in DBMS

Compared to the user-implemented models, the implementation of topology structure in LaserScan Radius Topology[100] is much more extensive, it is a 'complete' implementation of topology with support for linear networks and planar topology, including updates, insertions and deletions. All required topological references are stored explicitly: the winged edge representation in the edge-to-edge table makes up just a small part of the complete system (see Figure 8.4). Topological primitives are stored in the NODE, EDGE and FACE tables while faces are only stored by references to edges. A number of reference tables are used to store various types of topological references. The TOPO table is the link between the features and the topological structures. Topology is organised in 'manifolds'. Associated with each manifold and with the system as a whole are some metadata and error tables. Before topologically structuring data in Radius Topology, the user can specify rules in order to control the way the structuring works, such as snap tolerances, which features/primitives are moved and which stay while snapping, etc.

To realise geometry from a topologically structured dataset Radius offers a 'get_geom' function that is equivalent to the 'return_polygon' function of the non-

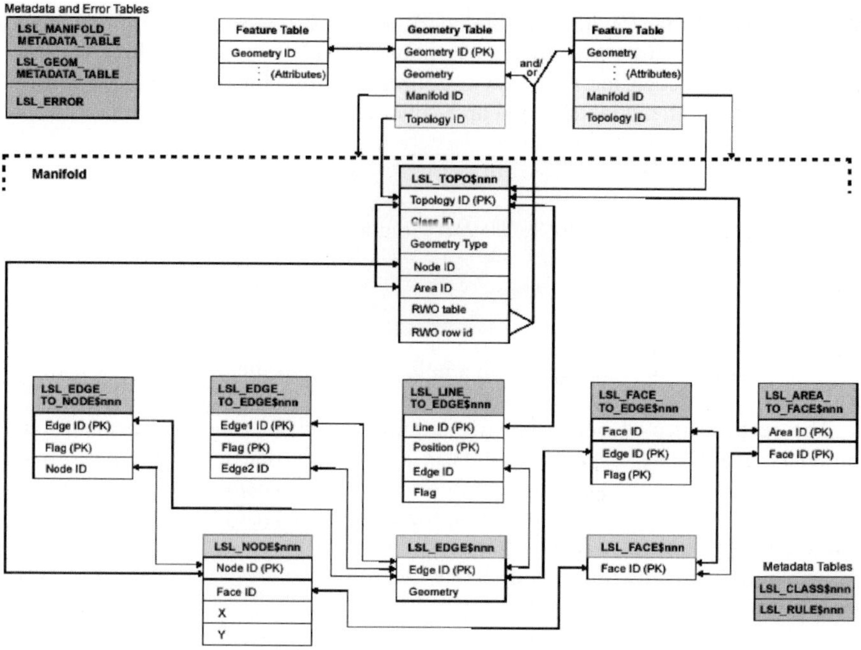

Figure 8.4. Radius Topology database tables (version 1.0), taken from LaserScan[100]

commercial implementations as described above. Most users however choose not to use this function, but instead store a copy of the geometry explicitly. This increases the storage requirements, but it means that there is no performance penalty when accessing geometries (e.g., for display or geometric queries) since the geometry is instantly available and does not have to be computed. The use of database triggers in the Radius Topology architecture ensures that the geometries and their topological representation are always synchronised. Additionally support for topological querying (containment, adjacency, connectivity, overlap) is available by means of a topo_relate operator.

Louwesma et al. (2003) describe a performance test in which topology structure of Laserscan Radius Topology was compared to the geometrical primitive of Oracle Spatial 9i. In the topology case less points are stored by avoiding storing 'common' boundaries twice. However disk space requirements where much bigger in the topology case due to the increased number of topology primitives and references between them compared to the number of area features and the way geometry is implemented in Oracle Spatial, small objects have relatively much overhead. The total storage requirement for topology is intended for references, id's and associated indexes that are required for the Radius topology structure. The storage requirement will probably be more favourable towards topology in the case of smaller scale data and data with a relatively high number of intermediate points in the boundaries.

From the tests described in (Louwesma et al. 2003) it can be concluded that performance of geometrical querying on a data set topologically structured with Radius Topology is slower. This is due to the cost of computing the geometries on-the-fly from the topological information. This occurs when geometries are not stored explicitly alongside the topology. For this reason users often store the geometries explicitly as described above.

Oracle Spatial 10g also supports 2D topology[101]. The basic topology elements in an Oracle topology are nodes, edges and faces. A node is represented by a point and can be used to model an isolated point feature or edges. Every node has a coordinate pair associated with it to describe the spatial location of the node. An edge is bounded by a start- and an end-node and has a coordinate string describing the spatial representation. Each edge can consist of multiple vertices, represented by linear as well as circular arc strings. As each edge is directed, it is possible to determine which faces are located at the left and right hand side of the edge. A face is represented by a polygon (that can be reconstructed from the several edge strings) and has references to a directed edge on its outer and (if any) inner boundaries. Each topology has a universal face that contains all other nodes, edges and faces in the topology.

Penninga and colleagues (Penninga 2004) tested the support of topology with two data sets from the Netherlands cadastre. The tests have shown that the current implementation does not completely avoid redundant data storage. The geometry is stored both in node and edge tables. However, as long as the user uses the supplied tools for data editing instead of directly updating the node, edge and face tables, data consistency can be efficiently maintained. In general Oracle topology can be considered a very suitable solution for the average user. The expert user might need to edit directly on the node, edge and face tables, as it is much quicker. More experiments are needed to explore the offered functionality.

8.2.2.4 Non-commercial 3D topological structures in DBMS

In 3D, there is no consensus on a single topological structure. Different topological structures can be defined depending on the number of primitives to maintain, and also the number and nature of relationships to explicitly store. The problems of defining 3D topological structures are relatively many compared to 2D. Due to the large amounts of data and higher complexity, one data structure representing a specific topological structure, which is appropriate for a certain application, may not be easy to serve another application. Unfortunately, 2D topological structures are not directly extendable to 3D. 2D structures are mostly built around the properties of an edge. One edge has exactly two neighbouring nodes (begin and end) and exactly two neighbouring faces (left and right). This property is not true in 3D space. An edge can have more than two neighbouring faces, i.e., the order of the faces has to be specified. An extensive study on 3D topological structures is presented in Zlatanova et al. (2004).

One of these 3D structures (i.e., the Simplified Spatial Model, Zlatanova 2000) has been used to test 3D topological implementations in object-relational DBMS. In this structure, edges are not explicitly stored due to performance considerations (Coors 2003). The role of the edge (=boundary) in 2D is now overtaken by the face (=boundary). Nodes describe faces and faces describe bodies. The 1D primitive as part of a body, is not explicitly stored (see Figure 8.5).

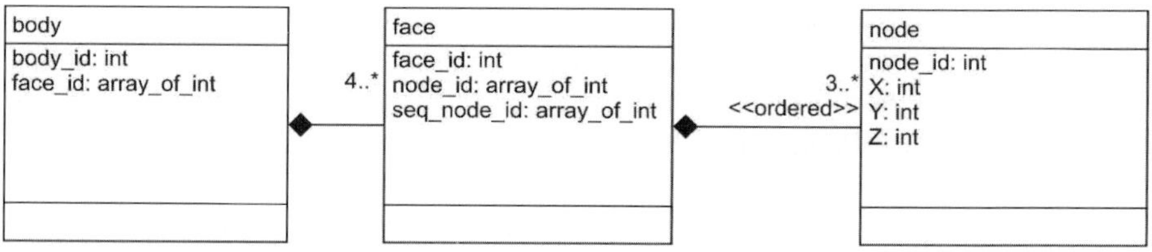

Figure 8.5. UML class diagram of Simplified Spatial Model (Zlatanova 2000).

This 3D topological structure can be implemented in various ways in a relational DBMS. The first straightforward approach is the relational implementation. The conceptual model can be converted directly into a relational data model. For each object (node, face, and body) a separate relational table is created. The NODE table contains the identifier of the node and the three coordinates. The FACE table contains the id of the face, a column denoting the anti clockwise order of the nodes in a face and the id's of nodes that the face consists of. A BODY table contains references to the id's of faces it consists of. Since the relationship between a face and constituting nodes is one-to-many (1:m), multiple rows represent one face. Theoretically, it is possible to have a separate column for each node but this approach is beneficial only in case the faces are triangles. Such an approach is reported by Coors (2003).

Another possibility is the object-relational implementation. The list of id's referring to lower-dimensional objects (faces, nodes) is stored in a newly defined object of type 'variable array' or 'nested table'. Such a new object type can consequently be stored in a single column. This means that the number of rows in the object table is reduced to the actual number of the higher dimensional object (body, face). Object relational implementation is a two-step procedure, i.e., creating objects (ADTs) and creating tables with columns of the created data types.

Similarly to the 2D case, the major disadvantage is that the DBMS is not aware of the spatial object. Spatial operations and spatial indexing offered by spatial DBMS cannot be used. To be able to use the spatial operations of DBMS, a function similar to 'return-polygon', as mentioned above for the 2D case, has been defined (Zlatanova 2001). This function realises the geometry of the 3D spatial objects, based on the topological tables and returns a 3D geometrical primitive as described before, i.e., a polyhedron consisting of flat polygons defined in 3D.

In general, regardless the type of topological structure used, the implementations have similar disadvantages as the ones specified for 2D user-implemented topological structures, i.e., the performance is relatively slow for certain queries. Operations on the topological model have to be further developed, to be able to use DBMS spatial operations, realisation has to be first performed. However, simple SQL statements comparing the identifiers of constituent primitives can perform a large number of topological operations. Section 8.3.3 describes a case study that illustrates the power of topological models in spatial analysis.

8.3 Spatial analysis in DBMSs

Spatial functionality is related to operations that are performed on spatial objects as very often no distinction is made between spatial and thematic components. In this chapter we will concentrate on the part of spatial functionality that is related to the spatial component. The most important aspect in the spatial domain is the framework for detecting spatial relationships. Amongst the different approaches (topology, distance, order, etc.) GIS society has largely accepted topology. OGC has adopted three topological frameworks: Boolean set of operations, Egenhofer operators and Clementini operators. Using these frameworks a large number of operators can be developed. However, the question 'who is responsible for the implementations' (front-end applications or spatial DBMS) is still open and even extensively discussed. Although several studies have shown that it is better to perform spatial operations close to the data, GIS vendors might not be willing to give up spatial analyses. In addition since spatial analyses in GISs have a long history, at least in the short-term feature spatial analyses in GIS front-ends will show better performance.

In the introduction (section 1) the new generation GIS architecture was made clear, in which all spatial data is maintained (and recognised as such) at DBMS level, without the intervention of middle-ware software. The functionality concerning spatial analyses that DBMS has to offer in such an architecture depends very much on the scope and constraints of spatial analysis: what is spatial analysis and what is spatial analysis in the DBMS context. DBMSs are essential in applications in which large amounts of large-scale geo-data need to be maintained and managed, such as cadastral data or spatial data used in municipalities. In general we can say that GIS functionalities that are not specific to a certain application belong in the DBMS and not in GIS (or CAD) front-ends. Examples are the spatial functions that examine the topological relationships between spatial objects. Arguments for this are the logical consistency of the data, better performance and better maintenance of the data, since unnecessary transport and conversions of data between DBMSs and GIS front-ends prone to errors can be avoided. On the other hand spatial functions that are specific for certain applications should be implemented in front-ends.

Here we consider the spatial analyses at DBMS-level that should be present to support the new generation GIS architecture. We will also give an overview on the spatial analyses that are currently supported in DBMSs. In the discussion (section 8.6) we will refine the statement on where spatial analyses should fit within the optimal architecture based on (past) trends both in computing technology, as well as in GIS.

8.3.1 Which operations in DBMS?

DBMS has not been designed to manage spatial objects, but does have a strictly defined functionality based on relational algebra and calculus (Ramakrishnan and Gehrke 2003). In principle, three generic operations are distinguished in the database literature (e.g., Tsichritzis and Lochovsky 1982): insert (add new data), delete (remove data from the database) and update (change existing data). A similar set of generic operations (but more elaborated) has to be available for spatial data. The operations related to introducing a new element, deleting and updating an existing one have to be extended with respect to the structure used, i.e., geometry or topology. Examples of such operations can be:

- operations to organise the data according to the used topological structure, i.e., operations for planarity, convexity and discontinuity as they are defined in the model.
- operators for consistency check: validation of the objects (e.g., polygon closed, body closed), node-on-line, node-on-face, node-in-body, line-on-face, line-in-body, intersection of lines, face-on-face, intersection of faces, face-in-body.

In addition to generic operations, DBMS offers a set of more elaborated operations known as selection, navigation and specialisation:

- selection: retrieve operation under a particular condition;
- navigation: describe the process of travelling through the database, following explicit paths from one record to the next in the search for some required piece of data;
- specialisation: complex operation that allows a new object to be created on the basis of existing ones.

Also these three generic operations have to be available for spatial data in a spatial context.

Selection in a spatial context is an operation that allows a number of data to be identified on the basis of three properties, i.e., logical position (first, last, prior), value of the content and/or relationship. An example of spatial selection operation is 'select all the buildings that are located next to the park'. Furthermore, three basic groups of spatial selection operations should be offered at database level (Zlatanova et al. 2004):

- metric operations: selection operations, requiring computations of geometric properties e.g., compute distance, volume, area, length, centre of gravity. The metric operations need coordinates of the spatial objects and the result is always quantitative.
- proximity operations: These are selection operations related to spatial location, e.g., objects in a certain area/volume, field of view. Position operations largely benefit of the spatial index offered by DBMS to restrict the search.
- relationship (topology) operations: selection operations based on spatial relationships. These operators will be dependent on the framework for detecting relationships. If topology is considered, the abstract specifications recommend three frameworks for implementation as was described before. Depending on the implementations (framework and model), topology operations are the ones expected to perform better on a topological model.

Besides selection, navigation and specialisation operations in a spatial context have to be also considered at DBMS level. Navigation is an operation that permits a logical path to be followed on the basis of a selection. Examples of spatial navigation operations are route planning (e.g., multiple topology operations 'meet'), shortest path (multiple topology operation 'meet' and multiple metric operation 'distance') and intervisibility. The navigation (in the database sense of the word, as it was explained above) should not be equated with spatial navigation (in the shortest-path sense). The latter is more of an analysis process using the data from the database and other external algorithms not standard within SQL or the DBMS. However, spatial navigation operations use the inbuilt navigation capabilities of the DBMS.

Examples of spatial specialisations are buffer, convex hull, union of objects and all types of generalisations. While navigation might be based only on topological operations, specialisations need in most of the cases the coordinates of objects. However, additional information retrieved on the basis of proximity or topology operations may be of use for some specialisations, e.g., union or generalisation.

In contrast to the group of selection operations, specialisation and especially navigation in the spatial domain can be very complex and time consuming. If they are performed at a DBMS level on the server, the performance can decrease drastically.

Furthermore, such complex operations may not be needed for all kind of applications. Therefore, these operations could be considered for implementation by the front-end.

8.3.2 Spatial operations currently offered by DBMSs

The spatial functionality presently offered by mainstream DBMS covers in general lines the scope of operations discussed in the previous sub-section. We use Oracle Spatial 9i to illustrate the possibilities of spatial analysis in DBMSs.

At database level, spatial operations can be defined utilising either geometrical or topological representations. It should be noticed (again) that the operations presently offered by the mainstream DBMS vendors (an exception is Oracle Spatial 10g) are built on the geometrical model due to lack of topology maintenance and therefore they work directly on the geometry (no geometrical realisation is needed). Oracle Spatial 9i supports the three groups of selection operations, i.e., topology operations, a variety of metric operations, proximity operations as well as simple specialisation operations. Another class of specialisation operations returns an aggregate of a collection of geometries. These are not defined within OGC. Tables 8.1 and 8.2 show examples of spatial functions implemented by Oracle Spatial and the equivalent OGC functions.

OGC	Oracle
Equals	EQUAL
Disjoint	DISJOINT
Intersects	ANYINTERACT
Touches	TOUCH
Crosses	OVERLAPBDYDISJOINT
Within	INSIDE
Contains	CONTAINS
Overlaps	OVERLAPBDYINTERSECT

Table 8.1. Topological operations in the DBMS according to Implementation Specifications of OGC and Oracle implementations

Unary operations

OGC	Oracle
Convexhull	SDO CONVEXHULL
Area	SDO AREA
Buffer	SDO BUFFER
Centroid	SDO CENTROID
Length	SDO LENGTH
Boundary	SDO MBR

Binary operations

OGC	Oracle
Distance	SDO DISTANCE
Intersection	SDO INTERSECTION
Union	SDO UNION
Difference	SDO DIFFERENCE
Symdifference	SDO XOR

Table 8.2. Metric and position operations according to Implementation Specifications of OGC and Oracle implementations

Although not that common, spatial analysis on topological structure is also available, e.g., Laser-Scan Radius Topology, Oracle Spatial 10g, but the knowledge is still limited. One of the most important operations, i.e., the realisation of geometry was already discussed in Section 8.2. The complexity of the functions considerably varies with respect to the different implementations. For example, the geometry (coordinates) of a body can be extracted by only one SQL statement in the case of relational implementation if the geometry is maintained explicitly, but an embedded script is required if the body is represented as 'variable array' of polygons.

In principle, all the topology operations have to perform better on the topological model than on geometry alternatives. On the other hand, some operations (compute area, distance, etc.) on topological structured data will be slower than on geometrical primitives since it requires querying and joining different relational tables, which is also discussed in (Hoel et al. 2003). Another explanation for the better performance of these spatial operations on the geometrical model is the internal optimisations provided by the DBMSs and the possibility to apply spatial indexes.

8.4 Topology or geometry

The statement that topological querying (queries that only require explicitly stored relationships) is much faster than in the situation where only simple geometry is available, was tested in a case study described in section 8.4.1. In section 8.4.2 it is discussed what to prefer: topologically structured data or geometrically structured data.

8.4.1 Case study

To illustrate the power of topological structure in performing relationship operations, we did an experiment in Oracle Spatial 9i, on a dataset, which is a selection of the cadastral database of the Netherlands. The test data set contains 1,788,019 parcels and 5,599,089 boundaries. The topological structure used for this dataset is described in section 8.2.2.4.

The query that we use in this experiment is to find all adjacent parcels to the parcel with object identifier 6862 (see Figure 8.6). The query was performed on both a topologically structured dataset and a geometrically structured dataset. The geometries of the parcels were therefore stored explicitly in the geometrical primitive of Oracle Spatial by means of the return_polygon function (section 8.2.2.2). A spatial index was built on the geometry-column to speed up spatial analyses.

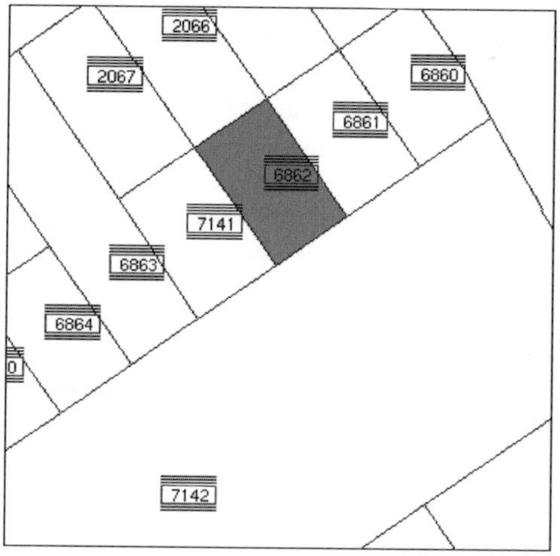

Figure 8.6. Data set used to perform the test query: 'find all parcels adjacent to parcel 6862'.

For the data set described by geometrical primitives, the query to find all adjacent parcels is given below using a 'subselect' structure, in which the polygons of parcels are stored in the table 'parcels_geom' in the column named 'shape'.

```
select object_id from parcels_geom where sdo_relate(shape,
(select shape from parcels_geom where object_id=6862),
 'MASK=TOUCH') = 'TRUE';
```

The query uses the built spatial index.

The query finds all parcels that have a 'touch' relationship with parcel '6862' using the spatial operator 'sdo_relate' which is implemented on geometrical primitives. The result is:

```
OBJECT_id
-------------
 7142
 2067
 2066
 7141
 2065
 6862
 6861
7 rows selected.

Elapsed: 00:00:22.05
```

In the topologically structured data set, all adjacent parcels to parcel '6862' can be found when all the boundaries are selected that have the specific parcel on the left or right side. The next step is to find the parcel that is located on the other side of the selected boundaries:

```
select l_obj_id, r_obj_id from boundary where
r_obj_id=6862 or l_obj_id=6862;
```

The result is:

```
L_OBJ_id          R_OBJ_id
--------------             ----------- -----
 2066             6862
 6862             7141
 6861             6862
 6862             7142

Elapsed: 00:00:00.01
```

The same test was performed for parcel '7142' with 28 adjacent parcels. The processing time for this second query was 22.56 seconds for the geometrical query and 00.01 seconds for the topological query. The queries were repeated a number of times which resulted in processing times of the same order every time. These examples show that the topological query is indeed faster on a topologically structured data set than on data set described with the geometrical primitives.

There is another conclusion that can be drawn from the first query: the results differ. The topological query does not give parcels '2067' and '2065' as a result since these parcels touch parcel '6862' only at a point and are therefore not seen as adjacent parcels from the topological point of view as defined in the winged-edge structure. The result set in spatial analyses using topological structure is therefore dependent of the topological structure implemented. The geometrical query does find parcels '2067' (neighbour on the right of parcel '2066') and '2065' as adjacent parcels since they do touch parcel '6862', even if it is only at a point. The geometrical query could be further specified by adding the condition that boundaries of two parcels should also overlap.

Although this case study is related to 2D data, a 3D comparison case study could even show bigger differences, as geometrical queries in 3D require more complex algorithms than 2D functions. This has a big influence on the computational complexity of topological queries that work on 3D geometrical primitives compared to topological queries that work on topologically structured data.

8.4.2 Discussion

The question whether or not to manage topology in DBMS is still challenging. Extensive argumentation for the need to organise the topology at DBMS level is given in Oosterom et al. 2002a. Some of the advantages are listed bellow:

- it avoids redundant storage and is therefore more compact than a geometrical model
- it is easier to maintain the consistency of data after editing
- it is more efficient during the visualisation in some types of front-ends, because less data has to be read from disk and transferred to clients
- it is the natural data model for certain applications; e.g., during surveying an edge is collected together with attributes belonging to a boundary
- it is more efficient for certain query operations, e.g., find neighbours

On the other hand geometrical queries on topological structure are much slower than on geometrical primitives, since a geometrical realisation is always required, by which several tables need to be queried that contain the lower dimensional objects. The same is true for visualisation of spatial data. Another problem of topologically structured data is the required storage capacity compared to the storage capacity needed for the geometrical primitives. Every row in the tables defining the topological structure has its overhead, and the references require a lot of storage capacity. An advantage of the topological structure is that topology structure management can be used in storage (maintaining consistency), data management and retrieval of data.

In principle, we believe that only one model, i.e., topological model, should be sufficient to manage spatial data in DBMS. However since performance of geometrical queries as well as of visualisation favours the geometrical model, which will be even more apparent in 3D, it could be argued that a DBMS that support both models simultaneously would be the most appropriate solution. In this case spatial objects are maintained in geometrical primitives and topological structure simultaneously and triggers are defined between the two representations to keep the two different representations up-to-date. Of course this solutions undermines the argument that topology structure avoids redundancy of the data. Another disadvantage is that the storage of spatial data becomes even less efficient compared to the topologically structured data.

8.5 3D and Geo-DBMS

The need for 3D information is rapidly increasing. Many examples of applications that have a growing interest in 3D information have been cited (Oosterom et al. 2002b; Oosterom et al. 2002c). Traditionally, the military applications were the first to look for 3D solutions and provided the first elaborated systems for 3D visualisation and simulation (Lindstrom et al. 1996). Nowadays more and more civil applications need the third dimension. In this section first examples of the growing need for 3D geo-information are given (section 8.5.1). In the second subsection (8.5.2) the question is addressed whether current DBMS technology is appropriate to meet the growing need for 3D geo-information.

8.5.1 Growing need for 3D information

The growing need for 3D geo-information is felt in different disciplines:

Applications in urban areas

- Urban planning is one of most demanding areas pushing 3D developers to provided fast modelling approaches, extended visualisation, interaction tools, and elaborated spatial functionality (Nebiker 2003; Shi et al. 2003). The influence of new buildings and infrastructure on the existing environment can be best visualised in 3D environments using virtual reality or augmented reality environments such as in Figure 8.7, which is important for presentation to citizens. In addition, 3D visualisations of planned infrastructure and underground constructions enables providing more insight into the vertical planning of regions.
- Cadastres traditionally register property rights to real estate on 2D parcels since the individualisation of land started with a subdivision of land using 2D boundaries. In today's world there is growing pressure on land that has led to stratified property (property units on top of and engaging each other). Cadastres throughout the world are confronted with the challenge how to

register and visualise stratified properties. This requires an extension of the cadastral map in 3D (Figure 8.8) (Oosterom and Lemmen 2003).

- Pipelines and tunnels can be better protected against damage when their 3D location can be visualised in the real world (Roberts et al. 2002). For example, knowledge about the location of cables and pipelines can avoid damage during excavation. Based on knowledge of the location of constructions precisely defined restrictions can be imposed on the owners of the surface land from doing anything that could damage the underground construction.
- Location-based services (LBS) for shopping, tourism, rescue operations etc. is another area, where the use of 3D visualisation and most probably 3D GIS is rapidly increasing (Coors 2002; Höllerer et al. 1999)

Applications in non-urban areas

- Road, railway, canal construction and maintenance benefits largely by 3D environments (Bresters 2003).
- Landscape modelling seeks specific 3D tools for interactive design and simulation (Lammeren et al.[102]; Blaschke and Tiede 2003).
- In telecommunications the decision on the locations of antennas requires 3D analyses to obtain information on the area that can be covered and on the costs of using the specific location.

3D Spatial analyses

- Maintaining 3D information on real-world objects enables the management of 3D characteristics of buildings, e.g., calculating the volume of buildings (for tax purposes) or dictating a maximum construction height and depth.
- 3D geo-information can serve as input for spatial modelling such as modelling noise levels (Kluijver and Stoter 2003) and risk modelling for buildings when a tunnel is drilled (Netzel and Kaalberg 1999).
- Geological applications require 3D analysis, e.g., finding fractures or salt domes, computing volumes of repositories, etc. (Wees et al. 2002).

Environmental management

- Knowledge about 3D characteristics of natural processes can be used to impose limitations and obligations, e.g., in case of noise control, odour nuisance and safety measures.
- In order to predict the consequences of bursting of dikes (flooding), a good terrain model is needed together with 3D software (Werner 2002; Zipf 2004).
- Zoning plans that have to regulate different types of land use on top of each other. An example of a zoning plan that had to deal with 3D information is

the 'Noord-Zuid lijn' in Amsterdam. In Amsterdam a metro-tunnel is drilled from north to south, the 'Noord-Zuidlijn'. A zoning plan was needed in which the use of a tunnel below other types of land use was guaranteed. The tunnel is planned partly below houses. Figure 9 shows part of the map that was produced for this zoning plan. It is a 2D map. The areas on the 2D map are encoded as 'multi-layers' and the 3D information (tunnel below houses) is added as a description in the legend and not as a 3D spatial description. Consequently, the zoning plan of the Noord-Zuidlijn does not include 3D spatial information (also not elsewhere in the zoning plan).

8.5.2 Towards 3D DBMSs

Developments in the area of 3D GIS are motivated by both a growing need for 3D information, as illustrated above, and technology developments. Concerning new technologies, significant progress has been observed in 3D data collection techniques and corresponding procedures for 3D object reconstruction. Computers (processors, memory, graphical cards and disk space devices) have become more efficient in processing large data sets. Elaborated tools and devices to display and interact with 3D data are already available on the market.

These developments pose the important question 'what is the readiness of spatial DBMSs for the third dimension'. The following sub-sections discuss this matter.

8.5.2.1 Other representations in DBMSs

The previous sections the readiness of geo-DBMS for boundary representations were discussed in detail. DBMS vendors still have not made the step to implement 3D data types in their geometrical models as was mentioned in section 8.2.1. Specifications for 3D features and consensus on a 3D topological structure have not been achieved as discussed in section 8.2.2.4. The current trend is to develop specific ad hoc solutions when using 3D geo-information instead of building a database for maintaining spatial objects. Non-commercial implementations of 3D GIS models can be found in (Cambray 1993; Oosterom et al. 1994; Rikkers et al. 1993; Wang and Gruen 2000).

Besides boundary representations, other approaches may appear also useful for 3D: Constructive Solid Geometry (CSG) and voxel representation (regular space subdivision). All approaches show advantages and disadvantages considering different criteria. The main advantage of boundary representations is that it is optimal for representing real-world objects. The boundary of real-world objects can be observed,

Figure 8.7. Visualisation of planned bridge, using a VR environment.

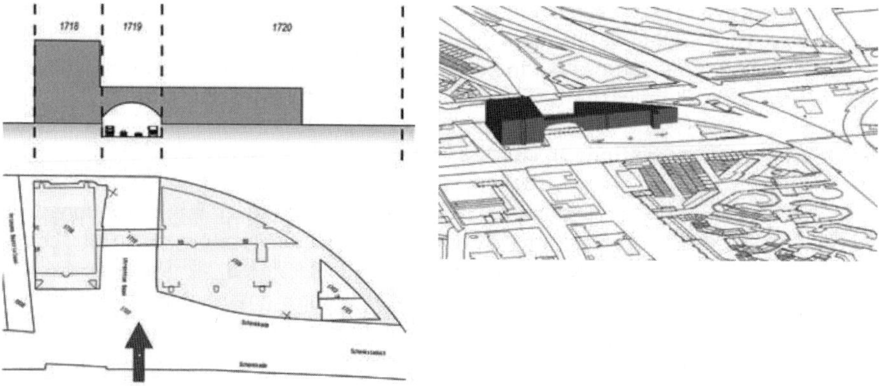

Figure 8.8. 3D cadastre example: the legal status of one building is established by means of establishing limited real rights (right of superficies, right of long lease) on three parcels (left), therefore the cadastral map does not reflect the real situation. 3D insight in the legal status requires extension of the cadastral map in 3D (right).

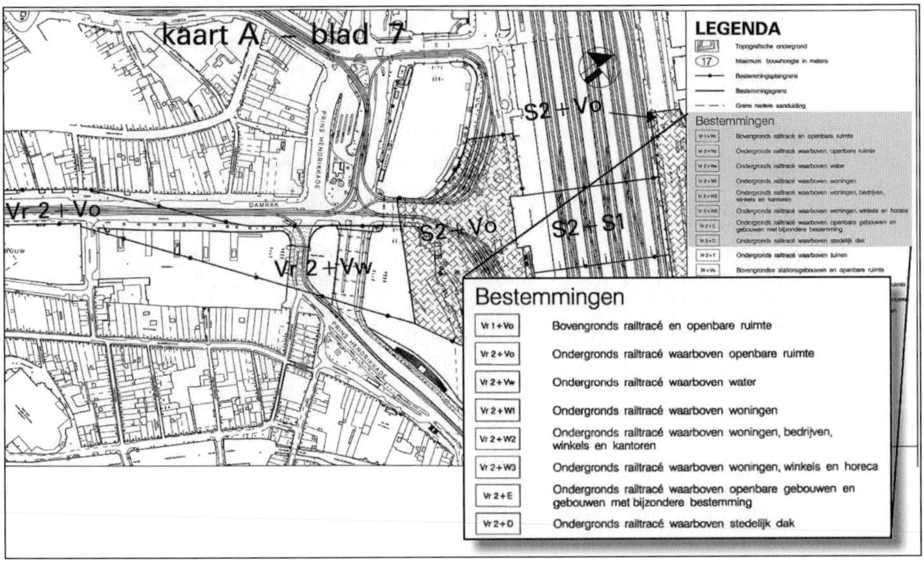

Figure 8.9. 3D zoning plan of metro tunnel (Noord-Zuid lijn) in Amsterdam. 'Ondergronds railtrace waarboven' means 'Subsurface metro line on which'.

measured and surveyed from properties that are visible, i.e., 'boundaries'. Furthermore most of the rendering engines are based on boundary representations, i.e., triangles. Unfortunately boundary representations are not unique and constrains (rules for modelling) may get very complex. Constraints of 3D objects are even more complex to deal with: are open-space objects possible, how to determine neighbours in 3D, how to ensure planarity of faces in 3D etc. Furthermore many large-scale real-world objects (trees, traffic signs, building ornaments, statues) or geological objects (surfaces, repositories, caves) may result in representations with unnecessarily high complexity. In such cases CSG or voxels might be much more appropriate. At certain stage Geo-DBMS have to open for other 3D representations.

8.5.2.3 3D visualisation of spatial features stored in DBMS

3D models usually deal with large data sets, requiring efficient hardware and software for visualisation. Different levels of detail (high detail when objects are close by and low detail when objects are further away) in a model improve efficiency of navigating through a model (Kofler 1998; Pasman and Jansen 2002). In new generation GIS architecture, the 3D data maintained in DBMS should be accessible by front-ends in a very efficient way. A study on the accessibility of 3D data organised in a DBMS by different front-ends is described in (Stoter and Zlatanova 2003). This study has clearly shown large possibilities for visualisation of 3D data organised in a DBMS, in this case Oracle Spatial 9i. However, to be able to query, identify and edit the 3D objects requires still some more research.

To improve performance, different representation of objects such as low-resolution geometries and imposters (image of object instead of geometry) can be either stored in the DBMS or created on the fly. The main problems of storing multiple presentations are fitting high detailed data to low level of detail and the redundant storage of representations.

A problem that comes with visualising 3D geo-data compared to 2D geo-data is readability of the data towards improved realism. To make a view realistic one can add, apart from traditional characteristics such as colour, illumination, shade, fog, textures, shadow, texture, and material to the geometry. Interacting in 3D environments, i.e., exploring 3D models also requires specific techniques. The new issue from a database point of view is the management of data, i.e., images, geo-referencing between images and geometry, etc., needed for realistic 3D visualisation and dynamics. The problem is well known and much discussed in 2D and traditional map production, i.e., if the geometry is described in a Digital Landscape Model, the visualisation parameters are then provided by the Digital Cartographic Model.

8.6 Outlines and further research

In this chapter we have discussed the responsibility of DBMS in new generation GIS architecture in which spatial features together with non-spatial features are maintained in an integrated DBMS environment and edited and visualised in all kinds of front-ends. The location where spatial analysis should fit in this architecture is not yet clear.

DBMSs have made the first step, they offer support and maintenance of spatial objects in geometrical models and some operations that allow spatial analysis of objects stored as geometrical primitives. The geometric operators offered as well as the possibility to use them in SELECT statements in different combinations form an extended set of tools for query and analysis. Still many issues related to the implemented data structuring and required operations have to be addressed. The geometrical model has been implemented but is still not complete. Real 3D geometric types are missing. Ad hoc solutions for representations of 3D objects can be found. Even some generic operations (edit, retrieve, etc.) are possible. However, validations of 3D objects, such as closures, and metric operations, such as volume, center, gravity, etc., need further development. Also visualisation of and navigation through 3D environments require additional attributes to be maintained in the DBMS compared to 2D applications. The lack of 3D support in DBMSs should be a point of attention both in research and in practice in the coming years as was illustrated by examples from practise in which a growing need for 3D information was met.

Now the 2D geometrical model has been implemented in mainstream DBMSs based on the OGC Implementation Specifications for SQL, the next step should be the implementation of the topological model and implementation of topology operators. Once implemented at a database level a large number of navigation operations can be implemented. The performance of the topological relationship operations also still needs improvements. Many DBMS are intensively working on topological models, which will definitely result in topology support at DBMS level in the coming few years. Laser-Scan Radius Topology and Topology-Network Model of Oracle Spatial (10g) are the first examples. A very tricky issue is the type of topology and dimensionality of the models. Current efforts are toward providing 2D topology that most probably will restrict the topology operators to 2D. Moreover, maintenance of many topological models appears unavoidable.

As was described in the introduction, DBMS plays an important role in the new generation GIS architecture. Does it mean that a DBMS will and should include all spatial analyses, including complex spatial analyses that have been optimised in GISs during decades? Does it mean that traditional GIS software (or extended with attribute maintenance CAD software) has to convert to a tool for import, visualisation, editing and exploration of spatial data?

Many spatial functionalities are (and probably will be) available only at the front-end and not at DBMS level (e.g., spatial analyses which are specific for certain domains and applications, tools for inserting new data, interaction tools for starting spatial analyses, visualisation tools). Also, too many operations performed at a DBMS level may lead to overloading of the server and affecting the performance of the DBMS. On the other hand, too few operations provided by DBMS will result in development of many functionalities by the front-end, i.e., duplication of development efforts and resources. This question is still challenging: which spatial operations should DBMSs take over? In principal, generic spatial functionalities that are not specific to a certain application belong in the DBMS and not in front-end applications. On the other hand, complex spatial functionalities that are specific for certain applications should be implemented within front-ends.

In this context we defined generic and supporting spatial operations, i.e., selection, navigation and specialisation. In contrast to the group of selection operations, specialisation and navigation in the spatial domain can be very complex and time consuming. If they are performed at a DBMS level (on the server), the performance can decrease drastically. Furthermore, such complex operations may not be needed for all kind of applications. Therefore, complex operations falling in the group of specialisation and navigation operations can be considered to be left for implementation by the front-end with respect to a particular application, while DBMSs have to support the more generic selection operations, i.e., metric, proximity and topology, and relatively simple specialisation and navigation operations.

Another relevant question in this discussion is whether spatial functionalities implemented in the DBMS will replace spatial functions that were originally built in GISs. GIS has become an important instrument in work-processes of companies and governmental offices. There has been a lot of money and effort invested by GIS vendors for selling their software and for giving support. They may not be willing to give up spatial analyses, i.e., the main part of GISs by which GISs will be reduced to a editing, visualisation and retrieving tool.

Chapter 9

Multimodal Interfaces for Representing and Accessing Geospatial Information

Reginald G. Golledge, Matthew T. Rice and R. Daniel Jacobson

9.1 Introduction

Multimodal interfaces have a great potential impact in our daily lives and in the education of students in all grades. In particular, they offer significant benefits for people who are disabled. The use of tactile, haptic, and auditory interfaces has a potential to make technology more universally accessible. To this extent it will mitigate the rapidly expanding digital divide between those who are able to use computers to access the Internet and web page information (i.e., those who are computer literate) and those who are not.

Information technology transformations are affecting how we communicate, how we store and access information, how we become healthier and receive more medical care, how we learn at different stages of our development, how business is conducted, how work is undertaken in order to produce income, how things are built or designed, how data is stored and managed, and how research is conducted. With the increasing emphasis on visualization as the main interface medium for computer based services, an ethical problem emerges regarding whether or not people who are visually impaired or who have other tactile, haptic, or auditory impairments that limit keyboard or mouse use should be increasingly disabled by the trend towards vision-based digital communication and information processing. We believe that such groups should not be shut out from the advantages offered by the use of this technology, just as we believe that multimodal interfaces will enrich the understanding of the computer-based input and output of information that is becoming a part of our everyday lives.

This emerging digital divide has been recognized by former President Clinton's Information Technology Advisory Committee[103]. Their slogan incorporates a theme of universal accessibility. More recently, President Bush's directive that "no child shall be left behind" provides the incentive for more search and exploration in the area of non-visual and multimodal interfaces. Progress towards either of these goals would be designed to remove age-related barriers to participation in society, would reduce language and literacy related barriers among different groups in society, would enhance global understanding, and would engage more people in the process of information exchanges. But, the transformation of the

Internet from a text-based medium to a robust multimedia environment has created a significant problem; this is the growing digital divide between the main user population and people with limited access to computers because their disabilities cannot deal with popular interfaces tied largely to mouse clicking and visualization (Waddell 1999). For example, the web has become the dominant Internet medium, and it accounts for about 60% of Internet traffic. Almost 75% of web traffic is image based (Shapiro and Varian 1999). By searching to find useful, satisfactory, or even optimal combinations of modalities that can be incorporated into various interfaces, it may be possible to create the hardware that will provide accessibility to a wide range of people at all ages, economic levels, social classes, and able-bodiedness. This goal is consistent with the mantra recited by Shneiderman (2002) and others: "The old computing is about what computers can do; the new computing is about what people can do" (p 2) Shneiderman describes his vision for new computing with three words: usable, universal, and useful. He seeks a goal of reliable, high-quality, usable, computing environments that are empowering and effective for all users and capable of supporting each user's personal needs. We suggest that the development of effective multimodal interfaces is a crucial part of achieving these goals.

9.1.1 Rationale for the Use of Multimodal Interfaces

Multimodal interfaces facilitate increased interpretation of scientific data through redundancy and augmentation (e.g., by providing some parallel alternative ways of looking at data or processed information). Their aim is to increase the reliability of data interpretation through redundancy of representation, to increase the number of data characteristics that can be identified and analyzed simultaneously, and to improve the navigation and search processes required for examining higher dimensional datasets. As opposed to pure repetition, redundancy means an identical display or the presentation of related information in different formats (e.g., representing text information as a map, a parallel commentary, or as a video). Wickens and Baker (1995) suggest three advantages for employing redundancy:

- the information flow from a computer to a learner is less vulnerable to shifts in attention of the user as the information is presented in different modalities
- learning theory demonstrates more long term memory storage of information when it has been presented in different forms
- information presented through different modalities allows a user to adapt to the format of information display that suits their own cognitive learning style or to which they have sensory access.

The benefits of adopting a multiple modal interface appear to be many and derive from research in psychology, computer science, information science and geography education. When making use of several different input channels (e.g., speech, touch, audition, vision), there is a positive probability that the information that reaches the brain is richer in detail. For example, in mathematics, formulae or where the QWERTY keyboard layout keys for the letters SDF and JKL are used to create 6-dot Braille, and ASDF and JKL are used to create 8-dot Braille symbols (see Figure 9.1). The Perkins Keyboard method for inputting Braille requires specialized software interpreters as well as specialized keyboard controllers that are capable of understanding simultaneous multiple keystrokes. As a device distinct from the traditional computer 100-104 key keyboard, the Perkins Brailler is a keyboard input device with nine keys that allow a blind individual to do simple Braille text entry and editing. The Perkins Brailler has a space bar in the middle, with three keys on either side associated with the six dots for Braille text, numbers, and punctuation. A backspace key and a line space key are typically above and to the side of the six Braille entry keys. The Perkins Brailler is often used with an embossing system that records the entered Braille characters on paper, but has also been modified to serve as a direct entry tool for a computer. Electronic versions of the 9-key Perkins keyboard are designed as a text entry tool for specialized word processing systems and are often combined with refreshable Braille displays consisting of a series of raised pins. These pins produce tactile Braille characters that can be felt with the fingertips, allowing a blind individual to check and modify text entered with the Perkins keyboard.

Figure 9.1. Braille Keyboard

Geospatial information input methods can be dichotomized as those associated with multi-dimensional spatial data displays (such as digitizing from maps and graphics) and those associated with linear text input. The traditional computer keyboard has, because of its nature and design, been almost exclusively associated with linear input of text and numerical data. For geospatial information input, the keyboard is an efficient and useful tool for entering coordinate data, specifying the attributes of geospatial data features, and for writing programs used in creating, editing, displaying, and storing geospatial data. Independent of the computer mouse, the keyboard can facilitate nearly every conceivable geospatial data interaction task that can be decomposed into a series of typed commands. Although nearly inconceivable with today's visual-centric Geographic Information Systems, a large portion of GIS user interaction during the 1970's and 1980's was accomplished entirely through the computer keyboard rather than with the computer mouse. Even in today's interface environments dominated by windows, icons, menus, and pointing devices, a large part of the customization of a GIS is still accomplished through programming and scripting done in the traditional method with a computer keyboard. Wood (1993) discusses the various differences between interacting with paper maps or visual depictions of maps on a computer screen, and interacting with the databases containing map information without the benefit of a visual depiction. Clearly, there are different cognitive burdens and tasks associated with each interaction methodology, and, for the blind or sight-impaired user, the necessity of non-visual interaction is a severe limiting factor in using GIS. Braille keyboards and other text input devices allow blind individuals to write scripts and execute GIS programs, but the inability to interact with the graphical display means that the role these individuals have in using a GIS is still relatively limited.

Computer keyboards are ideally suited for geospatial data input tasks that involve numeric and alphabetic text that can be type-written; namely, geospatial coordinates, object attribute data, formulae, annotation, metadata, labels, etc. However, some geospatial data input tasks are more efficient if the user interaction is guided by a 2-dimensional cartographic display or visualization, allowing the user to specify the location of features with a pointing device positioned on a 2-dimensional entry frame. These 2-dimensional input tasks usually involve gathering geospatial data from a paper map fixed to a digitizing tablet, or from a map that has been scanned and displayed on a computer screen (heads up digitizing). Geospatial data input in these cases is often a hybrid process involving the tracing and definition of geospatial objects with a mouse or digitizing puck, and the entry of attributes, labels, and annotation with a keyboard. Although the data input through the computer keyboard can contain spatial information such as coordinates, the input process is linear as opposed to the traditional map digitizing process where the user interaction is in a 2-dimensional plane and features can be defined through a spatial interaction process.

9.3.2 Haptic interfaces

Generally, a multimodal haptic interface is a computer controlled, motorized device held in the hand which provides information via haptic sensing of the user's hand and muscular sensors. While its predominant use is as an augmentation to vision, it is possible in some circumstances to substitute Haptics for other sensory modalities. Thus, touchable representations of screen objects can be created. These are a function of the user's motions and icon properties encountered by the cursor

For example, the MOOSE[104] was designed to provide blind computer users with access to graphical user interfaces (GUI). It is based on software that reinterprets a Microsoft Windows screen to the haptic senses. The MOOSE is essentially a two axis haptic interface which is capable of recognizing windows, buttons, slides, and pull down menus. They are represented as patches and lines of varying resistance. As compared to the usual hand driven mouse, the MOOSE can move under its own power and, therefore, make virtual objects apparent (e.g., makes them touchable). This facilitates location, identification, and, sometimes, manipulations of sensed features.

Apart from the MOOSE, other devices with a similar function include the Pantograph (Ramstein and Hayward 1994) and a version of a vibrotactile and force feedback mouse manufactured by Logitech using Immersion Corporation technology (Akamatsu et al. 1994). Perhaps the earliest version in this category was the Optacon (Linvill 1973), while still others have been developed by Eberhardt et al. (1993), Frisken-Gibson et al. (1987), Kelly and Salcudean (1994), Loubal (1992), and York (1989). In general, a haptic mouse operates as follows: the mouse ball moves the cursor across the screen and, when it lights on an icon, the haptic representation is felt and explored. For example, the edge of a window is a groove; a checkbox is either a repelling block or an attracting spring to initiate some action; and buttons can be activated from menus using force feedback or vibrotactile experiences. One thing that distinguishes the MOOSE from other devices is its capability for accessing a speech synthesizer going through a serial port for text output. This means that the MOOSE offers a combination of haptics and speech to provide a more powerful way to input and access onscreen information.

Kurze (1996) had examined the possibility for non-sighted persons to draw visual representations of three-dimensional objects using a computer tool. This uses the mental model approach to perception and assumes that the mental model is produced by touch rather than sight. He developed a device called the Tdraw Tool as a means to import drawings from non-sighted users. Objects were drawn on paper, placed on a digitizing tablet, and the nature of the lines/objects drawn were recorded via a coordinate structure and were supplemented with speech.

9.3.3 Sonification and Sound

Vision is, perhaps, a human's best sense for many of the spatial tasks associated with understanding and using geospatial data. There are other settings, however, where the abilities of the auditory sense equal or exceed the abilities of the visual sense. These settings are the primary focus for research in data sonification, which is defined as "the use of nonspeech audio to convey information"[105]. These activities, where the auditory sense is suggested to function as well as the other senses, includes preliminary orientation and feature extraction, i.e., 'telling the eyes where to look' (Gaver 1989; Wenzel et al. 1990; Wenzel and Foster 1990); temporal representation (Bly 1982; Buxton 1985), logarithmic data display (ibid.), representation of multi-dimensional data (ibid.; Krygier 1994; Fisher 1994b), and comparative judgments of distributional characteristics (Flowers and Hauer 1992). Scaletti and Craig (1991: 208) suggest that, "unlike the eyes, which have closeable lids, the ears are always open channels." The auditory sense's continuous information gathering nature is used in computer settings through auditory icons, that allow information to be communicated without direct visual attention from the user (Gaver 1989; Begault et al. 1996).

The auditory sense has several advantages and benefits as a method for inputting geospatial information. It is complimentary in how and what information it can communicate to a computer user, by broadening the human computer communication channel and taking advantage of unused 'bandwidth'. The biggest problem with using sound as an interface is its locational non-specificity (especially on the small screens used in handheld and wearable computing and communication devices). Localizing sound on a computer screen would require software capable of calibrating speaker output so that it appeared to come from specific places or locations on screen. Little successful research has been done on this problem; likewise, insufficient research has been done on the ability of humans to localize sound accurately enough to warrant the necessary expenditure to solve this screen-based auditory location problem.

Tannen[106] suggested an auditory mode for monitoring competing background processes. He suggested that, for web browsing, a non-speech, auditory information device would relieve the pressure on space required by other management processes. He suggests that file transfers, application processes, slow load monitoring, and the links could all be handled under this auditory display system. Ballas and Howard (1987) attempted to recover the essentials of environmental sound for presentation back to a user as they navigate through environments, and geographers and cartographers have attempted to map "soundscapes" as complements to traditional paper and ink maps.

9.3.4 PDA Wireless Technology, Cell phones: Stand Alone and Linked to Servers

The proliferation of small, powerful handheld computers and Personal Digital Assistants (PDAs) over the last decade has led to the development of specialized information services that operate within the limited settings provided by PDAs, cell phones, and handheld computer devices. These devices are by design small and streamlined. They lack many of the basic functions of a regular desktop computer, but share the capability of sending or receiving information from a remote location via an information exchange protocol and displaying this information in text form and graphic form (Figure 9.2). The information they can transmit and receive is limited, yet the most widespread information exchange protocol (Wireless Application Protocol) accommodates for this by providing a streamlined standard for efficiently exchanging information. WAP Forum Ltd.[107] proclaims: "Wireless Application Protocol is the de facto worldwide standard for providing Internet communications and advanced services on digital mobile phones, pagers, personal digital assistants and other wireless devices." Critics of wireless protocols like WAP cite the low bandwidth, low data exchange speeds, lack of client device processing power, and small client device screen size. At present, most of these criticisms seem to be valid.

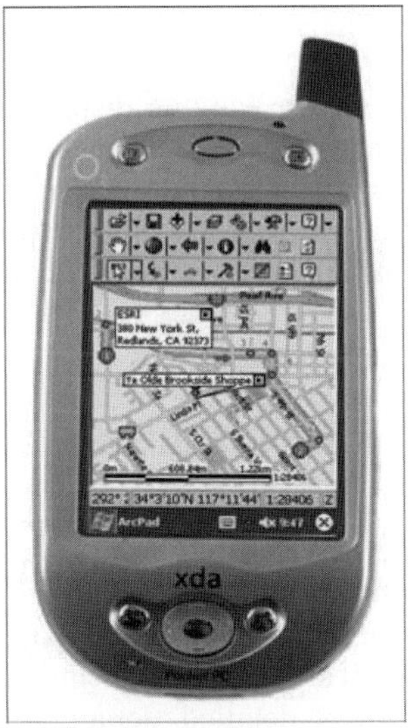

Figure 9.2. Personl Data Assistant (PDA)

One might ask, what do wireless technology and wireless computing devices offer in terms of digital geospatial information exchange? First, by their very nature, wireless devices offer an advantage to the consumer in terms of portability. This is a particularly keen advantage for exchanging geospatial information, because the portability of the device carries it into the real geographic world where geospatial information is used as a backdrop for navigation and travel, recreation, research, and analytical activities done in a field setting. Devices that do not utilize WAP or other wireless communication protocols are capable of storing and display geospatial information that is transferred onto the device through a wired connection to a desktop computer or server. Environmental Systems Research Institute (ESRI) announced in the Fall 2000 edition of ArcNews[108] that it intended to support the exchange of geospatial data through the WAP protocol using the ArcIMS and ArcPAD software, which is used for displaying and interacting with GIS data over the Internet (ArcIMS) and on Windows CE –based PDA devices (ArcPAD). Another major vendor of GIS software, Integraph Corporation, has announced plans to support the development of XML/GML (Extensible Markup Language/Geographic Markup Language) for web GIS and WML (Wireless Markup Language) for wireless GIS.

A variety of personalized household devices are now available for on-site real-time data entry and display. Voice Organizer v1.02 is a voice-driven scheduler that allows you to record an audio description of an "event" (an appointment, a task, or just a quick note to yourself) and have it played back to you at a specific date and time. Voice Organizer's voice interface allows you to step through the creation of an event, or listen to existing events, by pressing the Pocket PC's voice record button and speaking out different commands. However, every function can also be accessed with a tap of the stylus, allowing you to examine event titles and times quietly.

Voice Organizer's main screen lets one display a daily, weekly, or monthly list of events and planned activities. One can move to the previous or next day/week/month and scroll down the list of events and activities, all with voice commands. Each Voice Organizer screen displays the voice commands one can use with that screen. Commands are clearly marked with a special icon[109].

9.4 Output Devices

Much of the work to date on non-visual and multimodal interfaces has been specific: for example, using sine waves to interpret graphs; using force feedback to identify onscreen shapes; using audition to provide information stored in attribute tables in a GIS context but not represented on a visual display (e.g., Figure 9.3: a visual map of California counties that, when an individual clicks a cursor placed over a county, would provide a detailed demographic description of relevant county data[110]). Given this general trend, we offer specific context-relevant examples of how such interface have been used.

Figure 9.3. Map of California Counties with Cursor and Speech Bubble

9.4.1 Embossers

Geospatial information can be represented for use by those relying on touch using a Braille embosser. The TIGER Advantage is an embosser using variable height punching. The developers (ViewPlus Technologies) claim this gives increased flexibility and the potential to combine multi-color graphics with Braille for use in tactual documents. The device is claimed to be able to scan, draw, or import graphics of any format with any Windows application and, consequently, emboss it in great detail on the TIGER. It embosses Braille and graphics together and produces 3-D representations for video-tactile exploration. One important feature of the TIGER system is its capability to print text as DotsPlus Braille. Using a font of 36-point size, this can handle Greek letters, scientific symbols, and can compose equations. To do this, there is ancillary software called the Symbol TIGER screen font and the Math Type Equation Editor. These applications are displayed on TIGER screen fonts. DotsPlus (another ViewPlus Technology application) uses 8-dot Braille, capital letters, and Greek letters and can also be used for advanced math. A limitation is that the user must be able to interpret 8-dot Braille.

9.4.2 The Traditional Screen-Based Representation and Access Output Interface for Maps, Graphics, Diagrams, Using Mice and Joysticks

Geospatial information (in the form of maps and graphics) has become an integral part of modern life. Many individuals that begin a trip out of their familiar local area do so only after consulting a map or graphic of the route to their chosen destination. These destination maps are now commonly generated on a computer at a distant location, transmitted thousands of miles over the Internet, and displayed on a computer screen in the comfort of an individual's vehicle, home, or office. The traditional interaction method for these maps, graphics, and diagrams involve a web browser, a mouse, and a keyboard, which amounts to standard computer equipment located in a majority of homes, offices, libraries, and businesses. Decades ago, the same information was available only through printed paper maps sold or distributed by automobile clubs, gas stations, and bookstores. These paper maps still have a rightful place in the glove boxes of cars throughout the world, but over the next decade the spread of computing technology and in-vehicle navigation systems will cause these maps to be somewhat marginalized and be largely replaced by in-vehicle navigation systems and maps displayed on the screen of a Personal Digital Assistant (PDA). As with Internet maps, in-vehicle maps and PDA maps are displayed on an LCD screen and involve interaction via a pointing device, fingertip, or stylus. Internet maps are rapidly changing and evolving from static raster formats displayed with little interactivity on a computer display to dynamic vector-based formats that allow a great deal of interactivity (see Peterson 2003). Although map format and dissemination technology has changed over time and will continue to change in the future, the standard map display on a computer screen and interaction via a mouse or pointing device has been consistent over the last fifteen years. A recent survey of Internet map display techniques indicates the most common display size for Internet-based maps is between 400x400 pixels and 600x600 pixels. At this size, the potential for showing detail is generally poor relative to a printed paper map of the same size, but the immediacy and convenience of web accessible maps should be mitigating factors. The same study reported that less than fifty percent of surveyed Internet maps allow for resizing and recentering via zooming or panning, and an even smaller fraction allow for interactivity through web page form input or query.

In the near future, the computing hardware components associated with traditional input and output methods (i.e., CRT or LCD display, keyboard, and mouse) will likely remain unchanged. The methods for displaying maps and graphics will continue to be based on these traditional output devices. New interactive capabilities and new formats of computer maps will appear, as outlined in Peterson (2003), but the method for displaying and interacting with maps will remain much the same until major changes are made in computing hardware. Some of these major changes may be associated with wearable computing, mobile computing, and computers used for field research, where geospatial information reaches its most useful purpose in providing a frame of reference for data collection and analysis (Clarke 1999).

9.4.3 Making Non-Spatial Data into Spatial Representations: The Process of Spatialization

Spatialization is a process that takes data that may be acquired in any format at any scale from the nanoscale to the universal and then is transformed, using a spatial metaphor, into a perceptually-enabled representation. This can take place using real empirical data or by designing virtual environments in which data can be translated into spatialized information (e.g., such as in flight simulators). Pennock and Lantrip (1995) used a paradigm of ThemeScapes to implement a geographic metaphor by representing information as a topographical domain or virtual digital elevation model (DEM). The essential purpose of this experiment was to convey a global sense of data structure to the user. Dodge and Kitchin (2001) used the spatialization process to represent complicated spatial and non-spatial information flows in cyberspace in map form. Chalmers (1993) provided an overview of a variety of attempts to use a combination of geographic metaphors and virtual spaces to represent non-spatial information, and Parkes and Dear (1989) used "sound painting" to provide a virtual tactile and auditory representation of classic art such as the Mona Lisa for exploration by blind people.

One of the key components of the spatialization process is that it is designed to present information (i.e., to a computer or to a user) rather than just display data. Data can, of course, be machine processed and manipulated by software; information requires that a cognitive agent be involved. In other words, learning things from data involves cognition and, in the case of spatialization of data, involves spatial cognition for information gathering and learning. Thus, spatializing data is a way of unpacking and presenting information to a cognitive agent that deciphers the spatialization and provides an opportunity to turn it into information and knowledge for either personal or communicable use.

The process of spatialization is data-versatile. It can be used with metric, ordinal, or categorical/nominal data. This process can present data in one-dimensional (linear) form, two-dimensional (map or graphic form), or multi-dimensional form such as three- and four-dimensional arrays, clusters, scalings, and images. Data can be turned into information by representing it in Cartesian coordinate form, polar coordinate form, in real environmental spaces, or in psychological spaces.

Output representations can be in terms of location clusters, line graphs, connected networks, shapes and patterns in physical or virtual space, or by using a neural net paradigm. Statistical processes can be used to reduce data that might theoretically have hundreds or thousands of bits down to a number that can be conceptualized or visualized by the human mind (e.g., calculating means, variances, or factor analyzing large data sets). Some spatializations are constructed largely to represent interpoint distance properties, others are slanted more towards making direction and orientation explicit, and yet others use processes such as perceptual closure to present shapes or allow them to be inferred. Thus, spatializations can try to impose characteristics that provide a semblance of a physical space for the representation of data or they can be used to infer that data should be arranged in complex mathematical spaces or the ill-defined but powerful psychological spaces about which we still know very little.

The purpose of using metaphor is to embed information from one domain into another that is seen to be more powerful or familiar. Using a geography metaphor involves translating material from one (non-spatial) environment into a spatial one (Couclelis 1998). This is the essence of spatialization, although the translation from one domain to another is not limited to the domain of the earth's surface. Thus, spatialization is much more powerful than is implied in just using the geography metaphor. The geography metaphor seems to be constrained by scale and by physical structure, along with the human environment relations that are tied to that physical structure. However, spatial domains exist at many other scales, ranging from the micro space of the human brain or similar structures, to personal or body space of the "person in environment" relation, to geographic space that encompasses the obscured and distant environment that extends out to the extent of the entire surface of the earth, and gigantic space which extends beyond the earth into the solar system and beyond (Montello 1993).

Finally, spatialization has the capability of making complex things obvious, even when it is not possible to engage visualization. This occurs in the realm of the blind or vision-impaired where haptic or auditory spatializations are more important than visualizations. Spatialization is not bound to the visual senses. This means that it is a far more powerful metaphor than visualization. Traditionally, tactile maps and graphs have provided spatialized representations for those who lack vision, while the Braille alphabet provides a spatial-tactile map for "reading" text. Today, users can experience map and graphics information represented on a flat, two-dimensional computer screen using haptic concepts such as force feedback and vibrotactile experience (see next section).

9.4.4 Tactile Maps and Graphics

Tactile maps have been produced to supply geospatial information to people with visual impairment for more than half a century. Though somewhat uncommon, they can be found in public settings where locations of buildings, stairways, and exits are embossed into a metal or plastic surface. In combination with the raised lines that can be felt with the fingertip, Braille lettering is often used to label features or provide general information. Two common technologies are used to produce maps that, rather than being permanently mounted or fixed to a building or public display, can be carried and used portably by a sight impaired or blind individual. The first of these technologies involves vacuum forming, where reduced air pressure is used to draw a heat-softened sheet of polyvinyl chloride (PVC) over a mold. Vacuum formed maps can be mass-produced, and are considered advantageous in terms of consistency and variety of symbol design, and incorporation of visual and non-visual elements (Perkins 2001). The mold or 'master' used to create the vacuum formed maps requires construction by a skilled technician with specialized equipment via an often time consuming process. An alternative technology is the use of microcapsule paper, which is coated with a substance containing acrylic resin capsules containing alcohol. When exposed to a consistent heat source, the capsules explode, raising the surface of the paper. To create such a map, a black ink design is copied on the microcapsule paper which is then exposed to a heat source that causes a differential heating of the printed and non-printed areas and a subsequent raising of the surface underneath the black printed areas of the design. The advantage in using microcapsule paper is the ease with which maps can be created and reproduced, and its low cost per unit. No special equipment is required other than a consistent, uniform heat source that can be applied to the paper. The ICA Commission on Maps and Graphics for Blind and Visually Impaired People has reviewed the advantages and disadvantages of vacuum formed tactile maps and microcapsule paper maps as well as several other alternative technologies (Tatham 1991; Tatham 2001; Coulson et al. 1991; Perkins 2001; Siekierska et al. 2001). Several other experimental systems are being developed involving arrays of raised pins and specialized tactile hardware, including virtual reality equipment that can produce touch sensation, but these systems are not being used widely for the production of tactile maps.

Perkins (2001) reports that tactile map design can be taught to novices, and that the best results are obtained through an interactive map design process involving actual map users. Coulson et al. (1991) report that tactile map production can be automated, to some extent, through the use of Geographic Information Systems (GIS). They report on a study involving the use of ASCII symbols and standard cartographic symbols (MAP II symbols) for creating maps from a GIS, finding that the standard cartographic symbols were preferable over the ASCII symbols in terms of user understanding and correct interpretation. Coulsen et al. reported that experimental participants (who were blind) could use tactile maps to understand concepts of orientation, direction, and symbolization for elevation and land use categories. A major contribution from Coulson et al.'s work is the emphasis on the instant production of tactile maps from existing map databases stored in a GIS.

Coulson et al. argue that if tactile mapping is to be a viable means of communicating spatial information, maps need to be readily accessible rather than being produced through a painstaking and lengthy process. They continue by suggesting that a GIS contains the capability to generalize map information, data distributions, and scale automatically, to produce new datasets that are much more conducive to tactile representation.

If tactile maps are to be readily produced from a GIS as several researchers suggest, what effort should be made to standardize the symbolization? Tatham (1991) presents perhaps the most comprehensive set of suggestions for tactile symbolization guidelines for points, lines, and areas. These guidelines include suggestions based on previous experimentation by Tatham and other researchers, as well as theoretical considerations. Tatham suggests that point symbols be no smaller than 2mm and no larger than 10mm in width and height, that areas be no smaller than 13mm in width and height, that adjacent lines be spaced at least 5mm apart, that double lines be at least 13mm in length for a 3mm thick double line, and that radiant point symbols (i.e., stars and complex point symbols) be at least 4mm in width and height and no larger than 10mm. Other guidelines for symbolization related to areas and shape have been suggested by Ericksson (2001) and Clark and Clark (1994). Coulson et al. argue that it is highly desirable to standardize the symbolization and design of tactile maps, because the range of useful and distinguishable symbols for blind or visually-impaired individuals is a small subset of the available cartographic symbols, and the design process would be more effective if symbols were used. It should be pointed out that the goals of tactile map design are far different than the goals of visual map design that often involve issues of aesthetics and visual harmony. It is, perhaps, much more critical to create a well designed map for a blind individual, who may depend on this map for a larger percentage of information about his or her surroundings than would a visually capable individual who has many other sources of information about his or her surroundings. So Coulson et al.'s arguments in favor of a standardization of design for tactile maps are reasonably convincing. On the other hand, Tatham (2001) suggests that there is a difference between the standardization of design of map elements (akin to the effective design standardization of six dot Braille symbols), and the standardization of the use of such symbols (which is generally what is meant by those that suggest standardized design guidelines).

A new online tactile mapping capability has been developed at the Smith-Kettlewell Eye Institute in San Francisco[111]. The system uses a server-based GIS to produce the graphics for tactile maps ultimately produced by a Braille embosser, microcapsule paper, the ViewPlus TIGER embossers, and other tactile map or graphic producers. The developers' ultimate aim is to develop software that can handle any online request for geospatial information such as a street map of any place at scales ranging from neighborhood to citywide. For example, street maps would consist of line features that make up the street network, with Braille labels for street names[112]. The last 5 years have seen a resurgence in tactile map research in the academic community. Jonathon Rowell, the current chair of the ICA Commission, is conducting an extensive Tactile Map Usability survey through the Tactile Map Research Group at Anglia Polytechnic University in Cambridge, UK;

and is active in developing cognitive guidelines for tactile map design. Several other individuals and groups are actively researching tactile map design and tactile map use. These include Boguslaw Marek (Catholic University of Lublin, Lublin, Poland), Izabella Krauze-Tomczyk and Jolanta Korucka-Skorupa of the University of Warsaw (2001), Yvonne Eriksson (2001, Göteborg University and the Swedish Library of Talking Books and Braille) and Eva Siekierska (2001, Mapping Services Branch of Natural Resources Canada), and the UK's National Centre for Tactile Diagrams[113]. Researchers from Penn State University's GeoVISTA Center have contributed to the role of tactile and haptic mapping within visualization, including Amy Griffin (2001) and Frank Hardisty. Springsguth and Weber (University of Kiel – Multimedia Campus) have produced a variety of 3-D tactile relief maps of landmarks to incorporate blind-accessible tactile and haptic representations into VRML mapping environments[114] as a part of the MultiReader project whose goal is to produce multimodal interfaces for accessing information for all users, including print-disabled users.

9.4.5 Virtual Representation Based on Touch and Force Feedback

With the advent of virtual technology, the world of computer scientists attacked the difficult task of imitating the real world. In that real world, information is received and disseminated in a multidimensional space. Haptic interfaces have been developed as devices that allow users to interact with computers by receiving tactile feedback along with (at times) visual feedback. A haptic device achieves this feedback by applying a degree of opposing force to the user along the X, Y, and Z axis of a three-dimensional representation. It does so to orient users to the location and nature of objects in a virtual space. Whether embracing part of the body or only the tip of the finger, a haptic interface gives the user information about the nature of objects being thus presented.

To implement a haptic interface, researchers often begin with the creation of a haptic node in the virtual reality modeling language (VRML). This tells the interface how much relative force should be implemented when an object is touched. For example, to create variations in texture, the force being used to push the mouse must be compensated by an opposing force which can be less than, equal to, or greater than the original applied force. In devices such as the **PHANTOM** (Jansson 2001), the haptic interface allows the user of a virtual space to perceive physical characteristics of an object such as its solidity and its shape (Figure 9.4). This gives the illusion of actually touching the three-dimensional object in real space. The PHANTOM (developed by Massie and Salisbury 1994) is a device that measures the user's fingertip position and exerts precisely controlled force on the fingertip to enable the user to interact with and feel a variety of objects. It has been suggested that this would be an appropriate interface for control of remote manipulations such as are often found in robotics. To use the PHANTOM, a user inserts a fingertip into a thimble and can then sense an external force from the finger, creating illusions of interactions with solid objects (a stylus could be substituted for the thimble, so that users could feel the tip of the stylus touch virtual surfaces).

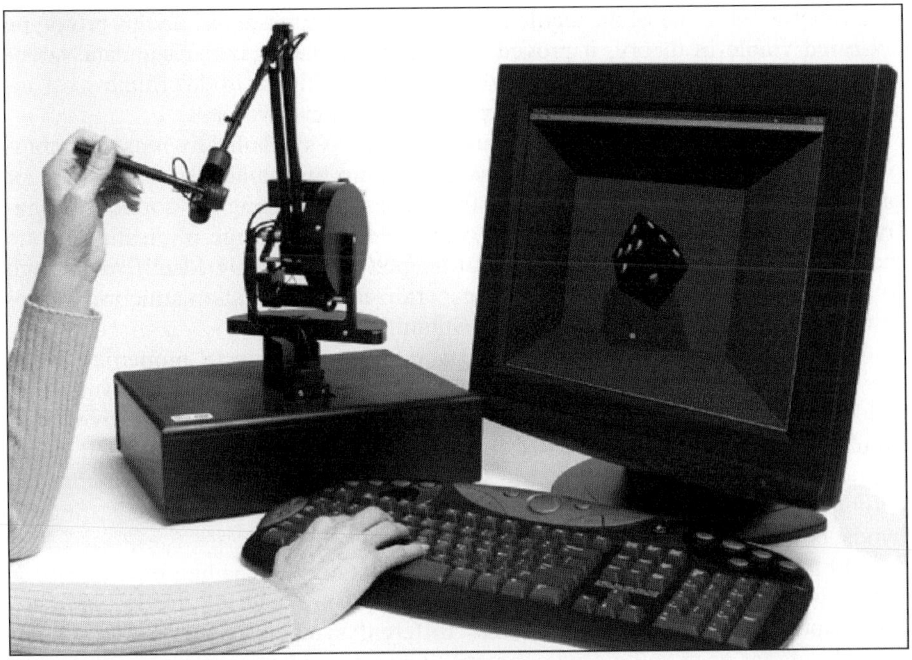

Figure 9.4. The Phantom

Hughes and Forrest[115] argued that a focus on touch mediated by vibrational or other stimulation or displacements on the skin represents the most comprehensible of all the touch interfaces. They claim that kinesthetic and haptic devices operate in a fairly limited way, with the user often having to wear special equipment in order to obtain the touch-based experience. They claim they are thus able to go beyond the traditional uses of a haptic interface which gives position by touching, as on touch-sensitive screens, or by pressure sensitive tablets or drawing tools, or by using force feedback which requires contact or weight in manipulating virtual objects. They argue that force feedback requires an input device that is mechanically linked to some anchorage or attached to the user and consequently limits the space in which it can operate (e.g., Magic Gloves). As an output device, however, sensing based on vibration or roughness gives immediate sensory perception via the mouse itself without the need for extra equipment. They also present arguments for the use of speaker pads attached to a mouse by which vibratory sounds can be emitted. Vibration thus, they argue, can be heard by sound or felt by touch. They have produced software that is useful in the GIS context. There are three options within the software: the first loads the image to be displayed on the screen, the second loads the image and stores it off screen in a bit map, and the third option allows the user to explore the screen bit map in tactile mode. In use, as a mouse is moved over the onscreen bit map, the corresponding common value in the off screen bit map is converted into a number which is interpreted by the system as a frequency of vibration of the tactile transducer on the mouse. While this prototype appeared viable in theory, it proved difficult to implement, because data values varied considerably from pixel to pixel and it was difficult to determine what information was being transmitted, except when the mouse was still.

Basic shapes are used on maps and graphics to symbolically represent complex real world objects. The process of reading and understanding a map or graphic involves correctly identifying the shapes and making the correct associative link to the real world objects. A major design challenge in creating haptic maps and graphics is determining what shapes can be easily identified through touch, and then forming the necessary generalized symbolization structure to represent the real world objects (Rice et al. Submitted).

2-D and 3-D features may be perceived as having material properties which are relevant to a class of features, not just to an individual one. Thus, density shading may be interpreted as a material property which can be perceived in a number of different places (e.g., to define countries, states, etc.). In such cases, haptic perception is macrogeometric, for the same pattern could be experienced and/or identified by exposing the object via various parts of the skin (e.g., fingers, hands; not just one finger) (Golledge et al. 2005).

In the haptic domain, roughness replaces 2-D density shading of a visualization (based on color, the grayscale or symbol structure). Using roughness to enable haptic perception usually invokes different symbols (e.g., dots, stippling, cross-hatching, open and closed geometric features, and so on (Clarke 2001; Andrews 1983). These 3-D symbols provide the appropriate surface "roughness" that facilitates haptic interpretation.

Salisbury and Srinivasan (1992) have an interesting historical summary of haptic interfaces up to the beginning of the last decade of the 20th Century. Hardwick et al. (1997) have discussed the problem of "feeling it as well as seeing it" (p 105). They suggested a haptic display within a gestural human computer interface for multimedia telematics services. Pennsylvania State University researcher Rajeev Sharma (2003) has been key in demonstrating the usefulness of multimodal interfaces that incorporate speech and gesture components, particularly for use in critical decision making applications. Related work by Penn State University GeoVISTA researchers Brewer (2002) and MacEachren et al. (2001), demonstrates the usefulness of multimodal visualization interfaces incorporating gesture within collaborative environments and emergency planning and response. The University of Washington's Human Interface Technology Laboratory (HITL or "hit lab") has also developed a substantial software toolkit for adding gesture recognition to interfaces. Their software toolkit, built with C/C++ and open source code, is now available as a commercial product. HITL has also contributed a significant amount of work toward hand motion recognition, which "provides functions for converting special hand motions into symbolic forms…to combine with other inputs such as speech recognition, static hand postures, etc."[116]. Wagner and Ludi (1999) suggested the potential of animated icons to assist interaction between the graphical user interface and the blind or vision-impaired user.

9.4.6 Auditory Interfaces and Virtual Environments

Seeing With Sound software claims that it is possible for people who are totally blind to "see" live camera views of their surrounding environment using their ears. This is different from echolocation or sonar; rather, it uses real visual input from a web cam. This software, called the "vOICe" has been developed to translate images that have been observed in real time into corresponding sounds. A blind traveler, for example, would wear studio headphones (a characteristic that many blind people will reject out of hand), and would carry a notebook pc equipped with a web cam to "hear" live views from their environment. Presumably, people hear the very same shapes and objects that sighted people see.

Sound has been demonstrated to be effective as a tool for simplifying the display of multivariate geographic data, such as raster imagery and associated uncertainty estimates (Fisher 1994b). Loomis and Soule (1996: 965) suggest that auditory displays have the potential for high temporal bandwidth, specialized temporal pattern processing, monitoring of spatial directions, and low sensory thresholds. Ramloll et al. (2000) have discussed the importance of making statistical information (specifically, line graphs) available and accessible for vision impaired individuals. They argue that a combination of touch and sound will provide the optimal combination for non-visual users to read line graphs.

Borrowing from Jacobson (1999: 108) and Kramer (1994: 9), the following summary lists the benefits of sonification:

- an auditory display's presence is generally as an augmentation and non-interfering enhancement to a visual display.
- an auditory display increases the available dimensionality of the representational space
- auditory displays have superior temporal resolution. Shorter duration events can be detected with auditory displays.
- auditory displays create user interest and engagement by decreasing learning time, reducing fatigue, and increasing enthusiasm.
- auditory displays have complementary pattern recognition capabilities by bringing new and different capacities to the detection of relationships in data.
- auditory displays provide strong inter-modal correlations, reinforcing experiences gained through visual or other senses.
- auditory displays enhance realism by adding immersive qualities and making virtual reality situations more realistic.
- auditory displays are synesthetic, i.e., they replace insufficient or inappropriate cues from other sensory channels.
- auditory displays enhance learning, by providing a presentation modality suited to many student's learning style.

Auditory interfaces include speech and sound, and are the focus of much current research attention in the computer and information sciences. Petrucci et al. (2000) suggested that information can be effectively communicated and integrated spatially, even for blind people, using a multidimensional auditory browser mechanism to replace non-visual, one-dimensional browsers (e.g., Apple graphic interfaces). Their technique, WebSound, uses earcons (auditory icons), text to speech conversion, auditory cues, and haptic feedback to communicate information held in a hierarchically organized spatial layout of a document. An audio and speech interface has also been developed by Lloyd and Schleppenbach (1997) to discuss ways to use speech and sound to allow blind or hearing impaired users to access complex, scientific instrumentation or other lab equipment. They argued that, while 67% of higher education facilities are available to mobility impaired individuals, only 35% are available to vision or hearing impaired people. Consequently, they present their LabVIEW interface with scientific equipment designed to allow the user to create virtual interfaces that represent alternative control mechanisms for scientific equipment. Brewster et al. (1993) discuss the importance of earcons as a method for providing navigational cues in a menu hierarchy. In this system, a hierarchical menu with four levels of 27 nodes is constructed with each node having a sound associated with it. Interpreting the sound allows the user to identify location in the menu. Subjects were correct over 80% of the time in experiments aimed at evaluating this capacity.

Location-based services appear to be a rapidly expanding and acceptable concept. Many of these are tied to wireless transmissions. In essence, they allow a traveler to stop, interface with a nearby server, and obtain verbalized locational information about the services offered in the immediate vicinity. The interface is usually a PDA, or a Smart cell phone, but could also be a Personal PC or other type of wearable computer. The possibility of developing this device depends on two things: a) the location of remotely accessed servers and b) the development of a very large comprehensive geospatial database that provides detailed information on the occupants and functions at specific locations. Most of these developments are likely to take place in commercial areas in large cities where some type of advertising or general information is necessary (e.g., to find the occupants of the 70th floor on a multistory building). Location-based services appear to be closely tied to the development of Geographic Information Systems (GIS).

Parkes and Dear (1989) developed NOMAD—a tactile/auditory information system designed to provide access to maps, graphics, and other visual displays. This originally was developed by Parkes for the use of a coauthor of this chapter (Golledge) when he lost his sight in 1984. The device improved dramatically with the help of software expert Dear, and was eventually made available commercially to the international blind community. This was the first on-screen tactile/auditory mode for accessing electronically displayed data. It did this using a touch sensitive pad on which a tactile image was laid. The pad was coded with a 1 cm. grid so that, if a finger or stylus depressed a given place on the pad, its coordinate location was recorded and whatever information was tagged for that location was spoken to the user. NOMAD was, however, a single-user device that was static. New information required removing the original tactile map and replacing it with another, then activating the 3-level menu of options again to access image information. While NOMAD clearly was a major contribution in terms of accessing geospatial information without sight, Jacobson (1998) suggested a more versatile mode for accessing data that is—in the emergent age of technology—online. This mode, called "Haptic Soundscapes," also combined touch and sound. The authors of this chapter have conducted basic and experimental research at the University of California Santa Barbara, Florida State University, and the University of Calgary for the last seven years. In this system, touch is available through a haptic or vibrotactile mouse; sound is available via speech or tonal variation. Geospatial information portrayed in, say, a choropleth map, would be sensed using a vibrotactile mouse that enables detection of roughness. Different density, for example, would be portrayed by varied roughness. Isoline maps could be likewise defined, or (since they often represent continuous data) could be represented by tonal variation along a sliding scale. This combination of software and hardware has opened up the Internet and its myriad of image-dominated web pages to those without sight[117].

9.4.7 Accessing Geospatial Information without Sight

One way of improving potential interactions between vision-impaired or blind people and their environments is to give them an in-place power of interpreting the environment by carrying all of the information that is needed around with them. This is the essence of systems suggested by Loomis[118] and Collins (1985) and elaborated further in devices such as the UCSB Personal Guidance Systems produced by multidisciplinary researchers at the University of California Santa Barbara (Golledge et al. 1991; Golledge et al. 1998; Loomis et al.1998, 2001). Other similar types of personalized guidance devices include MoBIC (Petric et al. 1996), a Smart Cell Phone (Makino et al. 1996; GPS-Talk [119], and an increasing number of other experimenters who have reported progress on various web pages. Of these, the UCSB PGS System, the earliest version of such systems, is elaborated here.

Developed as a succession of test beds since 1990, the UCSB PGS System has three dominant modular components. The first is a locator module. This consists of a Global Positioning System (GPS) carried in a shoulder bag or small backpack. Antennae for the system are mounted on the shoulder of the traveler or on the backpack itself. A compass is worn on the head (e.g., on the head clasp section of a pair of earphones) or on the torso. A compass is essential for obtaining precise directions. The GPS unit accesses locator information from the NAVSTAR Satellite system approximately every second and thereby records a trace of the traveler's path as one moves through an environment.

The second module is a spatial database contained in a small microcomputer or laptop. The database can be a commercial one of a large area (e.g., the TIGER files of a large city or part of a city), or can be specifically constructed for a particular local area (e.g., a college campus). The digitized local area map is usually incorporated as part of a minimal GIS which includes selected functionalities such as path or route selection (often shortest path), buffering, corridoring, and other necessary functionalities. The GPS signal (obtained in latitude and longitude format) is downloaded to the coordinate system used in the digitized map. A cursor in the computer system identifies where on the map the GPS places the traveler at any point in time. Via a speech interface or a keypad interface, a destination is input to the computer. A path selection algorithm is then accessed, and information is relayed to the traveler concerning the direction and distance of their destination. Given the go-ahead to plot a path, the computer plots a path to the destination, labeling each choice point consecutively. Choice points usually refer to places where decisions have to be made (e.g., turn left, turn right, cross street, etc.). As a traveler moves through the environment and is tracked by the GPS, the cursor similarly tracks the traveler's movement through the spatial database. Surrounding the cursor is a previously defined buffer of a specific size. As the traveler moves through the environment and the cursor moves through the digitized map, the buffer moves with the cursor and impinges on nearby features located in the database. These features are identified and communicated to the traveler.

The third module is the output interface. This has been specifically designed by Professor Jack Loomis, Psychologist at UCSB (Loomis et al. 1990). The interface is an auditory virtual system. Here, as the traveler moves through the real world environment and the cursor moves through the database, information on nearby features that are found to lie within the buffer are communicated by speech to the real world traveler. In this particular system, the interface is via stereophonic sound, communicated via headphones to the traveler. The information, however, appears to be coming from locations in the real world rather than just a message in the head. Loomis et al. (1990) defined the mathematics of a system that would allow sound to be projected outside the head and, in fact, to appear as if it was coming from a real world location at some distance away. Thus, if the traveler heard, via the headphones, "transit stop, 90 feet," he could point directly to the real world transit site and realize his distance from it. The location information given would appear to be coming from the real world location of the transit stop. So, as an individual negotiates an environment, in effect the environment talks to the individual, identifying what it is and where it is in a virtual, auditory system. This allows the traveler, even without sight, to build a mental map of the geospatial layout of the environment through which the traveler is passing. The information so obtained for vision-impaired or blind travelers is significantly greater than they could possibly obtain by walking unassisted through the same environment. In fact, very little of the locational and functional information obtained via the PGS could be obtained by an independent sight-impaired traveler without asking passersby or a travel companion.

A somewhat different version of this type of device that provides geospatial information to a real time traveler is that being developed by the Sendero Group and Pulse Data Corporation[120]. This device provides a Braille and Speak keyboard for input and output of information. It also uses street network databases, commercially available, and provides specific information on what street a person is currently traveling and what adjacent cross streets are nearby. Since it does not include a compass, directional information can only be obtained by exploration, but the device does appear to be a substantial advance over existing technology.

9.5 Where Do We Go from Here?

What still remains is to produce a general interface architecture that provides the user with the option of choosing different modalities or combinations of modalities depending on the task being considered. This is what is really implied by developing a universal interface. Such a universal interface does not exist at this time but individualized and specific research appears to be converging towards such an end.

Trends in the computer industry have been towards wearable computers and small wearable handheld devices such as PDAs and pocket PCs. These come along with very versatile cell phones providing new interfaces that are being well-integrated into everyday life. However, they have limited success when dealing with geospatial information systems. This is partly because of the output interfaces, which are generally restricted to small LCD screens. Pictures, maps, or graphics that have to be represented on these screens have to be extraordinarily simplified. They are not well suited to the complex multimodal representations needed to allow presentation of the depth of information that is often required for geospatial information processing and interpretation. There appears to be no solution to the screen size problem at this time, although virtual retinal displays (such as those being developed by the University of Washington's Human Interface Technology Laboratory) show some promise for those users who have at least a capability for low vision[121].

One of the more significant uses for both input and output of geospatial data is in the area of real time field-based data collection and representation. In the expanding area of activity analysis in transportation modeling, PC and hand-held computer-based survey designs are beginning to replace traditional paper and pencil or telephone surveying and sampling procedures. With the PC-based data collection device, a subject's input in the morning may propose a specific activity schedule and modify it each night based on recall or based on trip records produced by GPS units placed in the main household vehicle (Lee 2001; Zhou and Golledge 2004). Zhou and Golledge, for example, have modified a personal PC with a speech input card and wireless connection to a nearby server that allows individuals to input a daily activity schedule to the computer, and, as they move through their daily activities, they are tracked by a GPS unit. The GPS downloads their location into a spatial database carried in the computer's memory. If there is not a coincidence between the GPS recorded position and the expected position from the daily activity schedule, the traveler is queried by the computer to find what type of departure has occurred and what the specific reasons for the departure might be. The user is given a menu of possible reasons, ranging from changing environmental conditions to decisions to change, delete, or substitute activities. This real time data collection prompt obviates the problems of later recalling what may have been done during the day at the end of the day, as is the case with other existing desktop, PC-based, electronic data collection systems. The purpose, of course, is to fill in the data that is commonly missing from traditional surveys because of unwillingness to provide information, inability to recall the specifics of trips that were made during the day, and a tendency to omit multipurpose or short duration trips from reported schedules of activities.

The importance of computer gaming as a dramatically growing commercial business has encouraged exploration with multimodal interfaces to obtain more personal involvement with games. Haptic mice, vibrating mice, auditory interfaces, tonal variations, animations, and combinations of speech, keyboard, or roller-based mouse interactions have been developed and incorporated into gaming scenarios. Likewise, the use of color, virtual environments, and avatars (lifelike human representations on screen that use gesture as well as speech to give directions, to provide information, or to point out solutions) are all part and parcel of the gaming scenario. At this time, there is substantial research activity involving attempts to determine the extent that these innovations can help in data input, manipulation, representation, and output—particularly in the geospatial domain. The common sentiment is that multimodal interfaces provide richer and more lifelike interactions, whether it be with a dynamic gaming scenario or in terms of data representation and analysis. Consequently, the more general term "perceptualization" appears to be becoming more and more accepted, along with the traditional single modal term, visualization. Future developments in the input and output interface area, however, will depend on substantial developments in other fields such as nanotechnology (e.g., chip design and storage capacity), biotechnology (better comprehension of the way that interfaces can complement or supplement the different human senses), information technology (including database structure and management and data mining activities), and cognitive technology (particularly understanding the various ways that humans can interact with multimodal interface designs) (Golledge 2004). As advances in each of these NBIC areas develop, so too will we see changes in interface architecture that should, in the long run, substantially benefit the entire geospatial information processing and display domain.

Acknowledgements

Funding for this project was provided by NSF/PPD Grant # HRD-0099261, *Transcending the Digital Divide*

Chapter 10

Wayfinding with mobile devices: decision support for the mobile citizen

Sabine Timpf

10.1 The mobile citizen

People find their way every day: from home to work, from work to a restaurant, from the restaurant to a sports club and back home. Most people do wayfinding (Arthur and Passini 1992; Golledge 1999) very well within their neighborhood and less well in areas they have been to only once. When traveling into unfamiliar territory, people tend to prepare themselves using maps or by requesting instructions. The wayfinding task itself takes place much more intensively in the unfamiliar than in the familiar environments. In an unfamiliar environment we do more planning, more tracking, and we need to be more alert.

Sometimes we might wish for someone reliable murmuring into our ear how to find our way. This someone might in the future be a some*thing* such as our mobile phone, our personal digital assistant or our portable CD player. The only requirements for these devices are to know their own location and to have access to a location service. Recent technological developments such as Smart Phones, wearable computing devices, or Personal Digital Assistants (PDA) with integrated Global Positioning System (GPS) have made it feasible for a human person to carry their own personal navigation and information system with them.

The mobile citizen uses mobile location aware devices to let her find the way, gather and present relevant information, reserve hotels and restaurants, point out landmarks or tourist attractions, and manage the daily routing task.

10.1.1 What exactly is a location service?

Location services deliver information pertinent to the task, the time, and the location at hand. They help their recipients to make decisions based on local information instead of guesses or recall. They are set up as web services, i.e., "interoperable, self-contained, self-describing, modular components that can communicate with each other over the Web services platform" (Peng and Tsou 2003, p.640; see also Chapter 5 in this book).

> A location service is the delivery of location dependent information to the user of a mobile and location-aware device. The service is personalized and brings additional value to the user.

In the broadest sense a location service is any service or application that extends spatial information processing, or geographic information system capabilities, to end users via the Internet and/or wireless networks (Koeppel [122]).

In this broad sense, a mobile Geographic Information System (GIS) also qualifies as location service. Mobile GIS in combination with a localization technology allows the field worker to access information, databases and even programs at headquarters, thus always accessing the most recent and locally relevant information. At the same time the field worker can send results back to the office without having to travel there.

Location services emerged because of a convergence of multiple technologies including geographic information systems, Internet, wireless communications, location determination, and portable devices. They have been given a boost by the creation of the Open GIS Consortium (OGC) in 1994. OGC works with government, private industry, and academia to create open and extensible software application programming interfaces for geographic information systems and other mainstream technologies[25].

10.1.2 Why would we need a location service?

Simply knowing where you are, or how far you are from someone or something, is typically not valuable by itself. However, relating location to other pertinent information such as time or purpose of the trip gives it meaning and value. In general location services provide information in order to enable an informed choice and thus reduce the number of potential choices a traveler faces. Such a location service acts as a decision-support system. In addition location services reduce uncertainty about the environment and satisfy human's curiosity about surrounding space.

There may be many reasons to request a location service while wayfinding: we may be lost and need help, we would like to confirm a navigation decision, we need to change or adapt our wayfinding plan, we would like to acquire specific location information, we need more information about local transportation, we need information otherwise invisible or irretrievable while on the road, and lastly we might need help in dealing with an information overload (e.g., too many choices).

Other potentially valuable services for the traveler are: rain alert, weather service, yellow pages, shop/restaurant/hotel location and prices/menu, traffic updates (accidents, congestions, estimated travel times), entertainment information, tourist information, medical services, location based billing, tracking and routing, emergency response, and personal route planning.

10.1.3 The value of a location service

A user attributes value to a location service if he or she gains something from it. The main task in wayfinding is reaching the destination, but the value of a location service is heightened by presenting additional services that augment and/or complement the wayfinding task (e.g., get a new haircut before going to a meeting). As a rule, the value of a service is entirely user and situation dependent.

The value of a location service can lie in the provision of up-to-date information, i.e., information that is definitely newer than information provided in the form of maps or books. The value can be determined by the effort it spares the user, i.e., it retrieves and integrates information from many different sources or it anticipates information needs (see section 10.2). The value can lie in the solving of more than one location problem. This solution requires that more than one location problem is explicitly or implicitly posed to the location service, e.g., through the provision of a task list. The modeling of task lists and sequences is discussed in section 10.2. An automatic alert in case of medical or other emergency is probably invaluable to the traveler. The personal profile and other context information is the topic discussed in section 10.4. Location services take many forms and provide value in a diversity of ways, but the common denominator is the spatial data handling capability that links location to other types of data.

10.1.4 Research Questions and Structure of chapter

The remainder of this chapter is organized as follows: in section 10.2, the process of wayfinding is examined in detail. Section 10.3gives an overview of necessary technologies for wayfinding services, presents the different types of potential location services identified so far, and shows some examples of location services. Section 10.4 contains ideas to solve the current main research questions pertaining to the modeling of context including personalization. Section 10.5 provides conclusions.

10.2 The process of Wayfinding

Wayfinding is the part of the navigation process that deals with information processing (Blades 1991, Montello 2005). Information processing typically takes place in the head of the traveler and is triggered, inspired, or influenced by the environment in which the navigation takes place. Locomotion is the part of navigation that deals with the actual steering of movement, e.g., obstacle avoidance, adaptation to step heights etc. In this article we will concentrate on the wayfinding part of human navigation.

Navigation = Wayfinding + Locomotion

According to Norman (Norman 1988) we use knowledge in our head and knowledge in the world to carry out our tasks. This is especially true for the task of navigation, where we use a "map in the head" (Kuipers 1982) and feedback from the environment. In the case of location services a third information source comes into play: the mobile device or the knowledge in the pocket (see Figure 10.1).

Figure 10.1. Types of knowledge in navigation

Usually this mobile 'device' is a map of the area or a list of instructions. Recently this mobile device truly is a device, i.e., an information appliance such as a personal digital assistant (PDA) or a smartphone (see also section 10.3). Information appliances are mostly helpful in the wayfinding or information processing part of navigation. With some adaptations they can also be used for the actual steering of a vehicle, such as a robot or a wheelchair (Lankenau and Röfer 2000). The details presented in this chapter can also be applied to robot navigation or to automated cars, in which a location service would have to deal with both the wayfinding and the locomotion part of navigation.

What does it mean that we have three different types of knowledge at our disposal when carrying out the activity of wayfinding? In my opinion, it means that each of the three types of knowledge can substitute the other two, this means if all knowledge needed for the wayfinding can be found in the world, then we do not need additional knowledge in the pocket or in the head. If knowledge needed for the wayfinding is available in the head (e.g., in the form of a cognitive map), then we do not need additional knowledge in the world or in the pocket. If all our needed information can be found in the pocket, then we do not need more information in the head or in the world. Note that we are talking about the knowledge needed for wayfinding. We will still need operational knowledge about how to deal with the knowledge presented to us, i.e., in the first case reading signs and following sidewalks, in the second case matching our cognitive map to the real world, and in the third case following the instructions of the pocket device.

Wayfinding consists of three main actions or tasks: planning, tracking, and assessing (Infopolis2 1999). *Planning* answers the question 'where do I need to go?', *tracking* deals with the question 'where am I compared to the plan?', and *assessing* argues 'how good has my travel plan and subsequent execution been?'. Planning mostly takes place before the trip, tracking goes on during the trip, and assessing is the main task for the last part of the trip or for after the trip. In addition orientation plays a major role. The difference between orientation and tracking is that orientation answers the question as to where I am, whereas tracking compares the current location to the planned location.

Activity	Wayfinding: get from place A to place B		
Tasks	Planning	Tracking	Assessing
Operations	Information gathering, find routes, determine constraints, determine complexity, produce instructions	Orienting, track location, compare to plan, orient yourself	Compare needed to planned time, assess instructions, determine complexity of route

Table 10.1. Activity Model of Wayfinding, derived from Infopolis2

The three main tasks have been identified for the specific purpose of multimodal travel using a public transportation network. However, this differentiation is also valid for other means of transportation, such as traveling in highway networks or traveling by bike. Within the framework of activity theory (Kaptelinin et al. 1999; Nardi 1996; Timpf 2003) the relationship between those tasks and their corresponding operations is one of parthood: finding suitable routes is part_of the task of *Planning*, comparing the current space-time location to the planned one is part_of *Tracking*, and comparing the planned to the taken path is part_of *Assessing* (see Table 10.1).

In the remainder of this section we will only deal with the planning part of wayfinding. The tasks of tracking and assessing within the wayfinding activity are part of on-going and future work and deal also with the locomotion aspect of navigation.

10.2.1 Activity model of the Planning task

A case study about the usage of maps for trip planning on the Internet (Faby 2004) identified three sub-tasks: general information gathering, specific route planning, and detailed planning of the trip. Those three stages are at the level of tasks in the framework in Table 10.2. Timpf et al. (Timpf et al. 1992) proposed the tasks planning, instruction giving and driving for the activity of navigating interstate highways. Instruction giving is a specific type of operation within the planning task, whereas driving is part of the locomotion activity and thus not represented in the activity framework below.

Activity	Planning: plan a trip from A to B		
Tasks	Information gathering	Route planning	Detailed planning
Operations	Determine location of goal compared to start, assess if a path exists taking into account the trip constraints (time, finances, participants)	Determine optimal route, determine complexity, take constraints into account Result: coarse plan	Produce instructions, check if route is feasible within constraints Result: detailed plan

Table 10.2. Activity Model of "Planning a route"

The purpose of the planning activity is to derive an action plan from the minimal information on start and goal location. In addition, constraints influence the planning, such as time frame, finances, schedules etc. These are parts of the detailed planning process. The result of the planning tasks is a hierarchical action plan.

The **hierarchical action plan** details the spatio-temporal actions that a traveler needs to carry out in order to find a way from a start to a goal location. The plan depends on the complexity of the route (Heye et al. 2003), the aims and abilities of the traveler, and her familiarity with the environment. It corresponds in most cases to the detailed route instructions derived in the route planning and the subsequent detailed planning process (Timpf et al. 1992). However, depending on the user's requirements, the instruction can be more or less coarse. The complexity of the route determines the number of action levels. Each level of actions represents in fact a progressive abstraction of more detailed actions (Stell and Worboys 1999). Actions start at a certain time and place in the navigation process and end again at a specific time and place. Those space-time places are called decision points. They correspond to places where the correctness of the route may be reaffirmed or an alternate solution might be chosen.

In Table 10.3 three action levels are given for our wayfinding example: essential actions, actions, and operations. The distinction between different levels of detail for the wayfinding description provides us with a way to make distinctions between different user needs: the traveler that knows the area, has been to the city before, but does not know the route will be satisfied with essential actions or actions. The traveler new to the city but experienced in traveling will perhaps be content with action information. The inexperienced traveler new to the city will probably need information at the operations level. To identify possible obstacles and because redundant information might be the only possible one an impaired traveler will always need information at the operations level, with additional information on elevators, where the doors will open at which stop (on the same or the opposite side), if the bus leaving now is a low-floor bus, etc. Dewey has produced a detailed overview of the requirements for navigating in a wheelchair in a European city center (Dewey 2001).

Activity	Coarse and detailed Planning of the route from Informatikum to Hamburg main station				
Essential Action		Take bus 181 to Hagen-becks Tier-park		Take U2 to Hauptbahnhof	
Action	1	2	3	4	5
	Walk from Informati-kum to bus stop 181	Take Bus 181 to station Hagenbecks Tierpark	Walk from bus stop to platform of subway	Take U2 from Hagenbecks Tierpark to Haupbahnhof	Walk from sub-way station U2 Hauptbahnhof Nord to Haupt-bahnhof front entrance
Operation	Walk out of building	Get into bus 181	Walk towards U2 station	Get into sub-way	Take stairs at front of subway train
	Leave the campus	Buy ticket	Take stairs down to the platforms		Turn right at top
	Walk straight along Vogt-Koelln Strasse	Ride (4 stops) until bus stop Hagenbecks Tierpark	Walk on right platform with trains in di-rection Wandsbek-Gartenstadt or Barmbek	Ride (9 sta-tions or 20 min) until subway sta-tion Haupt-bahnhof Nord	Take next stairs to top
	Cross the next street	Get off bus		Get off sub-way	Turn right and walk 10 meters
	Turn left				
	Walk to bus stop				

Table 10.3. Hierarchical Action Plan

The hierarchical wayfinding graph (Figure 10.2) encodes the knowledge about activity, actions, and operations at the same time as it stores knowledge about the routes at different levels. Each level shows the route at a different granularity, i.e., at a different action level (Timpf 2002). Nodes are places (e.g., Informatikum, Hagenbecks Tierpark) and depending on the granularity level also specific locations (e.g., bus stop at Hagenbecks Tierpark, top of stairs). Edges connect nodes and thus carry information on the action or operation to be performed. Each edge at a level corresponds to exactly one action or operation, but may correspond to more actions on the level below. In addition, temporal and spatial information can be added to the edges.

Action plans are important for two reasons: firstly, the multiple granularities allow to adapt the presented information to the user knowledge, and secondly, the optimal presentation time for information delivery can be determined through definition of decision points and thus locations. It is important to note that the information presentation should take place just *before* a new action is to be taken by the traveler, i.e., proactively. The absolute time interval, which denotes *before* is

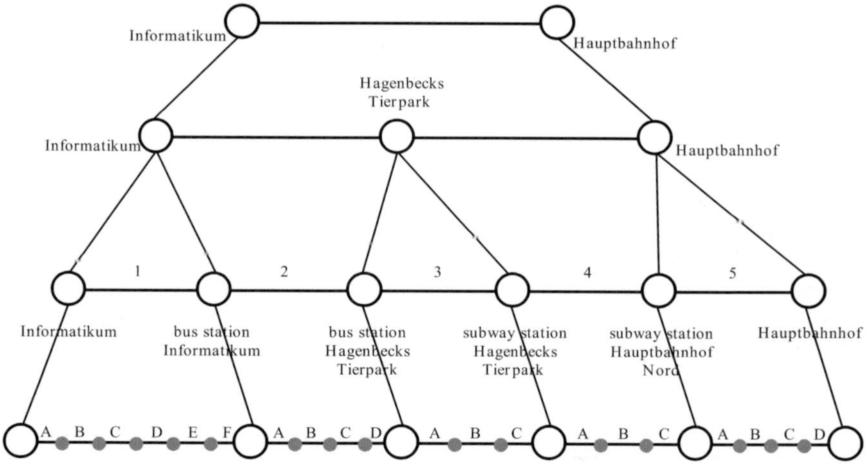

Figure 10.2. Hierarchical wayfinding graph including action plan

dependent on several factors (this is an open list): speed of travel, complexity of the route, number of potential interactions (may be dependent on the time of day, e.g., rush hour), number of alternatives, complexity of the choice, and potential disability of the person.

10.2.2 Information processing and information integration for the Planning Task

When looking in detail at the information needed to carry out the action plan it becomes clear that each of these actions needs a specific type of information or data and that these types of information are at different levels of granularity. One way to deal with different types of information and with different levels of granularity is to try to integrate them in a coherent and complex information system. This is what is done today when each transportation means provides its own database and information system. However, these complex systems are very unwieldy when a traveler needs to change means of transportation in a multimodal trip and thus must deal with many of these systems. In addition traveler's information needs extend beyond the information already existing in these systems. The additional value needed to make Location Based Services a success lies in the anticipation of user needs and in the provision of information required to fulfill all of the user's tasks.

The new paradigm of GI-Services (cf. Chapter 5 in this book) within a distributed multi-agent computing environment allows for a better suited approach: once the route is known and the action plan determined, agents retrieve information about parts of the route and present this information to the traveler using an information collage.

The term information collage (Figure 10.3) is derived from the term cognitive collage coined by Tversky (Tversky 1993). Tversky argues that spatial information is not always pictorial and thus cannot be represented in a cognitive map. She proposes the notion of *cognitive collage* to account for the fact that spatial information collected and accumulated over the years is very diverse, has different formats, different levels of detail, shows different aspects or point of views and even allows for erroneous, contradictory, and incomplete information. She also states the very personal nature of a cognitive collage that included recollections of places and memories, even overlapping (temporally or otherwise) information. In addition the metaphor is most appropriate for knowledge about environments that are not well known and thus have not yet been integrated into a survey representation (Siegel and White 1975).

Information collage = collection of information fragments about a specific route retrieved by an information system

We are interested in providing a user with a collection of information pieces or fragments associated with a route the user has to travel. An *information collage* is the collection of all the pieces of information that an information system could retrieve at a certain point in time as they pertain to the route, including information on the route itself. Each piece of information is called an information fragment. Instead of trying to integrate all the information fragments and make them consistent, we associate each fragment with the point of the route it corresponds to. These points may be places, locations or time points during an action or operation. Information fragments may be graphical or textual. The information collage is the physical manifestation of the plan such as it comes out of the Planning task. The traveler may use the information collage to comprehend the plan but also to track his/her progress.

We image mobile devices or electronic paper maps that can be 'loaded' with a specific route at information kiosks or via the internet. With the help of visualization techniques like magic lens (Stone et al. 1994) and fisheye views (Furnas 1986) only that part of the route is enlarged that is needed, whereas in the background the collage as a whole is visible. In addition, the map can be updated with newer information (like traffic information or times for the next train or bus).

Another application for the information collage is the integration of information by an information broker on the web. In a scenario, where location services are managed by information brokers the broker has to integrate information coming from different sources. The metaphor collage provides a good structure for (at least partial) integration or ordering of heterogeneous information based on the task of the broker. The information can be spatial (route), temporal (duration of trip), financial (ticketing), social (dangerous at night), etc. The integration has to be primarily spatio-temporal for wayfinding, but the other information may serve as constraints or additional decision criteria for alternative routes. The collage metaphor provides a very good structure into which information fragments can be integrated and ordered such that they form a whole.

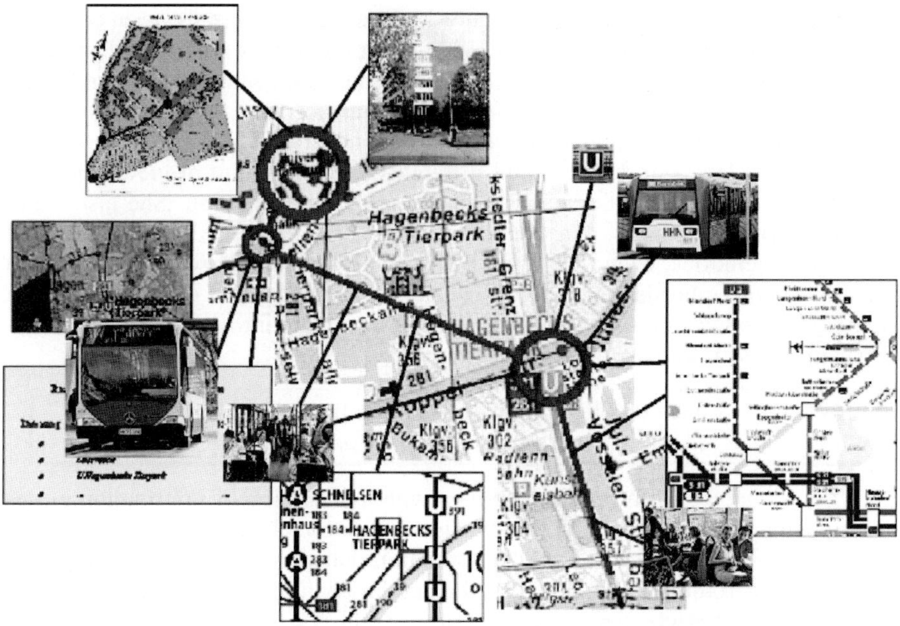

Figure 10.3. Example for an information collage: the route is represented using a graph, appropriate graphical and textual information is attached to each edge and node in the graph.

10.2.3 Information Needs and Behavioral Settings

In the last section we have seen that all available information *can* be incorporated into an information structure for presentation to the user. However, it is not yet clear which information *should* be presented and how to select this information. In order to determine which information should be presented, we need to look again at the knowledge repartition as described in the knowledge triangle (Figure 10.1.) Knowledge may already be present in the traveler's mind, other knowledge will be provided by the environment, and only the missing knowledge/information should be provided using the information appliance.

Settings (Schoggen 1989) play an important role in determining the importance of an information fragment for the traveler. By settings we mean the situation or setup encountered at a specific moment in space-time during the trip. The setting determines which potential actions we perceive and which behavioral schemata we expect. For example one specific setting is the location in front of the ticket machine, where we are expected to find out which ticket we need and how to pay for it. A similar, but complementary, setting with the same goal is standing in front of the ticket counter and explaining the person behind the counter where we want to go and what ticket we need. The end result is the same but the behavioral patterns are different for each setting.

The Infopolis2 project (Infopolis2 1999) determined the following situations (similar to what we call settings) important for a trip in the public transportation system: pre trip at home, on trip private vehicle (including walking), on trip first stop, on trip on board, on trip intermediate point, and end trip private vehicle (see Table 10.4).

Setting	***Remarks***	***Task***
Pre-trip at home		planning
On-trip private vehicle	first leg of trip	tracking
On-trip first stop	access public transport system	tracking
On-trip on board		tracking
On-trip intermediate point	changing modes	tracking
End-trip private vehicle	last leg of trip	assessing

Table 10.4. Settings for a multimodal trip in public transportation

Please note that settings also determine the types of objects potentially needed for a specific action, they determine ontologies (Kuhn 2001; Timpf 2002) for specific settings and tasks. For example, for the setting "on-trip first stop=train station" and the activity "board train" the important objects are platform, train, wagon, stairs, elevators, departure signs, and clock. The important actions are "locate correct platform", "walk to platform", and "enter train". Deriving the objects needed for a specific setting will give an important clue as to which information should be presented to the user in which setting. This is currently work in progress.

10.3 Wayfinding Services

A wayfinding service is the delivery of location dependent information to a mobile and location-aware device. There are four different types of wayfinding services we have identified so far: type 1 asks where something is, type 2 asks what there is, type 3 asks how x is here or there, and type 4 asks for route directions. These services can be implemented as push or as pull services. In Europe there is a tendency to prefer pull services, i.e., the user requests the service.

10.3.1 Technical prerequisites for wayfinding services

The main players in the GIS market have proposed to use mobile GIS for wayfinding services (e.g., ArcPad (ESRI), IntelliWhere (Intergraph), or mapXtreme (MapInfo)). The problem with this approach in my opinion is that they require a user to learn about GIS, which most users do not want to do. The power of GIS should be hidden from the user and should be suspected only in the quality of the answers. This requires a careful orchestrating of user guidance and an intelligent modeling of the potential applications.

There are several **technical prerequisites** for wayfinding services to work. The first one is the availability of a device that can determine its location, communicate via a network, and features graphical and auditory output and input capabilities. The second prerequisite is the existence of a communication network suited for high bandwidth. The third prerequisite is a distributed, interoperable service architecture as for example promoted by the OGC (Buehler and McKee 1998). Each of the three prerequisites will be discussed briefly.

Figure 10.4. Pocket PC and Smartphone used for wayfinding services

The **devices** currently used for location services are car radios, laptops mounted in the car, personal digital assistants (PDAs) or the newest generation of mobile phones (see Figure 10.4). Those have either an integrated localization component (car navigation systems), access an internal or external GPS receiver (PDA), or use a network based method to determine location (mobile phones). In addition devices with digital compasses and/or accelerometers are getting on the market, allowing to determine the direction in which the device is headed or pointed. So-called wearable devices (e.g., e-sleeve (Randell and Muller 2002)) are still in the experimental stage but can also be used for location services.

The devices differ mostly in their technical details such as size and resolution of display and input and output modalities. The differences play a role in the presentation of the information, but do not affect the general principles presented in this chapter.

The **location** of a device can be determined through a number of ways. We will just mention the different approaches here. For detailed introductions see for example (Hjelm 2002) or (D'Roza and Bilchev 2004).

A widely used method for determining location is the Global Positioning System (GPS). GPS is based on the measurement of the travel time of a signal to at least four satellites. GPS works very well at sea, in the air, and in rural areas. In urban areas the necessary satellite number and constellation often cannot be achieved. GPS does not work indoors, because the signal to the satellites would be too weak. Differential and assisted GPS both present solutions to this particular problem.

The second most used method is a network based positioning such as GSM or UMTS. Those methods return a cell id as location. Extensions to this method also exist in the form of measuring signal strength, timing advance, time of arrival or enhanced observed time difference.

The third method relies on the known position of LAN transceivers or dedicated beacons. These microlocation networks are for example installed in museums: when the device gets within the range of a specific beacon, the information is released.

The last method is often used when no positioning method is available or reliable: the user simply inputs the relevant location, such as geographic location, addresses, postal codes or other statistical area identifiers.

Other methods are for example a sensing floor, electronic tags on the doorjambs or other sensing equipment built into the environment (Hightower and Borriello 2001a,b).

The preferred architecture for location services is the **distributed service architecture** using multiple agents (Zipf and Aras 2002). GIS can bring its power into play if each of the used functions and procedures in wayfinding is available as a separate service. This has the advantage of working very fast, because only the necessary service is performed. Users of a location service do not want to deal with a powerful yet complex software package. For more information about distributed service architectures refer to (cf. Chapter 5).

The **wireless communication networks** needed for location services must enable the sending of other types of data than voice data, e.g., streaming data. The so-called third generation (3G) mobile networks possess enough bandwidth to enable these services. In addition to the 'normal' network, this network must also enable access to the Internet, usually through provider services (Braithwaite and Scott 2004).

10.3.2 Types of Wayfinding Services

Type 1: Where is?

The first type of service is the answer to the question "Where am I?". A slight variation of this question is "Where is a *specific object/subject*?", where *specific object* can be the Eiffel tower or my car and the *subject* could be my friend Eric. Another slight variation is the question "Where is *object type*?", where *object type* can be a bank, a river, a bus stop. This question always implies near here or nearest. So it is equal to the question type "Where is the nearest of my friends/bank?". In most cases the answers to these questions are given in the form of points on a map background. Another answer could be voice information on the neighborhood or area where the device is located. The solution with the map admirably solves the problem of finding the right level of detail for the answer. However, for blind people this solution is not feasible.

Type 2: What is there?
This second type of service provides answers to questions such as "What is there?" It returns a description of the object (building, sign, street) based on the inferred location of the object. For this question a pointing device is required for further identification of the object in question. The object can also be an area, such as a forest or a lake. A variation of this type is the question "What x is here", where x could be plants, wildlife, or public transportation. The answer to this type of question can be given in the form of a single phrase or as thematic map, depending on the complexity of the underlying system. As stated above, a complementary representation must be found for disabled people.

Type 3: How is?
The third type of service answers the question "How is x here/there?", where x can be the weather, traffic conditions, pollution, bird diversity or avalanche hazard. During navigation, this service pertains to the current location, whereas the service can also require the input of a location e.g., for planning purposes. The answer to this type of question is usually given as a phrase with a qualification and does not need a map representation.

Type 4: Path?
The fourth type of service answers the questions "How do I get from here to there?" or "How did I get here?". This is typically what is understood as routing service or path planning. In addition to the principal question, boundary conditions such as shortest or least complicated can be stated. The answer to this question usually is a map showing the path to follow with points of interests or landmarks for orientation (Agrawala and Stolte 2001). Information contained in the map can be made presentable to blind people using the path as ordering structure.

10.3.3 Some examples

The types of location services presented above are usually combined within a specific application area. For example tourist information services require a combination of all four types of services: the user wants to know where he is, what he can see at this place, how x was or is at a certain place and how to get from one place to another. By contrast, the 911 emergency localization service requires only a single type of service, namely the where-am-I service to work. Another typical application area is car navigation systems, which are currently a combination of type 4, type 1, and type 3 services. The application area of museum guides has been popular in Italy. Museum guides typically use type 1, type 2, and type 4 location services.

Car navigation systems

Probably the best known and best developed example for location services is a car navigation system. Car navigation systems have been under development for the last 30 years. Currently most car manufacturers serially produce cars with built-in navigation systems for cars in the upper middle prize range. Navigation systems can also be bought separately either for integration into the car radio (e.g., Alpine, Blaupunkt) or as mobile handheld devices with integrated or separate GPS receiver (e.g., TomTom, Magellan, Garmin, NavMan). Those mobile systems can be mounted on the headboard or on the windshield for better viewing. Most systems use voice messages in addition to arrows. Handheld devices produce their information in color maps or even 3D views.

Cyberguide

Cyberguide was one of the first implemented mobile guides (Abowd et al. 1996). It provided simple black and white maps and information services about predefined indoor and outdoor locations. The information provided was static and stored on the mobile device to be retrieved depending on the location of the device. Indoor positioning relied on infrared beacons, and GPS was used for out of doors.

BPN: Personal Situated Navigation Service

The BMW Personal Navigator combines several subsystems: a desktop event and route planner, a car navigation system, and a multi-modal in- and outdoor pedestrian navigation system. It also seamlessly integrates two- and three-dimensional maps (Krüger et al. 2004). BPN serves as a research framework although the prototype was developed in cooperation with BMW.

Deep Map

Deep Map is a prototype of a digital mobile tourist guide for the city of Heidelberg, based on agent-technology. It is part of a wearable computer system using a multi-modal and multi-lingual user interface. The natural language understanding was implemented for English, German and Japanese (Malaka and Zipf 2000). Deep Map is used as research framework to deal with context-dependence of information and personalization issues.

HIPS

The HIPS project was concerned with the development of a value-added service for museum visitors. The resulting Hippie system is conceptualized as a museum guide, which provides additional information on the exhibition, on each art work, and on the way to a specific exhibition area. The mobile device senses infrared beacons installed near art works and at the entrances to each exhibition hall. The system logs the visitor's movement and suggests other interesting exhibits augmenting them with background information (Oppermann and Specht 1999).

Lol@

The local location assistant Lol@ (Popischil et al. 2002) is a mobile tourist guide for the city of Vienna based on maps. It has been designed for the next generation of mobile phones (i.e., UMTS). Lol@ can dynamically generate maps and annotate them with labels and icons; it also provides basic means to interact with the map through zooming and panning. It provides several modi: pre-trip planning, on-trip re-planning, on-trip information, trip reviewing. The system is currently in a test phase using real users.

T-Info, Mobiloco, Vodafone Live

Telecommunication providers in Europe offer specialized services for subscribers using localization though the cell information of the mobile phone. Those services include finding the nearest cash machine (Find it), nearest gas station, nearest coffee shop, friends in the same zone/cell (buddyFinder), or interesting people in the same zone/cell (DateFinder).

Current location services provide information on user request. However, intelligent location services should take into account the stage at which the user stands in the wayfinding process. It is a matter of research to define stages of wayfinding, which information is needed at which stage and how best to present this information to the traveler determining mode (map or language) and best time.

In all of the above presented example applications, context is the main determinant for the way information is selected and presented. Context is especially important if location services provide information in addition to the main application. Thus a short overview of the research issues in context modeling is warranted.

10.4 Modeling Context

The main research problem for location services is the modeling of context in order to provide personalized and adaptive services to the traveler. Context in the literature often means a type of location, e.g., home or office. However, the scope of context is much larger. This section restricts context to be the context of a query, i.e., those filters or variables needed to adapt the information content of the answer to the user.

The context of a query is (a) the location of the device, (b) the environmental setting in which the query is made, (c) the time of day, week, special times, (d) the activity the user is involved in, (e) personal information of the user, and (f) the device and its characteristics. It may be necessary to make a distinction between context that is needed before the query can be formulated, such as the location, and context that is needed to filter the answers, discarding unimportant information and reducing duplicate answers to a single one. In the following subsections we will primarily list research topics with the occasional reference to on-going work.

10.4.1 Location

For location-based service the main context used before the query is sent is the location of a person. The other components mentioned above rather play a filtering role, although additional information on settings or time of day might be important as well. One impediment to the acceptance of location services has been the use of coordinates or cell id as location information. However, location is much more than a couple or a triple of coordinates.

Location can be expressed in the form of a nearness relation, e.g., I am near the lake, or in the form of a neighborhood, e.g., I am walking around the harbor. It is necessary to provide each coordinate pair with additional ("hierarchical") information such as the general area, city, district, county, country etc. More research is required to be able to determine the needed level of granularity of location for the query.

10.4.2 Settings, environment

Settings have been mentioned before in combination with information needs. Settings describe objects or concepts and their usage within this setting, i.e., their function. Settings also contain behavioral schemata, which make the modeling of the wayfinding process easier. Settings in terms of context provide the location service with a filter for important concepts to give information about or important actions to be carried out. Settings let us anticipate the next move of the traveler and thus provide valuable information just before this move is likely to occur.

Besides the concept of settings, the concept of Lynch's elements of a city (Lynch 1960) is important in wayfinding. Nodes, edges (in the sense of barrier), districts, landmarks, and paths play a crucial role in orientation within a complex environment. These elements could be seen as a kind of functional setting. Landmarks aid orientation and help with memorizing the environment. How to define and automatically extract landmarks is part of current research (see e.g., Elias 2003; Winter 2003).

10.4.3 Time

Time plays an important context role in location services. For example, a different type of instruction (e.g., different landmarks) is produced dependent on day or night; Schedules of public transportation change with certain times in the day, using profiles such as the early morning profile, the rush hour profile or the weekend profile.

Empirical research (Heye 2002) has shown that travelers overestimate waiting time and underestimate time spent sitting in the bus. Waiting time supposedly is negative, whereas the potential of using the waiting time for e.g., getting a newspaper or a coffee converts the waiting time to positive time. This distinction is important for location services because it offers the value added service a traveler prefers.

10.4.4 Activity

Activity models and knowing the traveler's motive ("getting to a meeting", "exploring a new city") allow to anticipate the next action of the traveler or to provide the right kind of information at a decision point. It is equally important to determine which peripheral actions the traveler might be willing to carry out. When going to a doctor's appointment, the traveler will not be interested in getting a coffee along the way, but she might be interested in acquiring some magazines for the waiting time in the doctor's anteroom. In this sense, settings and activity are intricately linked.

As discussed above, some work has already been done in the area of modeling the activity of wayfinding (Timpf 2002), but more research needs to be carried out in modeling those peripheral actions. Within the context of location services, we are not aware of any system taking the current action and its super-ordinate activity into account.

10.4.5 User

Within the location services community quite a lot of thought has gone towards the modeling of the user or customer (Ralph and Searby 2004). User modeling or personalization efforts have promised the most economic benefits in the form of targeted marketing, customer relationship management and service integration.

A **user profile** depends on location: a profile for work is different from a profile for home, different while shopping, watching TV or playing sports. We should expect future mobile terminals to provide profile/location-dependent services (Bishr [121]). However, nobody wants to live in a Big Brother World (Orwell 1949) where all personal information is available to everybody.

Newbould and Collingridge (Newbould and Collingridge 2004, p.74) solve this problem by making everything private including information on location. They define the user profile as dynamic information (location, devices, and networks), role information (see below), context, and information reserved for service providers. Role information contains the following: (a) user interests, (b) preferences, (c) bookmarks, and (d) personal information. With this setup role information actually depends on location (e.g., at work, at home, at sports center).

The solution by Newbould and Collingridge precludes the use of push services and only allows pull services, except in emergencies. This is a concession to the **privacy** of the traveler, which cannot be violated. However this solution also makes the personal device truly personal and when lost, all virtual information about the individual is lost.

10.4.5 Device

The device itself is also part of the context in a location service. Screen size, resolution, memory size, and processor speed determine how fast and what type of information may be presented to the traveler. The problem of adaptive displays (Wasinger et al. 2003), or in more general terms resource adaptive systems (Baus et al. 2002) have been occupying the community extensively. One solution is a standard to describe the device's properties. A second solution is to use the Java Virtual Machine (JVM) for the application development and leave the adaptation to the device to the JVM. At this time consensus has not (yet) been reached.

10.5 Conclusions

This chapter first defines what location services are with respect to navigation, why we would be interested in producing location services, and how the value of such services could be determined.

The second section describes the main impediment to producing good location services: the lack of a formal model of human navigation. In this model the main attention must be paid to the wayfinding aspect of navigation. The planning stage of wayfinding has been well researched, but the tracking and assessing stages are part of on-going and future work. For location services we need a firm knowledge about the tracking phase and how we can help users with this task. Most users accumulate knowledge about special places and routes and location services need to take this existing knowledge into account. This requires that the device is capable of learning about the user, a research topic in artificial intelligence. An information collage was introduced as an adequate structure for route information, which can be retrieved by the traveler when needed.

The third section gives an overview on the technological requirements for wayfinding services to work. The technological basis for wayfinding services are a distributed service architecture, a communication network able to deal with huge data streams, a device capable of representing spatial and temporal information, and an accurate and reliable localization method. The technology for wayfinding services exists in principle although some aspects, especially those concerning the interoperability between services, are part of on-going research. There are four types of location services, which often are combined to provide interesting applications. This section also describes in short some location services. Most location services need to deal with context information to provide interesting services.

In the fourth section an overview of the research issue 'context' is given. The lack of good context models is the second impediment for location services. The primary context for a location service is the location. But this information is not sufficient to present tailored services to the traveler. Context means the setting or environment, the time, the activity the traveler is engaged in, the traveler herself, and finally the device used for presenting the location service. The capture of the 'right' context allows for interpreting the query beyond the obvious and thus makes the location service valuable to the traveler. Location services need to provide enough value to the user to overcome the privacy barrier, i.e., they need to build up trust on the part of the traveler. Only then will more people be willing to use and to pay for location services.

A third impediment not discussed in this chapter is that the underlying GIS-service may be too complicated for an untrained user. Their output will only be valued if the input is very simple. This presents an interesting research challenge: how can we package a very complex program with an easy-to-understand and easy-to-use interface?

A fourth impediment may be that for location services to realize their full potential, i.e., to combine many services and anticipate our needs, we would have to "turn digital" (Negroponte 1995). This means that our schedules, shopping list, preferences, birthday wishes, address lists, trip schedules etc. would need to be kept up to date and available to our personal device.

However, research in location services is tackling the above mentioned research problems and impediments and the last years have shown an increase in location applications. The most promising application areas seem to be tourist services, navigation services for pedestrians and/or cars, and the virtual home environment.

Acknowledgements

We gratefully acknowledge support from the Swiss National Science Fund under contract number 205120-101631/1 and support by the DAAD through the International Quality Network on Spatial Cognition. Thanks to Urs-Jakob Rüetschi and David Caduff for numerous discussions about this topic.

Chapter 11

Augmented Reality Visualization of Geospatial Data

Daniel Holweg and Ursula Kretschmer

11.1 Augmented Reality

The achievements of computer graphics have supported the developments of Virtual Reality, a technique where a virtual environment is built up to envisage artificial environments. Especially due to the developments in hardware in the last years not only visualization but also interaction with these environments has become possible. One example of user interaction is the navigation in these environments. The evolving of the Internet led to the distribution of these environments to personal computers from remote servers. Here, a standard is established: the virtual reality modeling language VRML.

Parallel to these developments more and more mobile devices that become smaller and more powerful, are used to access and explore digital knowledge. Digital cameras allow users to send current visual impressions of the world to other people. By adding text or audio information users can try to convey personal points of view in connection with an image of the reality.

An automatic connection of the reality to additional information that is offered to a user in an unknown environment would mean a huge improvement for many applications. The most convenient way for a moving user is to fade visual information directly in the viewing field of this person. This technique is called Augmented Reality (AR).

11.1.1 Applications

One of the first application areas of AR has been in the industrial sector. Underlying cables could be made visible by use of a transparent display. But also instructions about how to repair a machine can be given directly in the viewing field of a person with direct connection to the real machine (Klinker et al. 1998). Arrows and short video clips enhance the understanding. These systems run mainly in closed environments. This can also be said to an application area that has gained much interest in the last years: collaborative environments. By use of virtual information discussions in groups can be enhanced. An example is the ARTHUR project (Moeslund et al. 2004). Here, different persons from their point of view can see three-dimensional virtual models and interactions with this information are enabled by special interaction techniques that aim to be as unobtrusive as possible. For this reason a method for recognizing hand gestures is being developed. Then, the user will not need to use a mouse or a pen to modify or highlight the virtual information. Another area for AR is the area of medicine. One example is the project Medarpa (Schwald et al. 2002). Here, the navigation of instruments inside a patient's body by a physician can be visualized by the use of a transparent display mounted on a swivel arm.

In these closed environments the tracking of position and head orientation as well as the presentation of information is much more unproblematic then the same task in an outdoor environment. Here, the hardware should be carried as close as possible on the body, and it requires not to weigh too much as it should not hinder the user in walking. The military sector is regarded as an important application area for the development of lightweight, small tracking and presentation devices. For many other application areas the use of AR to visualize geospatial information in outdoor environments has just started. Most applications in this area are focusing on systems which deliver information to tourists.

11.1.2 Augmented Reality by use of geographic objects

The main application task of outdoor AR is the presentation of information to a user while moving through an unknown region. Important objects in the user's field of view can be pointed out. One of the first outdoor AR systems has been the Touring Machine of Columbia University in New York (Feiner 1999). Here, single buildings of the campus are detected by the system and links to additional information about these buildings can be obtained by the user. At the University of Southern Australia, an AR system based on military applications has been developed (Piekarski et al. 1999). Here, information of a Virtual Reality environment for military simulations is faded in the line of sight of the user together with the view of the real environment. At the Technical University of Graz, the three-dimensional city model of the town has been used to fade in information about the current position of the user (Ribo et al. 2002). Behringer has developed a system that detects peaks of mountains and names them automatically (Behringer 1999).

These examples show that AR systems help to detect automatically objects in sight of a person who needs information about his current environment. Conventional AR systems place the information in the field of vision of the user. These can be text labels, arrows or even three-dimensional objects for which specific solutions have been developed. At the University of Tokyo, Chen and Shibasaki integrated the presentation of three-dimensional objects out of a data base on site in a GIS (Chen and Shibasaki 1999). The system not only expands a GIS to the third dimension, it also reacts to the current position and field of vision of the user on site. In the Archeoguide project, Stricker developed a system to fade in ancient reconstructions of buildings of the historic site of Olympia (Stricker 2001). In the GEIST system AR techniques are used to develop a learning game for students dealing with the history of a town (Kretschmer et al. 2001).

In case of geographic information these systems have been developed parallel to developments of other mobile information systems. Especially mobile phones or PDAs (Personal Digital Assistants) can be used to present information to a user on site (see Figure 11.1) .

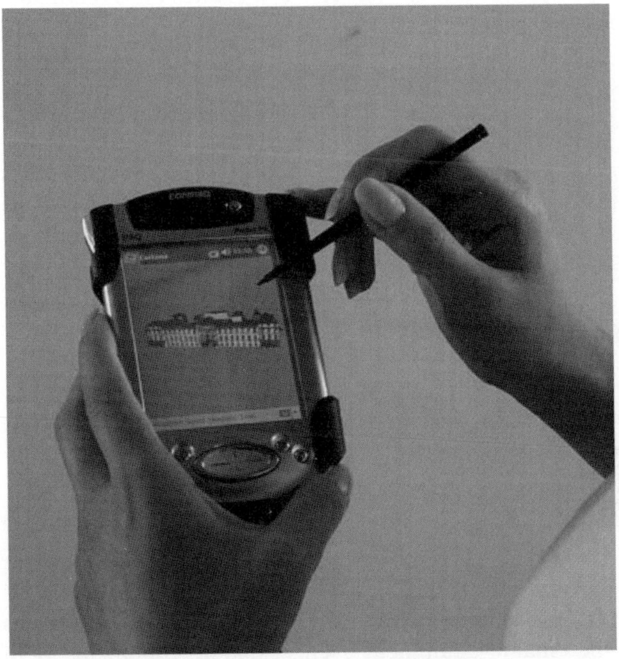

Figure 11.1. Visualisation of 3D Geometry as additional information on a PDA

Using such devices and additional sensors to calculate the position and orientation of the user, Location based Services (LBS) can be performed. Such Location based Services can be seen as a kind of augmentation of the surrounding reality. But AR itself means bringing the two kinds of information (reality and virtual information) into one presentation. Location based Services usually present their information separated from the surrounding reality.

11.1.3 GEIST: Outdoor Augmented Reality for Edutainment

An example of an outdoor AR system is the project GEIST, carried out by Fraunhofer IGD (Darmstadt), Centre for Computer Graphics (Darmstadt) and European Media Laboratory (Heidelberg). Within GEIST an outdoor AR information system with digital Storytelling to be used in urban environments has been created.

The story presented by GEIST takes place in Heidelberg during the 30 years' war. Users of the system are able to "meet" historical and fictional people that lived during that period. By interacting with these people and probably helping them with their problems, the student is taught on historical facts of the 30 years' war in general and especially on the situation of Heidelberg. The main purpose of GEIST is to create a lively and individual story for the users. Responsible for the story is the Story Engine, a kind of director that creates the overall story and out of this the single scenes.

Figure 11.2. GEIST-System in use

Figure 11.2 shows a user of the GEIST system at Heidelberg castle who is equipped with the "magic equipment". This "magic equipment" is the hardware necessary to explore the AR world of GEIST. Main parts of the equipment are a laptop, a wearable see-through display (here Sony Glasstron), and tracking sensors.

In GEIST a stage-based concept has been realized. Therefore those areas that are of some historical relevance have been identified as stages. Whenever a user, who walks around the city of Heidelberg, enters such a stage ghosts contact him via the headphones by asking the user to get in contact with their world, using the see-through-glasses. There he sees a scene consisting of ghosts (avatars), reconstructions of historical buildings and other virtual information overlapping the reality. The several scenes follow one storyline so the user gets an immersive and exciting impression of Heidelberg during the 30 years' war.

As described before, GEIST is driven by the idea of giving information to students in an entertaining way. The AR visualization system developed in GEIST certainly is not limited to such Edutainment applications and can be used even in less playful applications. Examples for such applications are AR information systems for tourists or the visualization of urban planning.

11.2 Requirements

Depending on the application of the AR system, specific tracking technologies are required. Knowing about the users' positions in the real world is necessary for AR visualizations like for any other LBS. In addition, it is necessary to know about the users' viewing direction to extend their view with virtual information. There may be some tasks where it is sufficient to track only the position or orientation of the user.

The accuracy of position and direction information depends on the kind of information that is used to augment the reality.

Highest accuracy is necessary if an existing building should be overlapped with a virtual model. In such cases even small inaccuracies can be seen by the user and will destroy the impression. Lower accuracy is sufficient if simple textual information (e.g., type of building) should be displayed in front of an object.

Especially in the area of LBS, there exist systems that detect position of a user once and offer additional information without updating these tracking results regularly. There are other systems where three-dimensional reconstructions of buildings are presented to the user. Avatars are placed in this environment, from where they explain details about the buildings to the user and even point to these. This requires very accurate tracking results as well as a high update rate.

The choice of the tracking technique depends mainly on the area where the system should run. Indoors, an accurate tracking based on sensors attached to the environment can be used, whereas outdoors no change of the environment is feasible and therefore hybrid systems that consist of several sensors are required. They are usually neither as accurate nor as reliable as devices applied to indoor AR systems.

11.2.1 Three-dimensional Geographic Data

The current interest in three-dimensional geographic data has emerged from the large number of three-dimensional city models that are nowadays available. Data for these models is gathered by photogrammetry or by laserscanning. The advantage of a photogrammetric evaluation is given by the higher resolution of the raw data as well as the higher accuracy of the processing of the data. Much of the data is still processed manually, i.e., an operator looks on the raw data, decides what lines belong to a building and digitizes it manually. A full automatic processing is not feasible at the moment due to a lack of stabilization and reliability. The first approaches in this direction can either not be applied to areas with buildings close to each other or the results have not been tested with a huge amount of buildings. This is the reason why semi automatic solutions have been developed and used. These solutions support the operator in either digitizing the data or in correcting the results automatically. As only semi automatic methods have been applied at the moment, processing of this data is more time consuming compared to laserscanning methods. The reason why this still emerging method has not managed to replace the generation of three-dimensional city models by use of photogrammetric methods is that the raw data from aerial laserscanning methods do not have a very high accuracy which effects the construction of three-dimensional city models. Frequently it requires additional data acquisition to get the same accuracy. The advantage of this method is that the processing can be done almost automatically. Especially for updating three dimensional data, these automatic processes can be applied. The result of both methods is the geometry of a three-dimensional city model, shown in Figure 11.3.

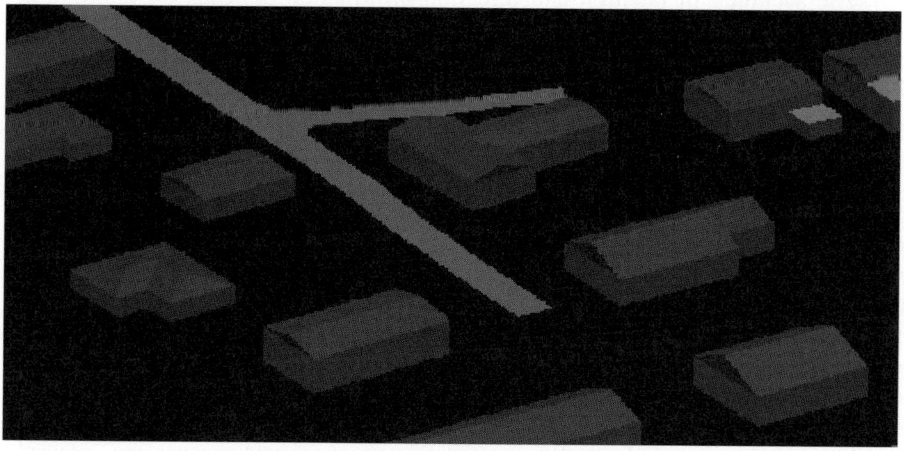

Figure 11.3. Results of the processing of photogrammetric data

The aerial images can be used to generate orthophotos to enhance the presentation of this data. These images with rather low resolution can only be used for applications where a huge area is presented to the user and for presentations which are seen from a long distance. Information delivered by mobile applications often consists of a single building that is presented to the user. In such cases, it is necessary to enhance the presentation of the buildings which can be done by gathering additional information. Methods of terrestrial photogrammetry help to get more details of the facade of an object. For a pure presentation of this data this is often too complex. The enhancement of the building by use of rectified photos helps to convey a real impression of the building. Another even very complex method to acquire the data for visualization is the manual modeling of buildings. For Virtual Reality environments this is the conventional method to process the three-dimensional models. For AR environments the modeler needs to adjust the results with the reality.

11.2.2 Sensors

As stated before, changing the environment to enable tracking for outdoor systems is impossible. This means, that magnetic and infrared tracking can not be applied to the system. These types of sensors require a transmitter of the signal with a known position within the environment. The user receives the signal from such a reference transmitter by a small device or vice versa. Network based positioning techniques like the use of the Wireless LAN are also restricted to limited areas, i.e., most often indoors.

Other network based approaches use the GSM network. These techniques can be applied to indoor as well as to outdoor areas. They enable coverage nearly everywhere even in tunnels or narrow streets. The accuracy is not high enough for all applications but in combination with more accurate tracking techniques these methods offer a first estimation of the position.

Most outdoor AR systems use GPS (Global Positioning System) to locate the user. The accuracy can vary between 100 m up to sub meter level, depending on the technique and on the kind of receiver that is used. One can distinguish between stand-alone and Differential GPS as a more precise application. GPS neither works indoors nor if the signals of less than four satellites can be received. Often this is not feasible in narrow valleys or narrow streets.

The techniques mentioned so far offer only information about the position of a person. Optical tracking can provide not only information about the position of a person but also about the direction the user looks at the moment. In laboratories and other environments that can be affected by the system developer, automatically detectable targets are used as a reference. Also, static cameras can be used.

In an outdoor environment, landmarks have to be detected by an optical tracking system. In comparison with a reference, information about position and orientation can be derived. By detecting features in the camera image and relating them to the correspondent features in the reference the transformation between the two information sources can be determined. This may lead to the determination of the position and orientation of the camera and therefore of the user. There exist different approaches: Behringer compared images of a camera with silhouettes derived from digital elevation models (Behringer 1999), Stricker used photos of the environment to compare them with a video stream of a camera (Stricker 2001), Chen and Shibasaki (Chen and Shibasaki 1999) as well as Ribo (Ribo et al. 2002) extracted information of a three dimensional city model to search for it in images of a video camera. These systems are then restricted to the availability of the reference. Additionally, they require much processing power that makes the usage of powerful laptops indispensable. If less accurate tracking results are required, GPS in combination with a small gyroscope is often sufficient.

11.2.3 Hardware

For both, audio and video information, special hardware is necessary to include virtual information to the user's cognition of the surrounding environment. For receiving audio information headphones or speakers are necessary. Using audio information for augmenting the reality is already in use within some applications. Especially in museums it is often possible to hire special equipment (headphones or speakers). Using this hardware the user gets information about the presented objects.

Augmenting the reality by vision requires hardware to integrate virtual information into the user's field of view. In general two ways can be seen for bringing together real and virtual information:

- Video AR: By using video AR, the reality is captured as a video stream that is used as background of an AR scene. Virtual information is projected in front of this background. The major advantage of this way of presenting an AR scene is that the current reality is captured and can be taken into account by the calculations. It is possible to use image processing algorithms to select information from the reality – captured as an image - and do the augmentation by using this information. Video AR can be displayed on any screen. In mobile environments a PDA can be used for the visualization of an AR scene.

- See-through AR: By using see-through AR, a see-through display (e.g., Sony Glasstron) is necessary. By using such displays the virtual information is presented on a semitransparent display so the user can see the reality as background of the AR scene. For such AR systems the reality has not to be captured but a connection of the user's viewing field to the real environment needs to be established.

Within the GEIST project see-through AR has been realized. In the following, we use GEIST as an example to present results and experiences we gained in this project by applying AR to visualize geospatial data.

11.3 Augmenting a users reality

Movement within the real environment is the basic kind of interface with the system in Location based Services. These services are able to react to the user's interaction by adapting the content as well as the kind of presentation of the information. Any kind of outdoor AR application can be seen as a Location Based Service. Contrary to usual Location based Services AR visualizations do not necessarily give the users information with respect to their current position by providing only virtual information, but they include virtual information into the real environment by enhancing the user's viewing field while he examines the location. Here, the existing and surrounding reality is augmented by virtual information so the user does not need to fuse the real environment and the information presented on a display. In AR visualizations the real environment and the additional data become one single presentation of information.

In general it might be useful using all human senses when augmenting the reality. As the currently available hardware does not allow giving information to all human senses for augmenting the reality, nowadays applications focus on the two most important and supported ones by available hardware. They are extending the reality by visible and audio information.

11.3.1 AR Visualization System

For visualizing virtual information in context of the surrounding reality an AR visualization system has been carried out during the project GEIST. This system manages the virtual information in a three-dimensional scene graph. The scene graph has to be created with respect to the real world. A directing engine does the composition of the objects that should be used for visualization. SceneML, a XML subset is used to describe this composition and the behaviors of the objects under several constrains. Figure 11.4 shows the general idea of the AR-Visualization System in GEIST.

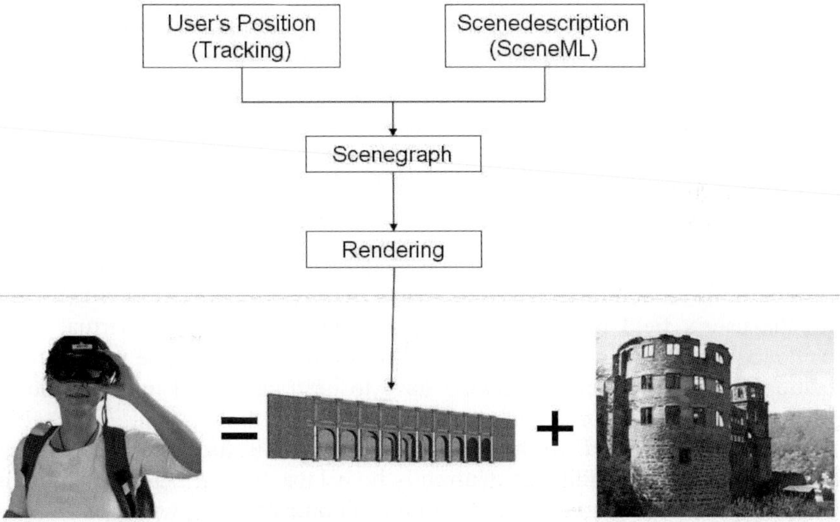

Figure 11.4. General Structure of the AR-Visualization System in GEIST

The visualization system converts this description to the three-dimensional scene graph and creates a view, which – faded into the user's field of view – overlaps with the surrounding reality. Therefore, georeferenced models of historical buildings are used as a frame of the virtual world. All other objects (e.g., avatars, objects of everyday life) are positioned with respect to these buildings. This enables the directing engine to create the virtual world without knowledge about the exact coordinates.

The positions of the buildings are stored in a three-dimensional database. Whenever a building has to be added to the scene its geometry is loaded from the database. To reduce the dimension of coordinates the object is transformed into a local coordinate system. This coordinate system is axially parallel to the superior coordinate system to reduce the transformation to a translation. The root of the local coordinate system is usually the centre of the area the AR-system works in. VRML is used as exchange format between the database and the AR visualization system.

All other objects like avatars and objects of everyday life do not have fix and known coordinates. Within the SceneML they are positioned relative (e.g., "in front of", "left of") to already included objects. When adding such an object to the scene graph the position coordinates within the local coordinate system have to be calculated by taking into account the user's position and the position of that object the new one is connected to.

Predefined animations can be used to make the avatars behave and react to the user's interaction. To make the avatars speak, Text-to-Speech is used. Even during a scene is rendered the directing engine can influence the action by adding new objects or specifying new behaviors for the scene. This allows the system to react to different behaviors of the user and present him more or less detailed information during a scene.

11.3.2 Rendering a scene

After a virtual world has been created as described above, a view of the virtual information is rendered. To make this view overlapping the reality when fading it into the user's field of view, the user's representation in the virtual world is positioned with respect to his real world position. Therefore, the results from the position and view direction tracking have to be transformed into the coordinate system of the virtual world.

The view generated for the user's representation in the virtual world overlaps the reality surrounding the user. Currently both Java 3D and OpenSG are in use as rendering platforms. Figure 11.5 shows an example of a real world extended by a virtual object.

The described process ensures the creation of an AR in general. Nonetheless in outdoor environments special constrains have to be taken into account. The most important one is allowing the real objects to be located in the area between the user and the virtual object. In this case, a virtual object that should be located behind a real one appears in front of it as it overlies reality. Figure 11.6 shows the basic configuration for this problem.

Several approaches for taking the occlusion aspect into account are known and in use with different AR applications. Being useful for AR visualization of geospatial information a model driven approach as described by (Klinker 2000) and (Breen et al. 1995) has been chosen. Using this method an accurate model of the real environment is necessary. The candidates for the occlusion are detected and their geometry is added into the scene graph. For rendering, the color of these objects is changed to black and texture information is removed. Within the AR visualization no information appears at the overlapping area and the virtual object appears behind the real one. So it improves the realistic impression of the AR scene.

Figure 11.5. Heidelberg castle extended by a reconstruction of the Ottheinrichsbau

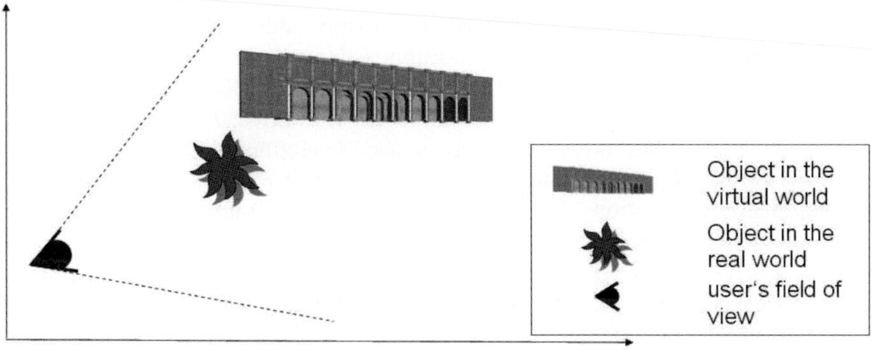

Figure 11.6. Basic configuration for occlusion of a virtual object

The geometry of the occluding objects added into the scene graph has to be at least of the same accuracy as the accuracy of the user tracking method. Usually this fits to the accuracy of geographic data and three-dimensional city models. So this data that is necessary for the video tracking can be used here again.

The necessary calculations are – in a first approach – usual GIS queries. It has to be calculated which objects are located within the cone / triangle coming from the user's position, his viewing angle and the distance of the most far away virtual object. For most applications this calculation can be done in two dimensions as objects are usually connected to the surface and the user's field of view is usually somehow parallel to it. Projecting the objects and the view cone onto the floor plan and doing the calculation with this data leads to sufficient results and allows it to work with well-known GIS tools.

A second important aspect to be taken into account is the brightness of the surrounding area and the location of the sun. First of all the brightness reduces the see-through impression for the user. It is necessary to bring the brightness of the virtual world somehow to the same value as the one of the reality. The second kind of influences the light gives to an AR visualization are shadows. While the user can see the shadows of the real objects, the virtual ones do not include them. To improve the realness of visualization it will be necessary to integrate a light-model and simulate shadows within the visualization.

11.3.3 Content

AR visualization of geospatial information can be seen as a next step within the development of geospatial visualization. It started with traditional maps and has gone to three-dimensional cartography within the last years. Three-dimensional cartography breaks the limitation of using the floor plan as a representation and only two dimensions for visualization. With AR visualization of geographic circumstances and traditional map content even the limitation of the available data pool breaks. The results of analyses based on geographical data or the spatial data itself can be visualized in context of the real environment so the user can detect errors or interdependencies with objects not yet included into the data structure.

The research within the GEIST project mainly addresses the visualization of detailed models of historic relevant buildings. Nonetheless AR visualization of geospatial data is not limited to this content and allows the presentation of any kind of information traditionally used in maps. Like virtual models of buildings even symbols or textual information can be used to augment the reality.

This kind of visualization improves the usability and intuitiveness of geographical visualizations. For those applications a high level of abstraction or a large area to be covered is not necessary.

11.4 Conclusion and Outlook

The chapter described the development and potential of AR visualization techniques for geospatial information. Currently this development is proceeded in a number of research projects. The technique today is especially used for entertainment and edutainment applications as well as information systems for tourists. Because of the new quality of integrating information into the real environment without abstracting it this kind of visualization will be useful especially for those users who have to make their decisions on recent data. For them a traditional map can be necessary to give an overview over a large area. AR visualization, that is limited to only small areas, allows a new quality of technical support for such decision makers and introduces a new kind of immersive information visualization.

Further research especially in the area of accurate tracking techniques will improve the most important problems of this kind of visualization. Nonetheless especially the dimension and weight of the hardware has to be reduced to ensure the acceptance of such visualization systems.

Chapter 12

Geo-ICT and Development – The Inverted Pyramid Syndrome

Satyaprakash, Ayon K.Tarafdar and Ravi Gupta

12.1 Geo-ICT and Development

In developing countries, the domain of development and of technology – more specifically geospatial and information & communications technologies – are conceptually complimentary and closely linked. However, cases of forming development strategies in developing regions on the platform of geospatial sciences or information and communication technologies (ICT) are rare. This chapter is an attempt to develop the argument that both these domains – technology and development – are very much similar in cognisance and perhaps need each other in more ways than many other research areas. The paper shall establish the strong need to amalgamate and internalise technology platforms in development processes of the developing world and demonstrate through the description of a pilot project how this may be taken upon.

12.1.1 Development: A Collective Responsibility

Development can be understood to be a process to decide for collective good and becomes helpful in distributing resources and opportunities in an equitable manner for a region in a time frame. Planning helps realise the aim to develop. It is a conscious (and often mandated) activity that enabled taking actions that lead to multiple aspects of development (human, economic, physical, etc.). Planning is done at various scales and levels and each type of planning is different than the other. The exercise of planning involves an intricate and complex mesh of multi-party involvement, consensus and conflict resolution. Ensuring planned development is very basic for any society and it does this through the different mechanisms. Considering the undoubted need for planning and development, we have to understand the premise that development focuses actually on collective rather than an individual well-being. This crucial premise makes it imperative to have a 'rationale' behind any planning approach. The purpose of having a rationale behind planning can be the following: firstly it gives planner a logical basis for development planning thereby legitimising his role in the society, secondly it shows function to a planner for his actions, and finally it acts as an ideological compass in his cognitive processes. One also should be clear of the fact that the purpose behind development does not have roots in crisis situations.

Now let us try to understand how rationality can seep into a development planning process. Having technology married to the decision-making mechanisms reduces chances of irrational judgement. However, it is important to understand that planning or development can very much be an independent phenomenon without a direct relation with technological know-how or its development for the same society. Technology development and presence of experts in certain fields (like geographic information sciences or GIS) may not be reflective of actual societal development of any community. Development, especially for developing countries, is an extremely complicated process that passes various corridors of power, decision-making exercises, political interests and bureaucracy. It may not necessarily be dependent on rationality. And hence what is practised in labs or academic departments as planning for development may not be what is done in the development authorities. One should keep in mind that since development planning adheres mainly to societal concerns, it is natural to expect interlaced interests and ambiguity. It is important to know the inherent complexities involved in development activities before understanding the role and effectiveness of technology.

If the intention is to establish a relation between technology and development, it is important to synergise the 'development of technology' with the process of organic development of underdeveloped regions and all planning activities related to it.

12.1.2 Understanding Geo-ICT

Geographical Information Systems is a strong tool, which stores spatial as well as non-spatial data digitally and establishes a link between these two. Resultantly it produces not just maps but an information system, which can retrieve, analyse and represent the stored data in desired ways. (Tarafdar and Krishna 2000) Geographic Information Technology (GIT) is the integration of mapping sciences and information technology. Although the use of such geospatial sciences in developed countries has been in almost all the spheres of development, it is the developing nations where this use has been minimal. With the emergence of mobile GIS technology and improvements in the desktop GIS, in terms of low cost and high performance hardware and software, GIS is increasingly being used for improving the efficiency in the delivery of public utilities and other infrastructure planning and management across the world. The development these technologies are happening fast and suitable enough to meet the demands of various scientific endeavours. Developing regions have also come up with extremely innovative cases and products. However, the question of having such tools as an internalised 'rational basis' of development strategies may not be evident. Most cases are piecemeal.

The chapter beyond here shall use the term Geo-ICT for amalgamating all references to geospatial sciences and the array of tools available these days in the field of ICT and IT in combination to geospatial sciences.

12.2 The Need for Rationality in Development

One of the reasons of considering at internalising Geo-ICT tools in development planning or activities is to enforce rationality and transparency. Rationales are attempts to correct imperfections of the production and distribution of economic and political subsystems of society. What makes an approach 'rational' or 'irrational' is a subjective matter when perceived individually. However when seen collectively (the typical case of development), it can be agreed upon that development planning is concerned essentially with collective rationality (Figure 12.1). Collective rationality can be of two nature – instrumental (functional) or substantive (Sager 1992).

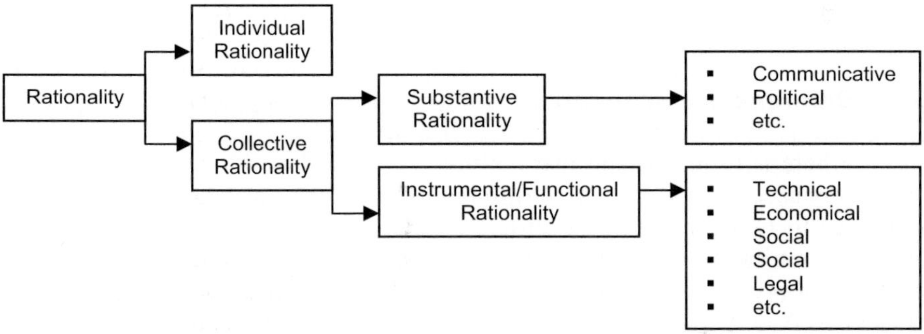

Figure 12.1. Types of Rationalities

A prime component of development serves to mitigate anomalies. Without going into the depths of the rationality theories it can be said in a nutshell that all the above types and forms of rationality emphasise the need to achieve different goals of planning in a manner that reduces the risk of arriving at non-conforming decisions.

The planning horizon is often strained or bounded due to two main sources of conflicts. These conflicts are more pronounced for developing nations. The first source comprises the organisations responsible for the functioning of a region. In every planning situation there are a number of organisations or institutions with varied responsibilities and power that eventually contribute to the overall development. Other than that, there are the end beneficiaries or the citizens of that area. Since development or planning serves to achieve multiple goals it encompasses most of these organisations. This often leads to conflicts among these organisations, institutional mismanagement and disharmony. The other source of conflict is the presence of a number of actors in the urban situation.

They comprise the beneficiaries in the society scaling from the common man to civil societies and business groups. The role of these actors in planning has risen over the years with increased access to education and birth of democracy.

Hence, it is essential to bring rationality to activities that affect multiple persons or a community. Hence development and rationality are terms that are in need of each other. Geo-ICT tools are facets of enhancing rationality.

12.3 Surveying Organisations, Planning Organisations and Users

The organisation of the different institutions contributing to surveying or planning or spatial data generation can be of vertical or horizontal arrangement or a mix of the two. Various sectors under the government in the form of ministries are formed at different levels (namely the national level, state level and municipal level). With different arrangements of the government power units there are 'different responses to urban realities'. This is a prime reason why Geo-ICT is still at large, in the domain of development planning in developing regions.

12.3.1 Differences in Mandates

The idea here is to understand that multiple issues and goals, different forms of organisational set-up and mandates, different resources and skills, varying types of limitations to rationality and the different nature of responses from each actor, together contribute different shades of complexity and unpredictability in development process. The other actors that add to the complexity are the citizens themselves. A city is an organic whole; it is a problem in 'organised complexity' where 'situations arise in which many quantities are arranged simultaneously and in subtle interconnected ways'. (Jacobs 1969). The citizens, comprise the individuals and different interest groups either in organised form of civil societies or in more formal co-operatives.

Table 12.1. Response to different complex development situations.
**adapted from Friedmann (1987, p 76-83)

	Goal of development	
Technology	Agreed	Not Agreed
Known	**A. BUREAUCRATIC MODEL** Policy response: pre-specify process and outcome	**C. PLURALISTIC POLITICS MODEL** Policy response: prescribe process, leaving outcomes open
Unknown	**B. SCIENTIFIC MODEL** Policy response: loosely prescribe outcomes, leaving process open	**D. WICKED PROBLEMS MODEL** Policy response: leave process and outcome open

In the case of India and most developing countries, these sectored corridors of power, especially for government bodies are arranged in a vertical fashion where each power unit functions under the overall guidelines set by the ones above. This reality and functioning paradigm gets eventually inculcated in the manner in which these government organisations interact beyond the scope of the state. In other words, spatial data or any data required for development purposes are generated, collected, maintained with high-end computing tools and scientific know-how mostly by a group of 'expert departments', where the data remains locked henceforth with the promise of reproduction on request. The main issue in such cases is the absence of knowledge of this rich repository of data and the issue of relevance of this data in local context. The case resembles the D segment or the wicked problems model. As per planning theories, this required an open process and outcome. Technology here need to evolve with the development strategies and not necessarily used as a tool to decide.

12.3.2 The Role of Geo-ICT in Development

For a long time, technologies like Remote Sensing, GIS, GPS, and Aerial Photography etc have been confined in research labs. Undoubtedly, these technologies have evolved to a large extent but the involvement of community in the progress and growth of these solutions is questionable. With very little application for social development and community use, social significance of these technologies have been very less and very few technologies and processes has been customized or adapted to meet the community needs. While the technology developers would argue that they have provided every possible tools and facilities to tackle complex problems and are developing them further everyday, the planners might still argue all these to be matters beyond their reach and relevance.

The following sections shall discuss a syndrome in this respect that is prevalent in developing countries that shall keep thwarting the amalgamation of Geo-ICT and development unless addressed categorically.

12.4 The Inverted Pyramid Syndrome

In South Asia, the existence of institutions that shine in spatial database generation and analysis are numerous. The understanding of Geo-ICT has evolved over the years in a differential perspective in Asia. Initially, it was an area of thorough research and got developed as a science. Its understanding as technology became prominent with the understanding of its varied applicability. However one has to translate and interpret these technologies beyond its attire of 'science/technology' unto the aspect of 'service' for most of these regions (Tarafdar et al. 2002). Perhaps this core lack of understanding results in the under-use or non-use of its potentiality. The rich knowledge remains as a subject of 'science' repeatedly discussed and researched in the 'islands of excellence'. Most of the relevant and rich database, application and research, are many a time not available with the right agencies. The figure below aggregates in general the case of developing situations.

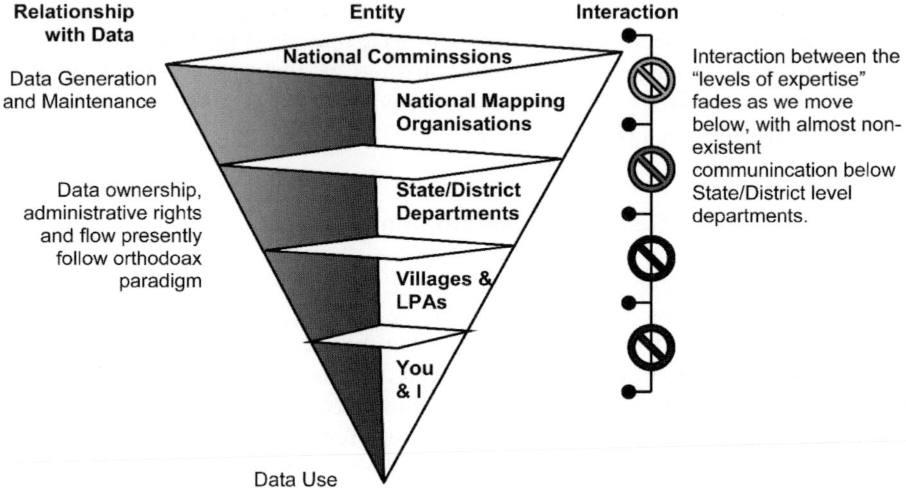

Figure 12.2. The inverted pyramid syndrome

The 'inverted pyramid syndrome' as depicted above is hence a conceptual reality that is inculcated in the functioning, set-up and mandates of state-owned organisations in 'nations of the south'. It is irrelevant to go into whether this is wrong or right. The above figure is a typical case for developing nations (often with a history of imperialist regime ruling for centuries). Most organisations in the developing world's spatial database domain are arranged in a top down perspective. There is a sense that policies taken at the top shall have its spread effect downwards. The agenda is set at the top level and followed up by various scientific and research organizations at central or state governmental levels. There is an assumption that the knowledge created from the data experimented with, shall have its own trickle down in the long run. On the supply side, the government makes huge investments. Many large national agencies in India (in surveying, geology, forestry, agriculture, water etc.) work independently of each other in the field of their region. Unfortunately, in spite of the huge amounts of information generated, the availability and usage of geographic knowledge becomes clogged within the government departments or research institutes themselves. The results of these high technology based studies are rarely available to the civil society organizations. There is also a clear assumption that the true bulk and wealth of information increases as one goes up towards the central nodal points. The ownership and decision making in terms of data to be generated, maintained and disseminated also follows orders from the top, especially in the spatial database domain. (Tarafdar et al. 2002) Most mapping agencies decide at the top what is to be done and the units at the village level simply implement it. This creates the ***inverted pyramid*** – with its assumed base (width of knowledge and responsibility) at the top and flow of every element (from data to decision) from top to down. In some cases, one might truly argue that policies need to be set at the top and decisions need to be made at the pinnacle. However, it is this inverted pyramid syndrome that is perhaps the stepping-stone behind amalgamating Geo-ICT and development.

12.4.1 Inverting the process of Geo-ICT evolution

Now let us try to reverse the inverted pyramid. Basing our rationale on the urban realities we need to reflect upon what can be the alternative approach to the present one. A strategy succeeds and becomes effective when it respects and acknowledges the context. In the context for developing countries the truth is that – communities and players at the grassroots are the best actors and stimulants in development. They have the real information base that is most related to the context and is of the minutest scale. This has been acknowledged globally through Local Agenda 21 and various national agendas that have shifted and decentralised planning and development related decision-making at the lowest rungs of administration. The role of the central units have been visualised in such scenarios as facilitators and planners at the national context that rises beyond the local and thinks beyond the nation.

However, if the functioning and mandates of departments and agencies working in the domain of spatial database include an essential clause within them that obliges them to inculcate their activities based on feedback from the grassroots, it might endorse the first step towards reversing the inverted pyramid syndrome.

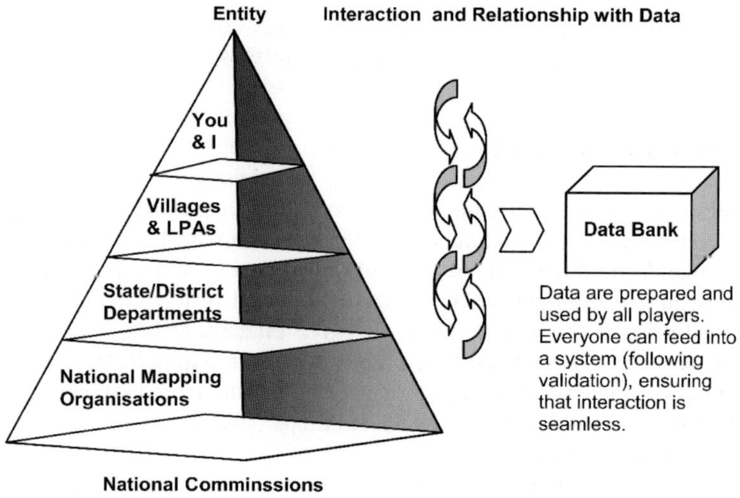

Figure 12.3. Reversing the inverted pyramid syndrome

We need to understand (from Figure 12.3) that data and information is the most at the grassroots level. The larger the scale, the more is the information. The same aspect also holds true for spatial database tools and policies. One needs to see this in the light of collaborative arrangement rather than bottom up. The figure above depicts a situation where layers cease to exist with respect to data generation or development of technology. Spatial data need to be a domain of all the players as long as they have the expertise. Anyone in need of spatial or aspatial data for development might consider developing it on its own. All such data available across various layers in the inverted pyramid syndrome need to be fed into a databank to be regulated and maintained by a mandated department. This databank shall be the seamless interface between demand and supply.

The notion of 'planning through collaboration' would relate to a condign approach that touches upon the communicative, social as well as instrumental rationality. Collaboration essentially is a form of collective rationality in its view where knowledge and understanding is produced through collaborative social learning processes. This is the first step towards putting technologies within the homes and offices at the local level. It lies heavily on communicative consensus since it means mutual understanding, co-ordination and formation of identities. It would call for societal transformation and inclusionary discourse where the multiple actors can go into argumentation in various forms, exerting their power and finally seek to negate their differences in possible outcomes.

The true raison d'être behind this concept is that such a process shall set the ball rolling in both directions. Grassroots will understand that it is their mandate to generate data for development and data actually lies with them within their locality. They will identify the data needs and fill the needs as per their levels of expertise. Agencies at a higher administrative level will leverage and move-in where expertise is lacking at the local level. The national level agencies will then need to shift their paradigm from prescription to enablement.

12.4.2 Need for Local Content

This situation that reverses the inverted pyramid syndrome, puts local content at the central of any discussion. Local content assumes a critical significance beyond any other mandates spatial available in the country. New or expert driven technologies are only meaningful if their content reflects local needs and conditions. The panel discussion on World Summit on Information Society recognised that ICT can be used to promote, distribute and create local content, moving from one-way media in which poor and marginalised people are information recipients to interactive media, in which they participate actively in its creation and dissemination.

Studies have demonstrated that the assumption that the supply of information created by the global network of ICTs will be sufficient to enhance livelihoods of the poor – as long as the poor have access – is fanciful and even unlikely. Encouraging collection, production, exchange and dissemination of relevant content in local languages can not only provide learning opportunities for the target population but can also provide an opportunity to the target local population to help shape the decisions being made in their community. In fact Internet access would be useful to the local people only if they are able to find/view local content of relevance.

Thus the concept of local content highlights the importance of moving beyond a quantitative and purely technology-driven approach towards people-centered and demand-driven solutions. In fact the benefits of producing unique local content can go beyond development. It can enhance ties within the community itself. It can define the image of the town/area/region, provide a sense of place, pride in common history and build a reputation for the community. By providing contents that connects directly with community, the local content website can forge critical ties with the people who may become dedicated viewers of such information. It will also widen the scope and viewers for Internet/websites use. However for local content to be relevant the content should be sufficiently focused or specific to be relevant to the local people.

12.4.3 The Issue of Accessibility

Access to geospatial information is vital for decision making for local, regional and national issues. Better-resourced (informed) communities can address their critical issues of social, environmental, and economic importance more effectively. However the common citizen even today has been kept away from the access to geospatial information. They have been denied the information they need to make meaningful choices. With reference to maps and GIS, there is loads of information to which the common citizen is still denied access. Some of the reasons why information access is still limited are as follows –

- The gathering/collection of geospatial information has also been very localized – mainly by the government department, leading to concentration of information in a few hands and depriving the community entirely from its access,
- The organizations or authorities that collect or generate the data have little or no contact with the community. As a result the data infrastructure produced by these authorities is rarely able to support community goals,
- Even within the department, data is organized in a format (and in different formats in different organizations) that is not easy to comprehend by other departments – let alone the community,
- Although in some cases sufficient quantity of data exists but existing data is not up-to-date and lacks necessary attributes (that can resolve local issues) making the data redundant for effective utilization,
- Except some isolated digital files most data is stored in traditional paper products. Mass storage of data online or wide area networking of information is still not considered – making information even more inaccessible.

As a result data and information is neither available to the community not is it able to meet the community goals. This has also led to a great disparity between those who have access to information and those who do not.

12.5 Moving Towards Pragmatism: Mapping the Neighbourhood

Let us briefly go through a project in India. This project generates local content at the local level. And the data generators (and sometimes assimilators) are the local people themselves, more specifically the students of the locality. The motive – local development. Even though the project propagators primarily aim to reverse the learning process and address alternative aspects of education, the fundamental element of the project that is relevant to the earlier discussions in this paper are brought forth in the following sections.

12.5.1 The Project Background

Mapping the Neighbourhood is a programme of CSDMS (Centre for Science, Development and Media Studies), an NGO based in Noida, India, has been sponsored by the Department of Science and Technology of the Government of India. The programme aims to bridge the digital divide prevalent in rural India, through participation of communities in making neighbourhood maps and using them for local development. The programme involves students from 20 schools across the Almora and Nainital district of Uttaranchal state in India. The project has attempted to facilitate community learning through neighbourhood maps that are grounded in ecological and social narrative with school children as the spearhead of knowledge creation.

The project has excited the students in scientific and technological innovations. They have become channels for the dissemination of scientific and technological knowledge. The focus is not just technology but also its innovative use for mapping the neighbourhood and its resources. This would build their professional capacity. The maps prepared through community-friendly-technology by the school children can then be fed into the ongoing planning and development process.

12.5.2 The Process

This neighbourhood mapping process not only involves generating community maps but also involves deliberative discussions as a framework for solving problems (Figure 12.4). With enhanced awareness about the community, the students are provided a ground where they can

- understand the complexities of issues their community faces;
- openly share their diverse perspectives and concerns; and
- identify solutions to problems

The project is attempted to bring high-end technology to the rudimentary level. School children are being initiated to use indigenously developed GIS software running on a Personal Digital Assistant (PDA) coupled with Global Positioning System (GPS). The data collected and map prepared by the PDA is transferred to a database. These maps, once finalized, are integrated with the data collected by the students and thematic maps are prepared which forms the basis of decision-making process.

Discussion forums are an integral part of neighbourhood mapping. The students discuss, from their personal experiences, issues that concern their locality and community. They proceed to examine multiple views and perspectives. In the end they understand the complexities of the issue and come to an informed opinion. Rather than responding to presentations or proposals by experts and advocates, the students exchange ideas on issues of common concern.

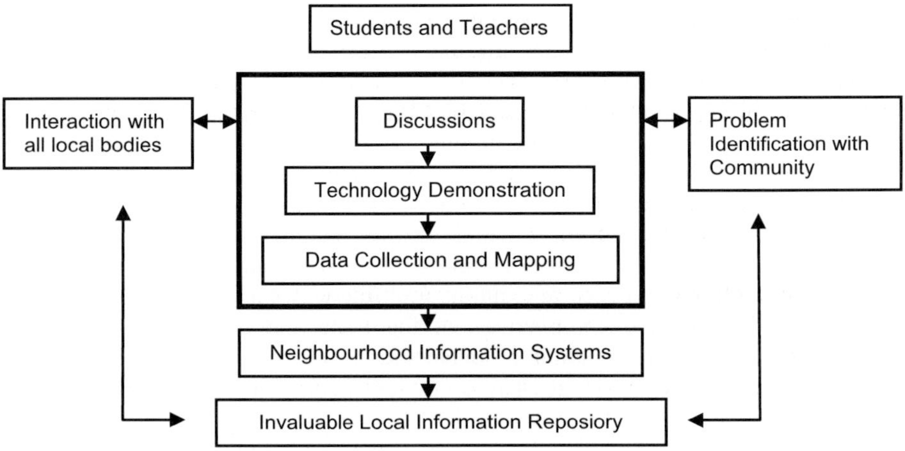

Figure 12.4. Process

In the same way, ordinary citizen from all sections of the community has been involved in open discussions about an issue of importance to them. After the conclusion of the discussion or study, results can be shared with the community, and action ideas may be considered and developed in an "action forum." The results of the forums can then be shared with local leaders. In addition, community members who have attended a forum may often decide to continue to work together to try to solve problems in their community. Thus neighbourhood maps created by the students with community assistance can help the community to make informed decisions about the best policies and most appropriate programs for the areas.

The students of most of the government schools in rural areas were found non-responsive and hesitant in the beginning of the project. However as the project proceeded and technology was introduced, the students became very inquisitive and excited. Most of the rural students involved in our project knew very little about the computers. Urban school students were more open and had in fact far more knowledge about the computers than their rural counterparts. The best part of it all was that they were all ready to learn. This was enough for us to start and introduce the new technology. The students, who were exposed to new technologies, which they had neither heard nor used before, received the exercise enthusiastically. They were thrilled to see their own school and its surroundings on maps in a computer. The use of ICT brought a new facet in the learning process.

12.5.3 The outcome

Under the project students are involved in collecting local information. Through this, they would not only be archiving their experiences of the project and their outputs (maps, information repository of villages collected through primary surveys) but they would also document facts about their town, history of development, local culture and other interesting facts. Thus considerable local content will be generated by the initiative of the students, which can be regularly updated by the students themselves and new content added. Local content that is in the process of being generated for Almora through this project are –

1. Information on demography (at the village level is available with the chief election office (voter's list) – this information can be utilised and made available to the common people.
2. There is very little information available on health and the available information is neither organised nor documented. The *anganbari* workers (social workers in health domain) also keep records of health of all children below 6 years. Information from the a*nganbari* workers can be combined with the information from primary health centers and the district hospital to form a health service network – something that is missing.

3. Communities in many hill regions have their own storehouse of indigenous or traditional knowledge that has been handed down by word of mouth from generation to generation. In Kumaon (project area), valuable information exists on indigenous biodiversity, traditional remedies and medicine, lost history and customs, innovative and indigenous farming techniques/practices, of indigenous crop etc. Few people outside the Kumaon are aware of such knowledge. The local people can document these facts in collaboration with academicians and researchers.
4. There is very little information available on the education system and the schools. There is absolutely no information available on the number of students and teachers in each school. Information on enrollment rate, pass-out rate and dropout rate is also not available for the common public. These schools have been graced by eminent academicians. Recording such information will provide inspiration for the other students.
5. Information on unique local features. One of the main sources of water in Almora is the *naulas* i.e., springs. But very few people outside Almora are aware that there was once over 200 *naulas* within the town itself, which are now on the verge of extinction and the block had another few hundred probably? Although the local newspaper does publish articles on the depletion of naula water due to misuse and increasing urbanisation, it has still not been able to generate mass movement or even a local movement to protect these naulas from fast disappearance.
6. There is also a considerable lack (almost non-existent), of information on the local culture, cuisine, local songs, customs etc.

These local contents /issues generated with the participation of local community in this project will help in collating the information about the region, which may be used for taking the right development strategies of the region. Since the community has collected the information, they will own it and further updating of the data could be done by them.

12.6 Acknowledging the Utility of Geo-ICT in Development

It may be attractive to assume, by many, that IT and GIS based spatial knowledge ought to be equitably distributed within communities for many of its 'far reaching positive effects', but this can be argued (McCall 2002). In most developing countries, collection of spatial as well as aspatial information is a challenge despite its effective role in problem solving. However, collection of geographic information requires appropriate GI technology. The decision support systems making use of socio-economic information and using interactive mapping technology, needs to be developed and the challenge lies in developing an appropriate GI technology for making spatial information readily available and easy to use for all communities. Decision making in the public domain becomes more effective when the decision-makers become more familiar with the geospatial tools, which enhances the scope of their decision-making capabilities. GIS can help to create database that can be universally accepted and in many cases it can modernize the overall public policy process-be it for planning or monitoring. ITC can have a huge impact in the way social services and amenities are delivered and managed. In developing countries ICT

- can provide transparency,
- can standardize procedures,
- can ease info exchange.

"Data are the fuel of GIS" - as somebody put it. Any meaningful application of GIS relies on the use of spatial data relevant to the theme. GIS applications, with their tendency to cover vast areas, require large amounts of data. Thus, the compilation of a database is usually the costliest and most time-consuming activity of any GIS project. It typically accounts for about 80 per cent of the total project cost, and it can take years to complete. Data are also the longest-living part of most operational GIS: hardware generations change every two or three years, software versions even more frequently, but the data are there to stay, and they even retain their value once an updated version exists, e.g., time series' analysis.

At this age when scientific tools and techniques are inseparable from our day-to-day life, it has become essential to address the problems in a scientific manner. Without an integrated methodology, identifying viable technological and institutional options for sustainable development in developing regions is not possible. For sustainable development, everyone is considered to be a user and a provider of information. This includes information, appropriately packaged with experience and knowledge. The need for information arises at all levels, from the decision-makers to the grass-root level individuals. In order to ensure that the decisions are based on correct information, we need to bridge the data gap between the decision makers and the implementers of the programme and make the information (read data) availability reliable. Although it is recognized that the data is necessary for sustainable development, it is not available for most of the regions in developing countries. Even if it is available for the developed regions in developing countries, it is not being put to maximum use for sustainable development. Under-developed regions neither have the data nor have any processes for the assimilation and collection of data, thus slowing the development of the region.

From this we can arrive at an argument that Geo-ICT applications have to rely on an organized infrastructure of spatial data if they are to become economically viable and deliver the results within a useful time frame. The advantage of an infrastructure is that one can use it without having to pay for the full cost of its establishment.

12.7 Implications

The above-discussed project, which presently involves 10 schools and about 250 students, is a piecemeal effort of significance. It is a multi-pronged approach that tests a number of elements on ground. If such an effort remains piecemeal, the idea of reversing the inverted pyramid doesn't comply. This exercise if seen to happen in a large and adequate number of schools, where the findings or the total investment becomes important enough for the local authorities to take notice and utilise the information – then the base of the pyramid reverses and becomes upright. Otherwise the syndrome remains.

Discourse and argumentation in such contexts needs to address the four key aspects as suggested by Forester (1989); comprehensibility (including all actors involved in development process and technology development – in argumentation), sincerity (allowing all opinions – of development strategy and technical solutions – to take shape and interact), legitimacy (discourse should be focussed upon the concerns of the locality and not just scientific argumentation) and accuracy (of information and depicting truthfulness).

Technology should be used as a tool for improving human life and thus access of every citizen to the benefits and advances of technologies is imperative. Facilitating the access of all sections of the population to information and public services through ICT would improve the communication with each citizen to increase individual's participation in the democratic process. This would support the establishment and management of geographic information systems and the formation of a knowledge-based human capital and digital democracy. In the new economy, only knowledge and information based development can simultaneously achieve the sustainability, equity and efficiency objectives of development. Thus for developing countries, use of adequate geographic information needs to be targeted and enhanced in sectors of public policy, human resource development, and public participation.

Chapter 13

Privacy Issues in Geographic Information Technologies

Alastair R. Beresford

13.1 Introduction

The concept of privacy varies widely between societies, across time and even between individuals. Therefore, providing a succinct definition of privacy is difficult because one person's concept of what constitutes privacy intrusion can be very different from another's. An extensive survey of personal privacy (Rotenberg and Laurant 2004) was carried out by Privacy International as part of the Global Internet Liberty Campaign. The original report is now revised and extended on a yearly basis and identifies four broad privacy categories:

Information privacy:
: protection of data containing personally-identifiable information; examples of personal data include medical records and bank statements.

Bodily privacy:
: protection of people from physical invasion; examples of bodily invasion include drug tests, cavity searches and genetic testing.

Privacy of communications:
: protection of all forms of communication from interception; examples of interception include monitoring telephone, email and written correspondence.

Territorial privacy:
: protection of domestic, work and public space from intrusion; examples of intrusion include search warrants, video surveillance and identity checks.

Gavison (1980) took an alternative approach and defined privacy in terms of three commonly asserted moral rights connected with privacy:

Solitude:
: the right to control access to a personal space; examples include undisturbed reading of a book in a park or enjoying a meal with family and friends at home.

Secrecy:
: the right to control the collection and onward distribution of sensitive personal information; example types of information include medical records, family photos and private communications.

Anonymity:

the right to remain anonymous and unnoticed in social interaction; examples include shopping for goods and services without identification and the ability to speak freely with anonymity.

In many traditional social structures, individuals experienced relatively little solitude, secrecy or anonymity; gossip could spread intimate personal details and actions of other members of the community quickly. Such information was rarely spread outside a community since the method of communication was often by word-of-mouth and the information transmitted would generally have little relevance to remote communities, except where it concerned a famous or wealthy individual. As most events were committed to memory rather than written down, past events were tempered by time since recent actions of an individual allowed the community to subconsciously re-evaluate an individual's previous behaviour.

At the dawn of the industrial era, large metropolitan areas developed and people gathered together in much greater numbers. This enabled the general populous to enjoy much greater levels of anonymity than was possible in a small and relatively closed community. During the same period the reading and writing skills of the population improved as better education became available to the masses. Writing enabled individuals to record and distribute an immutable record of current events; this, together with the development of the printing press and instantaneous photography, enabled invasion of the secrecy and solitude of an individual on a much larger scale than was previously possible.

Developments in science and technology over the last century continue to provide individuals, companies and governments with new and more invasive ways to intrude on the privacy of individuals and communities. Since the latter decades of the 20th century, computer systems have developed their capacity to collect, analyse and distribute sensitive information at an exponential rate. This, together with advances in medicine, transportation, finance and communication have resulted in the recording and preserving of many more private facts.

Globalisation has also accelerated the analysis and distribution of private data through the development of high-speed computer networks and the growing number of interoperability standards. Examples include the Internet Protocol, facilitating ubiquitous network connectivity, and standardised data storage formats and meta-formats such as XML.

There are many reasons for recording and distributing personal information. Most governments, while wildly varying in political constitution, now demand some personal information from citizens; examples include earnings, family make-up, religion and qualifications. The stated goal of data gathering is that of collective good for society; in many cases the benefit is clear, for example, the regulation of medical practitioners, but in some cases the societal benefits are less obvious.

Commercial organisations are primarily concerned with profit, yet companies are often much more invasive than their governmental counterparts. Traditionally, marketing, advertising or brand loyalty have been suggested as the major motivation, but more recently *price discrimination*–the act of charging individuals a personalised price based on the amount they are prepared to pay–has been suggested as a strong motivating factor.

Discriminatory pricing works best in markets with large, fixed, up-front costs and low marginal expenses. With the centralisation of services and reduced cost of communication and transportation, more industries fit this model. Price discrimination is not a new phenomenon: it was used extensively throughout the early development of the railways in the US and Britain until a customer backlash introduced extensive regulation (Odlyzko 2003). Traditionally its use has been limited by the lack of technology to perform detailed customer profiling, however recent developments in computer systems make such techniques relatively simple and cost effective.

Recent price discrimination examples include flights, computers and even DVDs. The region coding scheme on DVDs was (in part) motivated by the desire to enforce a price differential on discs sold in different areas of the globe. In September 2000, Amazon experimented with price discrimination on for customers buying otherwise identical DVDs; this practise was quickly discovered by Internet users and was widely reported in the press[123]. The situation had the Amazon community split–some thought the practise was fair, others abhorred it.

13.2 Specific GIS privacy issues

The development of the Geographic Information Systems (GISs) mirrors the technological development of computers and the globalisation of government and industry. Traditional GISs ran on mainframe computers with relatively little computational power and data were often entered manually by the users of the system. Desktop GISs widened the market and, with the development of computer networks and cheap, removable media, increased the ability of individuals to copy and distribute geographic information. Recent developments in GISs have seen further growth in distributed processing and real-time data sharing over the Internet (Peng and Tsou 2003).

The types of data stored in a GIS are not restricted to solely topological facts; rather, the aim of many GISs is to spatially index and cross-reference multiple datasets containing a wide variety of information, and much of this data may well be of a personal and private nature. Therefore, the potential for privacy invasion of spatially-indexed personal data can be at least as problematic as a generic computer database.

Due to the expense of mainframe computer systems, early GISs were predominantly used by governments and large corporations. Typical government uses include tax assessment, zoning and planning, environmental impact analysis and maintenance of transportation and utilities. Businesses with a direct involvement in the exploitation of natural resources have an obvious need for recording and processing spatially-indexed data, yet the spatial indexing of personal data has many other uses for commercial enterprise.

Geodemographic software systems are a particular type of GIS used to categorise individuals living within a particular geographical region with a common set of socio-economic attributes. Such databases are popular with a wide variety of businesses. For example, mass marketeers use demographic data to direct advertisements to specific regions and thereby reduce postage costs on sending adverts to areas deemed unlikely to be responsive to a particular product. Other businesses use location data to determine the best location for retail outlets or to offer a potential customer a personalised price for house and motor insurance.

Traditionally, companies producing geodemographic software believed the accuracy of the system depended on the size of the region associated with a particular demographic, the quality and variety of the data used to generate the demographic for each area, and the number of possible socio-economic attributes. Therefore, by reducing the size of the geographical region, a geodemographic system should select a more socially homogeneous group, and this, together with more detailed and accurate input data about the group, combined with a greater number of possible attributes, should produce more accurate geodemographics. There are, of course, limits to the accuracy of such an approach; nevertheless the continued existence and popularity of geodemographic systems in the business community demonstrates their utility.

Recent technical innovations such as GPS and mobile telephone location systems show the potential for dynamic geodemographics (Phillps and Curry 2003). With this technology, geodemographic analysts can use the *set* of locations an individual visits during the day rather than just their permanent home address. Proponents of geodemographics anticipated that this set of data will provide better insights into the socio-economic groups of individuals. This concept can, of course, be extended to allow fully fledged location-based services where advertisements or products are offered to potential customers based on their current geographic location.

Geodemographic software encourages the practise of "redlining", where business avoid offering services such as mortgages or insurance to geographic areas considered to be a poor economic risk. Such practises represent a form of discrimination, but may also be interpreted as form of *collective* or territorial privacy violation, since personal data is gathered from a community and used to generate a, perhaps unfair, or at least generalised viewpoint of the population.

Indigenous people have also expressed concerns over the use of GISs. Worries about GISs stem partly from a cultural desire for collective territorial privacy, where certain regions are considered sacred and private and as such should not be visible to outsiders (Madsen 1994). GISs can be used to determine regions rich in natural resources and therefore indigenous populations are concerned that recording such information may lead to pressure from governments and commerce for exploitation. The data does not even have to be collected locally, and this reduces the ability of the local population to control the access to, and dissemination of, information about their land. Satellite imaging systems integrated with GISs can be deployed remotely by governments and commercial organisations to intrude on territory without the consent of the local population.

GISs raise both information privacy and territorial privacy concerns. Protecting the set of movement data from a specific individual is a form of information privacy, whereas protecting the information concerning personal or collectively private spaces is an instance of territorial privacy.

13.3 Future directions of computing and GISs

The price of computers continues to fall dramatically, following the downward price trend which started almost as soon as computers entered mass production. Astoundingly, as the price of computing has dropped, the performance and portability of computers has increased immensely. Most people in richer nations no longer own just one computer, but a dozen. If the price, performance and portability trend continues for much longer, the average consumer will soon be able to afford a hundred or a thousand small and powerful computational devices. Many of today's computers operate independently, however as the deployment of wireless technologies takes hold, devices will soon start to inter-operate.

Weiser noticed these trends over ten years ago and used them to provide some predictions about computing in the 21st century. He coined the term *ubiquitous computing* (Weiser 1991) to describe a world in which computers "weave themselves into the fabric of everyday life until they are indistinguishable from it" and claimed that "the most profound technologies are those that disappear." He noted that other technologies have done this before, and envisioned computing becoming sufficiently embedded into our world that users would often interact with computers at a subconscious level, much like we do today with the written word on signs, billboards, books and magazines; "in essence...only when things disappear in this way are we freed to use them without thinking and so to focus beyond them on new goals."

Most individuals are incapable and unwilling to manage a thousand devices in a similar fashion to the control and configuration required for a typical home PC today. Therefore, if ownership of a thousand computers is going to be productive, methods will be needed to automate the vast majority of the tasks these new devices are designed to perform. At the same time, techniques for reducing the cognitive load placed on users through improvements in human-computer interaction will be required.

Context-aware computing (Schilit et al. 1994) or *sentient computing* (Hopper 2000) augments computers with sensors and actuators in order to achieve a better understanding of, and interaction with, the physical environment. Such systems collect sensor data to build a model of the environment or *world model* (Adly et al. 1997). Human-computer interaction can then be improved, since individuals can use their five senses to interact with the computer within their real environment, rather than through a screen, mouse and keyboard. A computer system can also make use of the rich and diverse set of sensory inputs to trigger actuators and therefore automate tasks on our behalf. Effective context-aware systems should be easier to use than the traditional computing infrastructure commonplace today and it will therefore enable the average individual to make use of a much larger number of computers than would otherwise be possible. In many respects context-aware computing provides a way to realise the goal of ubiquitous computing.

Researchers in the field of ubiquitous computing have struggled to provide a precise definition of context, yet, from a privacy perspective, it is important to consider what types of information may be required to enable the successful operation of a context-aware system. López de Ipiña describes some of the common attributes, which include identity, location, time, features of the natural environment, such as temperature, light level, air quality etc., physiology, activity, social interaction and nearby resources (López de Ipiña 2003). Dey and Abowd claim there exist four *primary* context types which characterise any particular situation of an individual or object (Dey and Abowd 2000):

Identity:
the identity of relevant entities.

Location:
the geographical position of relevant entities.

Activity:
the activity or activities being performed.

Time:
the time period at which the entities perform the activity.

All other types of context information are declared *secondary* because they require association with one or more pieces of primary context to be meaningful. For example, a phone number, address or date of birth is usually associated with an identity; similarly temperature information is only useful when combined with a location, time and possibly activity. It is usual for secondary context to require more than one piece of primary context in order to identify unambiguously a single piece of information; in database parlance, the primary contexts must be combined to form a composite key.

A large amount of research in the field of ubiquitous computing has focused on enabling location-awareness and location-based services (Hightower and Borriello 2001a,b). In many cases the scale of these research projects is small in comparison to both the geographical size and data richness of commercial GISs. However integrating serious GIS systems will become a necessary component of ubiquitous computing systems as the technology is transferred from research groups and into commercial deployment.

Providing computers with knowledge of the context of users means that computers will gather, collate and distribute much more personal information about individuals than they do today. Such personal information is often considered private. Therefore, there appears to be a fundamental conflict between personal and collective privacy and the needs of context-aware computing. The next section outlines some possible methods which can redress the privacy imbalance between the deployment and use of GISs. Particular focus is given to future issues which arise when real-time location information of individuals becomes available for use by GISs and ubiquitous computing systems.

13.4 Protecting privacy

Lessig (1999) describes the four basic constraints which regulate the behaviour of individuals: architecture, social customs, market and law. Each constraint poses a distinct but interdependent cost on the individual. Constraints can be complementary or opposing, yet all can ultimately be governed by the rule of law. Architecture and the market regulate behaviour before an action can take place: a locked door prevents unwanted entry and the cost of cigarettes deters consumption. Social customs and law regulate behaviour after the event: diners may scorn if you smoke while eating and the police may arrest you for trespassing on private property. Most individuals feel the *threat* of social customs or law before the action even though social customs or law cannot actually prevent the act, unlike architecture or market forces.

Law is able to directly regulate behaviour through statute, and indirectly by regulating architecture, social customs and the market; examples of indirect regulation include building codes, sex education in schools and cigarette taxation. Such indirect regulation does raise questions of transparency, particularly in the regulation of architecture. For example Lessig describes how Robert Moses commissioned the bridges constructed on Long Island to prevent the passage of buses–the primary transportation of African Americans–preventing the progress of public transport to the beaches (Lessig 1999).

There are occasions when breaking a regulation may be morally justifiable, for example, speeding to hospital with the seriously ill; law and social customs are good at adapting in this situation, but architecture and the market are not; for example, a vehicle's speed limiter always limits. Therefore architectural and market forces are arguably more powerful but less adaptable regulators.

This distinction has important implications for the regulation of privacy. Law alone is not sufficient as it is not the most efficient method to prevent unwanted intrusion into the private lives of individuals because: firstly, privacy is very subjective–what is acceptable to one person is unacceptable to another; and secondly, post-action retribution through law may not be able to provide any good remedy—"the cat may already be out of the bag." This does not mean law-makers provide no aid in protecting privacy; quite the opposite is true–law has a crucial role to play in controlling architecture, market forces and social customs to foster privacy.

13.4.1 Legal instruments

Almost all countries now recognise privacy as a fundamental right and have attempted to codify privacy in law. The first known piece of privacy legislation was the 1361 Justices of the Peace Act (England), which legislated for the arrest of eavesdroppers and stalkers. In 1890 Warren and Brandeis argued that citizens should have the right to prevent disclosure of personal information as a separate right to that of copyright law or of slander or libel (Warren and Brandeis 1890). More recently, the 1948 Universal Declaration of Human Rights (United Nations 1948, Article 12) states everyone has a right to privacy at home, with family and in correspondence. This declaration was reaffirmed in The European Convention on Human Rights (Council of Europe 1950, Article 8) and the International Covenant on Civil and Political Rights (United Nations 1966, Article 17).

Many other more recent pieces of legislation aim to protect citizens' privacy; some modern constitutions even go as far as expressing the specific rights of access and control of personal information. Lessig argues that we should go further than just protection and recognise (certain pieces of) information about individuals as "real property" in order to provide individuals with the ability to control the dissemination of their data (Lessig 1999, Chapter 11).

The legal protection for personal privacy differs markedly across the globe. Some countries, such as Europe, Australia, New Zealand and Canada, have a *comprehensive* regulatory model. For example, the European Union regulates the collection of and access to all *personally-identifiable* information. The 1995 Data Protection Directive (European Union 1995) harmonised data protection law across all member states and embodies Westin's principles of privacy (Westin 1970), namely openness and transparency, reasonable security, accountability, collection and limitation, data quality, usage limitation and individual participation.

Accountability, openness and transparency is ensured through the *supervisory authority* which has the power to investigate, intervene, temporarily or permanently stop processing and ban the use of, or order the erasure of data. Further, the supervisory authority has the power to initiate legal proceedings in cases where the directive has been violated. All companies and individuals collecting and processing personal data must notify the supervisory authority of their intention to process data.

The directive demands that personally identifiable data are, firstly, collected for explicit and legitimate purposes, and not further processed in a way incompatible with the original purposes; secondly, accurate and kept up-to-date–inaccurate data must be either erased or rectified; thirdly, relevant and not excessive with respect to the purpose of collection; and finally, only stored for as long as is necessary for the purposes for which the data was collected.

The directive also requires explicit consent or a demonstration of necessity of processing–this can include a legal or contractual requirement where the data subject is a party, or for the protection of vital interests of the data subject, for example, emergency recovery of medical records in the event of an accident. Furthermore, data subjects are also entitled to access their personally identifiable information "without constraint, at reasonable intervals and without excessive delay or expense."

The comprehensive model proposes an *architecture*, backed by legal punishment, to regulate and control data processing of personally identifiable information. It is worth noting that, at least in the case of the European model, regulation only applies to personally identifiable information, and as such does not necessarily provide protection against territorial privacy violations. In addition, the directive does not require consumer choice; as Langheinrich notes (Langheinrich 2001), notice and consent are of little good if consumers have no choice.

Other countries take a broadly *sectorial* approach to privacy legislation by enacting statues when specific privacy problems arise. Specific regulation often lags behind current developments in technology, and can leave vast areas of commerce insufficiently regulated. The United States, for example, does not have a comprehensive regulatory model, and the US constitution does not specifically mention privacy at all, although the Supreme Court has interpreted certain passages in the Bill of Rights to protect a person from government surveillance when they have a "reasonable expectation of privacy" and has recognised a right to anonymity in the context of free speech. In addition some states have incorporated a more general right to privacy into their own constitutions.

The US Federal Privacy Act of 1974 attempted to introduce some of Westin's principles to protect the privacy of citizens from invasion by the government; however, the act did not limit or control the behaviour of private industry. The effectiveness of the legislation has been limited by poor enforcement and a broad interpretation of "routine use" which permits data collected for one purpose to be reused for other similar purposes.

In 1977 the Supreme Court ruled that individuals do not have constitutional protection for private data once it has been transferred to third parties. To date, the federal government has not provided comprehensive legislation to regulate private industrial use of personal data. Instead it has enacted a series of laws to cover specific cases, with the result that, for example, video rental records are better protected than drug purchases.

Various industries have attempted self-regulation through codes of practice, however many of these bodies suffer from a lack of effective enforcement and therefore have less impact than primary legislation; furthermore, industry-wide consensus does not necessarily result in good consumer privacy, although the threat of government intervention for poorly performing industry codes of practise may have some positive effect.

13.4.2 Architecture

New technologies do not necessarily decrease the level of privacy experienced by its users–some technologies can be used to increase the level of privacy experienced by an individual or group. Technology is not privacy neutral: the details of the design and implementation, or *architecture*, of a computer system can have a profound effect on privacy experienced by its users. The previous section described how different countries have regulated the design and use of computer systems in processing personal information. This section describes some possible design strategies which can be used to minimise the privacy problems presented by GISs.

The first question to ask is: should the data be collected at all? There are numerous occasions when data collected for one purpose are subsequently used for many others. Such "mission creep" is explicitly forbidden in some territories, such as the European Union, where data collected for one purpose cannot be used for another. Unfortunately, there remains a strong desire to reuse information already collected and processed, and this motivates governments to pass exemptions or reinterpret data protection legislation. The same situation encourages companies to write overly broad usage requirements for the collected data, so that every conceivable data processing need is covered.

There are two common techniques which can be employed to reduce the privacy impact of storing personal data; both of these approaches are suitable for GISs. Firstly, *access control* can be used to restrict the flow of data to specific individuals or organisations; however configuring access control is difficult, and assessing the trustworthiness of the entities given access can also be problematic, especially in a large system. Secondly, *anonymisation* can be used to remove any personal identifiers from the dataset, thereby mitigating many information privacy concerns; however, difficulty arises in determining whether the dataset has been sufficiently anonymised and whether it is of any use after anonymisation. An overview of these two methods of privacy protection will now be considered in the next two sections.

13.4.3 Access control

Traditional database systems store relatively static information on a server which is usually run and managed by the company, government or individual who wishes to process the data. Therefore the data subject, be that an individual or a community, loses the ability to directly control the use of that information. The comprehensive model, and in particular the one promoted by the European Union, attempts to redress this imbalance by giving data subjects the ability to access and control the onward distribution of their data, even when it is no longer stored on a computer under their control.

Allowing data subjects to determine a policy which controls access to, and onward distribution of, their personal data is an example of multilateral security, where each data subject can each impose a set of their own, independent access controls on their portion of the overall dataset. Distributed GISs (see Chapter 5) make distribution of an access control policy harder, since the physical storage location and usage of personal data is distributed; there is no longer a single entity an individual can approach in order to access and correct personal data. Distributed systems also make the specification of policy itself inherently more difficult. Distributed systems are going to become increasingly commonplace in computing, and therefore this is a problem which must be tackled, yet the information privacy issues presented by highly distributed systems have received relatively little interest. One example use of multilateral security to control personal data in a distributed system is the British Medical Association model for electronic patient records (Anderson 2001); individual components of a patient record can be compartmentalised to restrict access to particularly sensitive information, such as HIV test results, from the majority of medical staff.

The specification of access control policies is likely to get increasingly difficult as the number of computers increase and data is distributed over increasingly heterogeneous computer systems and networks. The reduction in cost of computers and rise in the number of broadband Internet connections does, however, present a new possibility. In the near future, data subjects may be able to afford a computer and network connection which is powerful enough to store and distribute all their own personal data. Third-parties which request access to personal information may be forced to request access to the data as and when required. This, highly distributed, architecture, places the data subject back in control of their personal information. Corporations and governments may resist such an architecture however, since: firstly, the current relative speed differential between local and network access to data makes a centralised solution more efficient for the foreseeable future; secondly, distributed systems are harder to build and placing users in control of data hosting may make such system less reliable; and, finally, data subjects may take more interest in the manner in which their personal information is used and therefore limit its use in a more rigorous manner. Therefore primary legislation may be required to force governments and companies to offer this architectural option to their citizens and customers.

Context-aware computing anticipates an era when computing systems will need to gather, collate and distribute much greater volumes of information in order to function effectively. In particular, continuous streams of location data from a variety of sensor networks, such as GPS and mobile telephone location systems, will be used by computers to automate tasks and reduce the level of concentration required to interact with a computer system. In this environment, spatial indexing of data will be a crucial component, since location data is one of the four primary types of context, and will therefore be used to index lots of other datasets.

Protecting privacy for individuals who release a continuous stream of location events is much more challenging problem than protecting relatively static location data. Combining a detailed GIS containing static information together with a stream of location information of sub-metre accuracy can reveal a wealth of personal facts concerning the individual being tracked, including, for example, the clinics visited in a hospital or the genre of books borrowed from the local library. Uncontrolled access to multiple streams of location data from multiple individuals reveals details of the social and business connections which exist in a particular community. The remainder of the section explores some possible computer architectures which can be used to control the release of personal information whilst simultaneously permitting location-based applications to function.

The IETF GeoPriv Working Group define a location services architecture designed to protect location privacy of individuals who make use of location-based services. The requirements draft (Cuellar et al. 2004) defines four, logically distinct, service components, which are shown in Figure 13.1, as well as two data objects: location data owners specify *privacy rule* objects to describe the access control to be applied to every *location object*, which represents a location event associated with a particular identity.

The *location generator* service object determines the location of the entity being tracked and constructs a location object to describe the current location. The

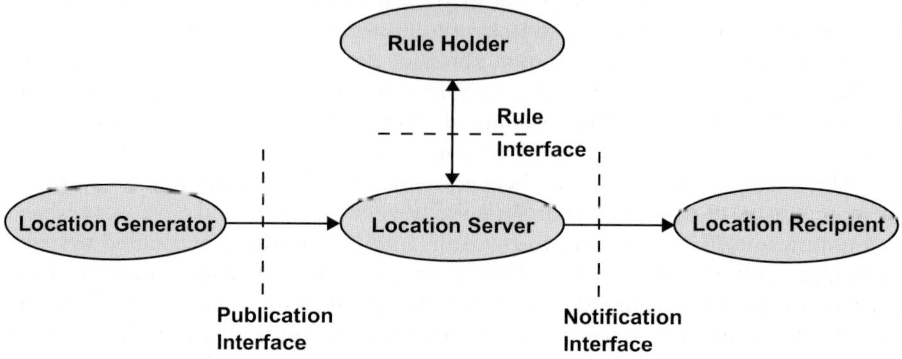

Figure 13.1. The four primary GeoPriv components.

location generator then publishes the location object to one or more *location servers* which receive location objects from, possibly multiple, location generators; the location server may receive subscriptions for location information from location recipients.

The *rule holder* service stores privacy rules on behalf of the *rule maker*. The rule maker is usually the owner of both the location device and the location information; there are exceptions however; for example parents may be rule makers of their children's location information and employees may be rule makers for corporate mobile telephones. The location server applies any privacy rules it learns from the rule holder to the location objects it receives and then notifies the location recipient of the location object as necessary.

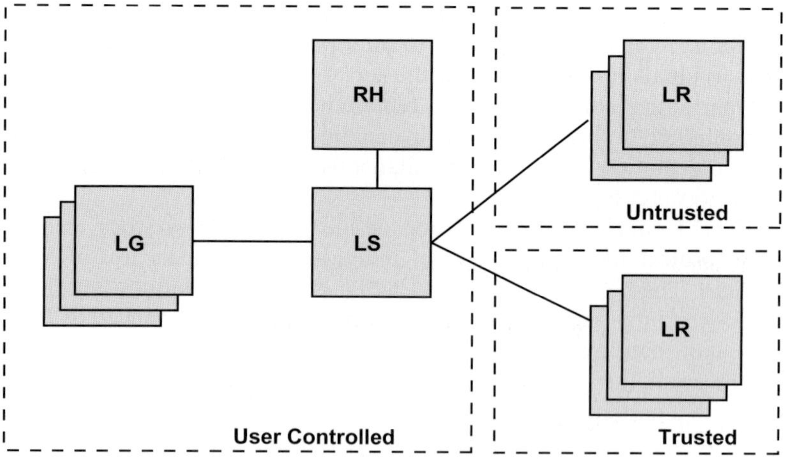

Figure 13.2. User-controlled model.

The rule maker may amend any of the parameters in the location object to increase the privacy of the target; common adjustments include: anonymising the target identifier by using pseudonyms (see next section); reducing the spatial resolution, for example converting 15 Bridge Street, Cambridge, England to City Centre, Cambridge, England; and reducing the temporal resolution of the data.

The location application can be stream-based and therefore request from the location server all events concerning a specific set of individuals, or event-based, and request from the location server information about any user located within a particular spatial region. An example stream-based application might locate a particular user on map together with the current location of local amenities and perhaps the location of friends; an event-based application might notify a user if they pass an historic building of interest or offer a reminder when they next meet up with a particular companion.

The ownership of, and trust placed in, each of these services can affect the level of location privacy offered by the architecture to the users of the system. User ownership of all the components in a location architecture reduces the need to trust third-parties and implement (potentially complex) privacy restrictions. User ownership of all components of the location architecture is not always possible, or indeed desirable; these issues are outlined next with the introduction of four possible architectures in which service components are owned and controlled by either: the data subject, a trusted third-party, or an untrusted third-party.

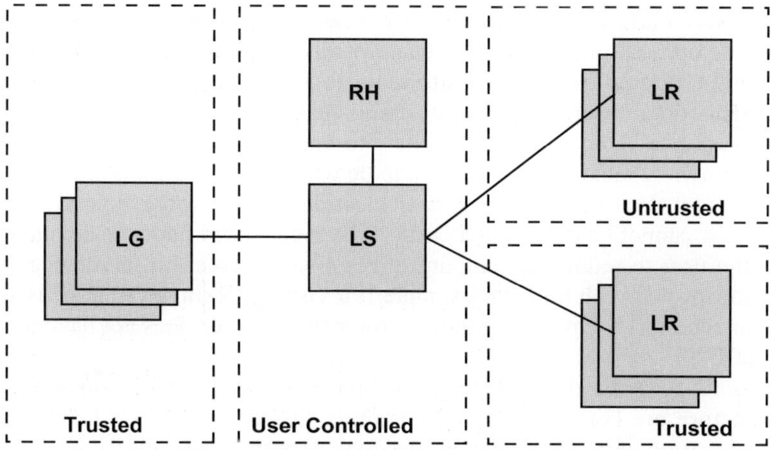

Figure 13.3. User-mediated model.

The *user-controlled* model is depicted in Figure 13.2. In this model the location generators are location systems which can calculate their location without revealing their presence to any infrastructure controlled by third-parties. Therefore, all location data initially resides solely with the data subject; this places the user in direct control of any release of location information. In addition the location server and rule holder are under the direct control of the user, and therefore the user can restrict access to location information to authenticated applications or (after anonymisation) to untrusted applications. In the case when location applications are also under user control, no trust is placed in third parties and therefore location privacy is assured if the software and hardware implementation is secure. Car navigation software fed from a local GPS receiver is one example of a user-controlled location service.

The user-controlled model is very suitable for personal location applications–the user is in direct control and, provided the hardware and software is well designed, nobody other than the user can possibly determine any personal location information. However, this model has two major drawbacks: firstly, users can (via applications) locate themselves quickly and efficiently, but determining their *context* (e.g., other users or objects nearby) requires the application to know about, be trusted with, and contact every other user's location server (an overhead a centralised location server is designed to prevent); and secondly, the model requires the user to carry around a device capable of deriving location information, storing the relevant GIS databases and executing applications. The device may need significant resources in order to store and process a lot of data.

The *user-mediated* model is depicted in Figure 13.3; in this model the user does not own the location generators, which may be controlled and managed by a central authority, but instead the user owns, and controls, only the rule holder and location server. The rule holder and location server could reside on a roaming device carried with the user, or remain on a home PC connected to the Internet. The optimal position of the location server and rule holder is dependant on the likely usage of the location information: a home PC is suited to location-aware applications used by people other than the owner of the location information (since its connectivity is likely to be better), whereas a PDA may be more convenient if location services are to be used by the roaming user.

This model does not support shared location information efficiently since a user's context cannot be determined easily. The model does have the advantage of enabling the user to reduce the amount of resources present on the device which must be transported with him; for example if a visual tracking system is used and application actuators are placed in the environment, the user does not have to carry any device at all.

The *third-party* model is depicted in Figure 4. All the main components of the architecture are controlled by one or more trusted entities on behalf of the user. The location server and rule holder may be run by a collective of friends and family, an independent service provider (e.g., mobile phone operator) or by an employer. In this model applications can be built to use contextual information based on location information from other users of the location server.

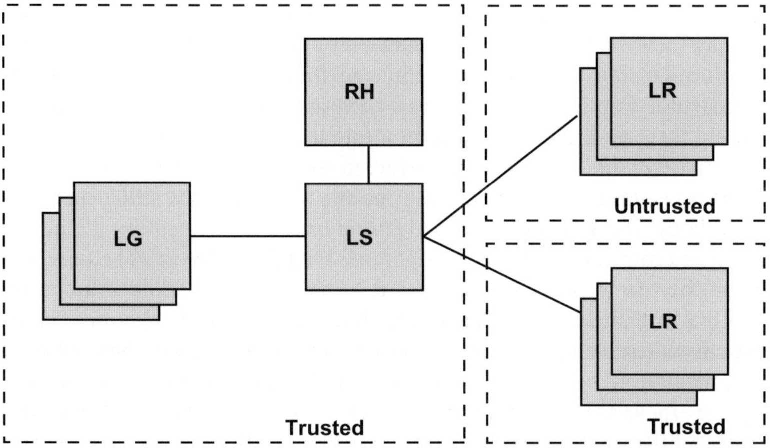

Figure 13.4. Third-party model.

In this model the user must trust the service provider to correctly configure the required software and hardware for the rule holder and location server. At the most fundamental level the user therefore relies on contract or privacy laws in order to ensure the correct level of location privacy is provided. Relying on legal protection ultimately requires a reliance on the government; in other words, if the government grant themselves (and possibly others) exemption from privacy legislation then the user may have to accept invasions in location privacy.

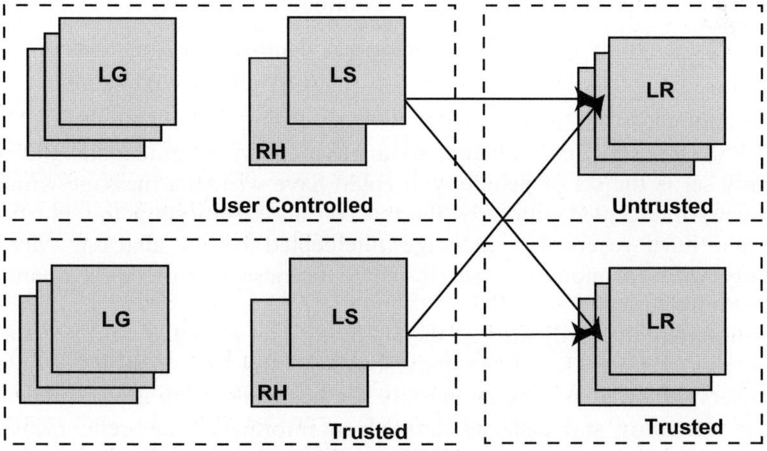

Figure 13.5. Hybrid model.

In some situations a *hybrid* model is a likely outcome where different location technologies are combined in an iterative fashion to build a hierarchy of location servers; see Figure 13.5. For example, some mobile phone users will soon have handsets with built-in GPS and WiFi and in this case GPS and WiFi location services may act as location generators; the user can distribute the resulting location information via the data interface of the mobile phone either directly to their friends and family (by running a location server and rule holder on the phone) or via their own user-controlled remote location server.

Irrespective of the distribution of location information determined from the GPS and WiFi location generators, the mobile operator can still determine (less accurate) location information from the phone using GSM location techniques and feed this information into the mobile operator's location server. The user may also have the ability to provide a subset of location information from the (more accurate) GPS and WiFi location generators either directly to the mobile operator or via their own user-controlled location server. In this model the distribution of location information can be viewed as a multilevel secure system (Anderson 2001) where location servers form a trust hierarchy and distribute location information adjusted by privacy primitives to applications or to other location servers with less security clearance.

13.4.5 Anonymisation

Anonymity is defined by the Oxford English Dictionary as "the state of being anonymous" which in turn is described as "nameless, having no name; of unknown name." Anonymity can be used in order to protect information privacy by ensuring any data released to an untrusted party cannot be associated with a real-world entity. Note that anonymisation techniques do not necessarily protect collective or territorial privacy since these datasets do not always identify individuals. Within the computer security community, Pfitzmann and Köhntopp promoted a more specific, quantitative, definition of anonymity:

> "anonymity is the state of being not identifiable within a set of subjects, the anonymity set."

The anonymity set is associated with an action or role performed by the members of the set. For example, in anonymous communications the sender anonymity set is the set of people who could have written a message which was subsequently intercepted; similarly, the receiver anonymity set is the set of people who could have received a message intercepted by an attacker. Measuring anonymity with an anonymity set is useful because it provides a quantifiable metric for anonymity.

Traditionally the cardinality of the anonymity set has been used as a measure of anonymity; a user who is "k-anonymous" or "has k-anonymity" is one of at least k users within a specific anonymity set associated with a particular action. Recently, Serjantov and Danezis defined an information theoretic measure of anonymity (Serjantov and Danezis 2002): each member a_i of the anonymity set A is assigned a probability equal to the (estimated) likelihood that member a_i performed the anonymous action. Shannon's entropy measure (Shannon 1948) can then be used to quantify the level of uncertainty or anonymity the members of the anonymity set achieve collectively with respect to their action.

A wide variety of organisations collect personal information from individuals in the form of a survey or census. Inference Control or Statistical Disclosure Control (Willenborg and de Waal 2001) is the discipline concerned with the modification of personal confidential data collected from individuals in order to prevent third-parties using the data (e.g., government departments, academic researchers and pharmaceutical companies) from correlating confidential information with specific individuals.

Traditionally data were provided to third-parties in the form of aggregated tables of values, but increasingly individual data values, or *microdata*, are used. Statistical disclosure occurs when data provided for statistical purposes is misused. It is often not sufficient to just remove explicit identifiers, since implicit information may also be hidden in the dataset. For example, a sociologist investigating urban deprivation may notice data concerning an (anonymous) 40-50 year old vicar, knows there is only one such respondent in the geographic area and knows his or her identity; the sociologist can then misuse the data to disclose other potentially sensitive personal data associated with the poorly anonymised record.

If access control could be used to facilitate location privacy it is important to ask the question: why bother with anonymity? There are a combination of factors which make anonymity-based techniques more attractive than traditional access control methodologies. First, users may accept anonymised location-aware services but shy away from services which require concrete identity. Typical real-world examples include the use of cash as an anonymous payment method for sports betting and the acceptance of false details as the norm when visiting a sexual health clinic.

Second, configuration of access rights is a difficult task for a skilled computer scientist; the situation may be nearly impossible for an average user. Role-based access control (Sandhu et al. 2000) and well configured defaults may mitigate the worst privacy worries, but these solutions will only partially solve the problem. Data subjects in today's GISs often cannot adequately control the initial and onward distribution of personal data. In the world of ubiquitous computing GISs become more wide-spread and simultaneously the datasets are augmented with large collections of dynamic location information; providing manageable access controls on this scale becomes even more intractable. Anonymity side-steps the configuration problem by denying an application access to any personally identifiable data, and therefore is more suited to ubiquitous computing applications than access control.

Third, if applications can be made to work with anonymous location sightings, then the applications themselves need not be trusted. Any reduction in the trusted computing base of a system increases our confidence in its correctness; in location-aware computing, removing applications from the trusted domain will increase our confidence in user privacy. Concerns about programmatic correctness alone (e.g., buffer overflow exploits) would justify this reduction in the trusted computing base since the vast majority of code will reside in the application layer. (Making applications untrusted means that an attacker cannot use, say, a buffer overflow exploit in a location-aware application to determine the location of a particular person).

Privacy worries extend beyond concerns about program correctness. Several mobile phone operators have already deployed systems which permit third-parties to connect to their networks and gain location information of users. Therefore trusted applications require trusted third-parties. This requires contractual agreements between mobile phone operators and third-party application providers. In addition, third-parties may have to adhere to data protection laws since they are handling personally identifiable information. This would not be the case, if anonymous location information were made available.

Whilst anonymisation is attractive from the perspective of the user, it can make writing location-aware applications harder. Some location-aware applications will function on anonymised location data, in other words, data with all direct identifiers removed. Uses of anonymised location sightings include service planning, (for example, how many users of bus route 12 are there on an average Monday morning in winter), environmental discovery and control of shared resources.

The main characteristic of a truly anonymous application is the one-directional flow of data from user to application; information flow in the other direction is difficult because the application does not know which user to communicate with. One approach is to broadcast reverse path messages to all users, but this technique will not scale to large anonymity sets. Applications which interact with users via devices which have an inherent user identity associated with them cannot provide an identifier for return messages without sacrificing privacy; for example a mobile telephone or home server cannot provide a telephone number or IP address because these are considered direct identifiers under many threat models. Therefore many interactive applications require an *unlinkable* pseudonym in order to communicate with the user.

Fine-grained location information of the type required for many location-aware services provides a very rich data set of user movement. Users often have one or more regions of space and time they predominantly occupy; such regions are called *home locations*. The term *simple home location* is used to refer to any home location which can be identified with a single location event (i.e., a single co-ordinate, containment or proximity measurement associated with a timestamp); conversely, a *complex home location* requires two or more location events to enable an attacker to identify an individual. Consider the following examples: the set of people sitting at a particular office desk for the majority of the working day; people found entering or exiting a family home at a particular time in the morning; the route (time, speed and direction) taken by someone to work.

Simply replacing user identity with a single pseudonym is not sufficient to protect privacy. If an attacker has unrestricted access to fine-grained location information, it is often possible to determine which underlying identity is represented by any given pseudonym since publicly available data can be used to correlate pseudonyms with underlying real-world identities. Large amounts of correlating data are available from a wide variety of GISs; examples include the electoral roll, company and university websites, phone books and poorly protected databases connected directly to the Internet.

If location data is recorded at sufficient resolution, a single static pseudonym does not protect the privacy of an individual. To demonstrate this, two weeks of location sightings from an indoor location system were recorded using the Bat

System. The Bat System (Ward et al. 1997) locates Bats, which are small wireless devices worn by people in a sentient office environment; position data for the Bats is recorded at 50 Hz with an accuracy of < 3 cm 95% of the time.

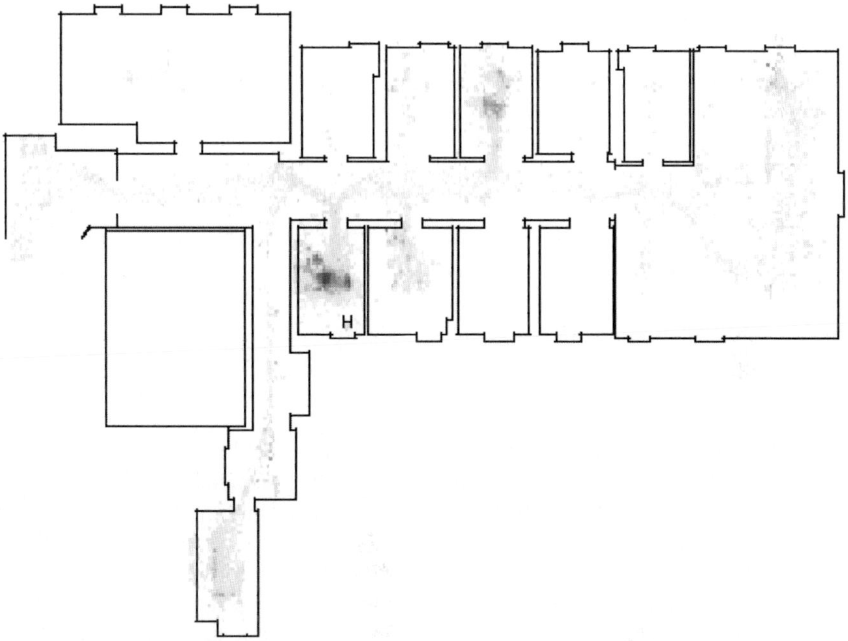

Figure 13.6. Knowledge of a user's primary desk location can be discovered by identifying the place where the Bat resides for the longest period of time; 'H' marks the user's home office.

Measuring the total time each pseudonym spent at every floor-plan position reveals that most occupants of a research lab spend the majority of time in a small contiguous area centred around their desk. These data can be extracted from the location information and correlated with known user desk positions to determine, with high certainty, the underlying user of any particular Bat; Figure 13.6 provides one such example.

Replacing user identity with a single, static pseudonym does not guarantee the user any level of anonymity (and therefore little privacy) if the location data associated with the pseudonym is of high temporal and spatial resolution. In this case, simply changing pseudonyms frequently may not be sufficient either: an attacker may be able to "follow the footsteps" of the underlying users simply by matching a disappearing pseudonym with the nearest (in both the spatial and temporal sense) newly appearing pseudonym.

Location-aware applications often require much less spatial and temporal resolution and coverage area than the underlying location system provides. This, of course, is the exact reason *why* location service platforms provide a reduction in network bandwidth and processing requirements when compared with providing location information of all users directly to applications.

In order to prevent an attacker from correctly linking two pseudonyms to the same underlying user, the spatial and temporal resolution of the location data must be reduced in order to introduce confusion about the possible mappings of pseudonyms (i.e., prevent the linking of pseudonymous "footsteps"). Resolution reduction can be applied globally by reducing the spatial and temporal accuracy of all the location data provided by the location server to the location applications. For applications that require high levels of accuracy but reduced coverage area, it may be possible to constrain the resolution reduction methods to regions which are of little interest to the application. Broadly speaking, global resolution reduction techniques are suitable for stream-based applications, whereas coverage reduction techniques are suitable for event-based applications.

In order to guarantee user privacy through pseudonymity we need to limit the amount of information presented to applications to ensure: firstly, that the information distributed with any given pseudonym is not sufficient to identify the underlying real-world entity with sufficient certainty; secondly, different pseudonyms cannot be combined together to form a single pseudonym with sufficient information to identify the underlying real-world entity. Provided these two statements hold, we can argue, that no personally identifiable information has been made available to the application and thus information privacy is assured. There are two possible methods of achieving this.

Reducing the spatial and temporal quality of location information associates multiple pseudonyms with the *same* piece of location information. For example, given three pseudonym-location-time tuples from a location stream as:

(*a*,*Room SN13*,14:52:23) (*b*,*Room SN17*,14:52:25) (*c*,*Room SN17*,14:52:29)

these could be modified to:

(*a*,*Floor 2*,14:52:30) (*b*,*Floor 2*,14:52:30) (*c*,*Floor 2*,14:52:30)

These modified location objects then have an anonymity set size of three. Therefore resolution reduction introduces uncertainty as to the underlying identity associated with each of the pseudonyms and removes all simple home locations from the dataset. Gruteser and Grunwald (Gruteser and Grunwald 2003) describe in detail a method of reducing the accuracy of location information in order to protect privacy through anonymity. Beresford provides an analysis of their approach and provides some additional improvements (Beresford 2004a).

Reducing the coverage area of location data given to particular applications can also be used to enable unlinkable pseudonyms. In the mix zone model (Beresford 2003). location applications specify a set of application zones for which they are interested in receiving location data. (This can be done, for example, as part of an event-based request from a location application to the location server.) A mix zone is defined as all the regions for which an application zone does not exist. Users entering a mix zone change their pseudonym, and therefore “mix” with other users who are simultaneously located within the same region. An attacker viewing the movements of users into and out of a mix zone is then uncertain of precisely which users went where. Providing a measure of the level of anonymity provided to the users of the system is more complex than with resolution reduction. This is because every pseudonym entering a mix zone does not have an equal probability of becoming every exiting pseudonym. Therefore an information-theoretic measure is more appropriate than simply measuring the cardinality of the set of users inside the mix zone (Beresford and Stajano 2004).

13.5 Conclusions

Technological developments have had a profound impact on the level of privacy experienced by individuals, and the rapid progress of computers in the 20th century has provided many new ways for governments, companies and individuals to invade both the collective and personal privacy of the individual.

Traditional GISs spatially index and cross-reference multiple datasets containing a wide variety of information and much of this data may be of a personal and private nature. Corporations and governments have used GIS technology to build geodemographic software systems in order to associate socio-economic attributes with particular geographic regions. Such databases may then be used for a wide variety of tasks, including urban planning and direct marketing; some individuals believe their use is discriminatory and invades the collective privacy of their community.

The ubiquitous computing revolution will change the way humans interact with computing systems; such systems which will gather, collate and distribute much more personal information than computers do today. Future GISs have a crucial role to play in supporting this new computer architecture because location data is a primary source of context information.

With the rapid decreases in cost and increases in performance, the average consumer is already able to store their personal information on a computer under their control; developments in broadband Internet connectivity will soon enable the consumer to share this data as part of a large-scale distributed system. This architecture presents individuals with the opportunity to manage their own personal data for the first time. However, integrating this decentralised architecture into location-aware applications is not always optimal, and some compromises must be made between protecting privacy and efficient use of location data in a ubiquitous computing environment.

Enabling privacy for GISs can be done through access control or anonymisation. Neither of these approach is perfect, and each has their advantages and drawbacks. Future GIS designers must consider the typical application usage scenarios and design the architecture of their computing infrastructure carefully in order to minimise privacy concerns.

Chapter 14

Frontiers of Geographic Information Technology

Sanjay Rana and Jayant Sharma

14.1 Outlook

In this chapter, we try to perceive the future of GI technologies from many directions ranging from governmental, economic, societal, and technological viewpoints. The aim behind presenting such a broad sweep of views is to link the nitty-gritty of modern science, technologies with the micro-macro politics that influence it. Like with any predictions, there is good news and worrying news in fact much like the two most recent predictions about the future of GI technology. Wegener and Messer (1996) predict the state of GIS in 2015 as being one of the following four possible scenarios:

- Trend Scenario – In this scenario, GIS and the related technologies are well integrated into education, society and consumer technology. This would be achieved by significant leaps in the technology and acceptance of GI technologies by governments.
- Market Scenario – This scenario predicts a segmentation of global industry into the western data user and the eastern data producer countries. Subcontractors would be swept away by major global ventures. While most communications would be geocoded, public wouldn't have access to the data they require.
- Big Brother Scenario – This scenario involves an ever present monitoring system for people's movements with the development of massive software systems. Despite the invasive observations, the public would be satisfied as they would feel safe about their living environments benefiting government significantly.
- Beyond-GIS Scenario – In this scenario, at first GIS would become obsolete and irrelevant to the society because of the shift in thinking towards more non-technological means of knowledge collections. However, it would come back in a new form as a much more personal medium for participatory planning.

The dramatic and to a large part cautious outlook by Wegener and Messer (1996) is not too far from the reality today. However, a more recent report by National Research Council (National Research Council 2003) on the future of GI technology in 2010 takes a mainly discipline oriented view of the future. In short,

NRC report highlights the state of the art and potentials in popular topics such as location-aware computing and Geovisualisation as these two themes present the most number of exciting research challenges. This type of exclusive thinking is common across the broad spectrum of technology experts. For example, the predictions about the future of technologies in general by 40 leading academic and industrial leaders (Applewhite 2004) revolved around their particular discipline. However, like Wegener and Messer (1996), we believe that technologies are not isolated parts of the society and hence the societal factors should also be taken into account by research bodies. The following sections present our perceptions about the future directions. In particular, we follow the "cause" and "effect" logic in presenting the ideas.

14.2 Good Predictions

14.2.1 Development of larger, heterogeneous, integrated systems

Despite the amazing leaps in the processor powers, most analysts now believe that desktop computers and workstations are not sufficient for all computing demands. The idea of dedicated clusters and super computers has also started to go out of fashion because it involves too much cost in development and maintenance on the part of stakeholders. The Grid Computing initiative, which seamlessly allows the usage of dispersed heterogeneous and multi-party computing resources, is being touted as *the* computing paradigm to form future applications. The other obvious reason for such a buzz of activity is the huge grants being given away by research councils all over the world. For example, the Engineering and Physics Research Council (EPSRC) in the UK allocated several million pounds on Grid Computing related projects in different broad programmes such as E-Science, and novel forms of computing such as Biologically Inspired Computing, and High Performance Computing. There is vast scope of using high-end computing in GIS, ranging from viewshed analysis to multi variate census statistics. The first GridGIS conference was organised in mid-2004 which brought together academic and industrial role players. The general opinion amongst the participants was one of enthusiasm and a feeling that the Grid Computing is still in a nascent state mainly because of the technological hurdles and lack of simple run-time environments. These issues will soon be resolved, however, making the utilisation of Grid Computing fairly simple. The onus would therefore lie on the GIS vendors to allow a seamless integration of an off the shelf GIS with a Grid Computing service. Geospatial Database vendors have already started to include Grid Computing capabilities e.g., Oracle 10g.

14.2.2 Commercial growth in the applications industry

The recent economic liberalisation of China, India and several East European Countries has added new markets for the GIS applications industry. The continued demands for the management of natural and urban environments are likely to yield major contracts, for the next decade or more. This is a lucrative market for large vendors, which was otherwise generally left for small governments and private contractors. However, due to the complex geo-political and historical monetary balances in these regions, the markets may not open up for everyone in the short run. Small overseas vendors may have problems in competing against the bias towards large local solutions providers.

14.2.3 Innovation in the miniaturisation of GI Technologies

Mapping interfaces have now been built for the handheld computers and mobile phones. Due to the lack of processing power and storage space, these software are presently limited to mainly mapping and field data collection. However, the inclusion of solid state hard disks with several gigabytes of memory, like the ones used in mp3 players, in handheld computers would lead to increased cache memory and processing speed.

Another potential for GI technologies lies with the miniaturisation of computing technologies, i.e., tiny computers generally used as sensors or to control mechanical devices. Some of these devices could require simple spatial analyses, e.g., rule-based guidance systems in robots. In order to speed up the processing, the spatial analyses could be built into the device. Such kind of hardware programming, common in computer graphics, is generally not done in GI technologies. This is changing, however, as evidenced in Intel's recent announcement of plans for an LBS chip[124].

14.2.4 Increase in interdisciplinary and industrial liaisons

Most of the large research councils are generally over flooded with applications for grants. Like any other discipline GI technology also faces inter-discipline competition. One fairly reliable way of observing the short-term future of any discipline is to follow the grant money trail. For instance, majority of GI technologies related grant awarded in the recent years have focussed on novel technologies such as Grid Computing, crime mapping, military applications, and geosimulations. It must be noted that these are our personal observations although an attempt was made to collect data on a discipline based funding trends but it was not so successful. It is quite difficult to provide detailed figures on the actual amount earmarked for such projects as most research councils tend to keep a scheme-based rather than discipline based record of grants. The nature of the research schemes has not always been strictly based on the desires of the disciplines but perhaps more from the viewpoint of urgent needs of society, and availability of industrial sponsorship. Nonetheless, there is an overall trend towards projects based on interdisciplinary, large groups, industrial sponsorships and clear deliverables to the society and industry. Hence, there will be less duplicate researches leading to better focussed science and scientists.

14.2.5 New trends in GI technology applications and data

Several new strands of GI technologies appear promising e.g.,

- Geographic Information collection continues to grow thanks to military spending. Database and imagery software, and storage vendors have simplified the creation, maintenance, and dissemination of huge digital imagery libraries. Collecting and using raster imagery in GI applications will be cheaper and more efficient and hence more widely used.
- Improvement in photogrammetric techniques and camera technologies would produce multi-spectral high resolution sensors and mobile sensor platforms.
- Innovations in immersive 3D visualisation has started to gather momentum with the launch of Holographic displays (Sullivan 2005).
- Unless there is a privacy related backlash, GPS or GPS-lite (as in RFID) will be ubiquitous. This will lead to innovative uses of location as context in information dissemination, retrieval, and discovery.
- Orientation sensors would become smaller, cheaper, better and available on many handheld computers.
- With Google purchasing Keyhole and developing maps.google.com and Google Earth, there is bound to be a significant new community of developers and applications incorporating such services. Therefore, there could be a Google effect like a MapQuest effect? However, it remains to be seen as to what forms they make take.

14.3 Worrying Predictions

14.3.1 Identity Crisis?

Many GI technologies and technologists are still tied up in the geography departments and have to compete with the institutional resources. However, there is still an unspoken divide between the more social/physical geography oriented patronage of geography and the recent GeoComputation oriented topics. One evidence for such a divide was the lack of dedicated platforms and discussion groups on GI technologies in major geography societies e.g., RGS (UK), IBG (UK), and AAG (US) only recently started to recognise the increasing relevance and significant value of the GI technology. Nonetheless, we doubt whether the rift will be fully bridged, because while the focus, and thus evolution, of social/physical geography topics have remained more or less same as they were four decades ago, the GeoComputation research on the other hand has undergone dramatic changes. The main contributing reason for such a contrast is the strong affinity of GI technologies to the overall IT industry. GI technologies evolve in correspondence to the developments taking place in general software and hardware industry and it has to because any IT-oriented topic can not afford to be left behind. Many wonders of GI Technologies such as database queries, data mining, immersive visualisation are clearly more computer science and electronics than geography. We wonder if GIScience can self-sustain itself so that it doesn't have to be an appendage to geography departments but be either a separate department or at least a part of some department with a dedicated affinity towards computer science and information technology. The lack of such a sympathetic research and development environment could lead to suffocation or stunted growth of the GI technologies. Sui (2004) hopes that GI technologies and in GIS in general will perhaps herald the "Third Culture" that will bridge the gap between the approaches.

However, with decreasing research funds there would be a need to develop successful business models, especially in the academic system, to promote research and development of GI technologies. The following organisation models are some we could think of, based on existing research groups:

Model 1: GIScience and GIS could exist as a group within or part of science departments with a focus on informatics and computing. This group will include faculty with diverse backgrounds and an interest in geospatial topics e.g., the GIScience research groups at the School of Computing of the University of Glamorgan, and School of Informatics of City University (London). These departments develop their own research agenda, run academic courses, are supported by typical departmental routes of money.

Model 2: Centres of excellence or specialisation could be established, in Universities, where a small expert group of GIS scientists caters to requests from, or collaborate with, other academic departments, external governmental and industrial organizations. Their staff would work in usual industrial environments. In other words, it would be a company owned by the University. These centres would be less likely to be involved in mainstream academic activities i.e., teaching. The funds would come mostly from external grants while the University provides access to space and library resources in exchange for GIS support i.e., the university acts as an incubator. We doubt if any example of this model matching all the above characteristics exists, but CASA at University College London, GeoICT at York University, GeoVista Laboratory at University of Pennsylvania, and Centre for Computational Geography at University of Leeds come close.

Model 3: GI technologies can be developed at the Research Laboratories that are solely funded by the government/industries/university. They are plenty examples of this type of organisations in the USA (JPL, LLNL, NCGIA, Bell Labs), Europe (IGN), and Asia (ETRI, Korea). They would have large funds and research programmes. As deliverables, they would apply for patents, and sell licenses to software and hardware technologies. However, their growth and focus is largely dependent on the individual goals and preferences of the researchers. This model is ideal for core research with blue-sky thinking.

14.3.2 Technological bottlenecks

Scalability of current GIS

Owing to the rapid digitisation of spatial datasets there is now a real crisis in analysing and storing large amounts (in petabytes) of spatial data. Large datasets initially started off in the form of earth observation data, e.g., satellite images, and have grown even bigger with the use of high resolution sensors (e.g., LIDAR), and duplicate collection of datasets (e.g., several satellites mapping the same part of earth with various sensors). The other types of data that have started to overwhelm are demographic data, e.g., census, and socio-economic profiles, as they change constantly and involve multiple variables. The typical response towards this issue has been to aggregate data, and where possible to resample/generalise the data to manageable chunks. The resultant loss of spatio-temporal variations in the data is not always recoverable. At present, the GI technologists rely upon data compression and large-capacity hardware storage. However, non-uniform practises for versioning and aggregating data create problems for data interoperability and spatio-temporal analyses.

Another limiting aspect of most commercial GIS is the inability to process very large spatial datasets. This is actually quite a complex problem and involves issues such as limitations of current computing hardware, operating systems, geospatial data file formats and clearly the GIS. This limitation means that large scale analyses such as various geosimulations (e.g., agent-based modelling of complex environments and weather), and data mining are generally avoided by researchers or processed offline via dedicated computers.

Lack of advanced topological operations

Although the desire to analyse three-dimensional geospatial datasets have been felt since the beginning of GIS, surprisingly there is not a single commercial GIS that incorporates 3D topological analysis. The ideas such as 3D GIS and Multi-dimensional GIS (Raper 2000) appear and seem to fade away. Multidimensional analyses and 3D topological queries are well established topics in computational geometry and computer science hence the lethargy amongst the big GIS vendors seems unfair to the users. The visualisation of 3D geospatial data, however, has had significant growth. There are several off the shelf products (e.g., Map3D by Autodesk, GRASS) and in-house prototypes that provide visualisation and manipulation of 3D geometry but as highlighted by Zlatanova and Stoter in Chapter 8, we are a long way away from the kind of advanced topological queries routinely done in 2D GIS applications.

Blind reliability on GIS

Since the beginning of GIS, while the packaging of useful spatial analysis function in software has made our task easier, at the same time it has isolated the end-user and GIS has now become a black box. The majority of the end users are unaware of the assumptions and limitations of the spatial analysis functions built into the

GIS. The GIS vendors are partly to blame for this too as they seldom provide details about the type and limitations of the algorithms implemented in the GIS. As a result, it is not uncommon to observe vast errors in spatial analyses when computed from two different GIS (Fisher 1993).

There is a also growing consensus amongst GI technologists that deterministic spatial interaction models are only as good as the data they are based upon and reliable under given constraints (e.g., given spatiotemporal extent). Alternative and unconventional approaches such as storytelling have been proposed as way of describe modelling (Guhathakurta 2002) .

Limitations of computing hardware

Finally, due to its nature GI technologies inherit part of the future of computing technologies. One of the biggest concerns in computing technologies is stated in the famous Moore's Law, named after Gordon Moore (Moore 1965). According to the Moore's Law, the number of transistors (an indication of processing power) doubles up every couple of years. The following list shows the phenomenal growth in the number of Intel transistors.

CPU	**Year of Introduction**	**Transistors**
4004	1971	2,250
8008	1972	2,500
8080	1974	5,000
8086	1978	29,000
286	1982	120,000
Intel 386™ processor	1985	275,000
Intel 486™ processor	1989	1,180,000
Intel® Pentium® processor	1993	3,100,000
Intel® Pentium® II processor	1997	7,500,000
Intel® Pentium® III processor	1999	24,000,000
Intel® Pentium® 4 processor	2000	42,000,000
Intel® Itanium® processor	2002	220,000,000
Intel® Itanium® 2 processor	2003	410,000,000

(source: Intel[125])

One of the reasons for the growth has been the improvements in the lithography tools to make the transistors smaller than 0.5 microns. This allows the packaging of more transistors in a processor. However, eventually the lithographic technique would hit the minimum molecular size below which it may not be possible to fabricate a silicon semiconductor. At present, the desire to have faster machine drives the computing hardware industry. The industry and investors are now banking upon wonderful ideas such as nano-technology and biological computing. However, early attempts at these futuristic technologies have not been so successful so as to assure the investors. We believe that the vast amount of commercial potential still left in the applications industry will soon be realized and result in a much brighter future.

14.3.3 Geopolitical and Socio-Economic Issues

An idea of how the society and political situations influence the developments in the GI technology was presented in the first chapter. There are a number of recent events that add new directions in the conventional growth pattern of the market. The first most is the influence of global economical shuffling being undertaken to drive the costs down, harmonise wealth across the world, and increase competition. For example, the popular policy of outsourcing back office development works such as digitising, beta-testing of software, and to a lesser extent new product development, to cheap and qualified labour market elsewhere is bound to have some effect on the market and research from the outsourced countries. It is non-trivial to make any sensible prediction about the effects without having an access to the business plans of the corporate bodies and the participating governments. However, one can hypothesize that if the outsourcing is a short-term solution to the long-term skills shortage, then the likely effects (whether good or bad) are also likely to tail off as soon as the skills shortage has been addressed or the demands have been met in some other way. If, however, the outsourcing plans are long-term, then the effects could be pretty much anything. Certain experts worry that long-term outsourcing is likely to cause a wealth and technology exodus from the outsourced country[126]. Others feel that the benefits reaped by the outsourcing in new economies will come back in terms of demands for high-quality products and specialised equipment (Jones 2004). Both arguments seem plausible under certain assumptions, and hence it's a confusing story overall. However, while the short-term effects are obvious, an analysis of the long-term effects is not so straight-forward at least to outsiders who are not party to monetary or strategic plans of the economic heavy weights such as G8, OPEC, WMF, and other such industrial bodies. It is conceivable that some effects are inevitable and being part of a strongly high-end or specialised application oriented industry, GI technologies are likely to be influenced as well.

Another social issue related to the GI technology, which seems somewhat clearer, is the one about privacy in Location Based Services (LBS). LBS would manifest in various forms ranging from short text messages sent to one's mobile promoting a new offer whenever one is near his or her favourite shop, or as trackers fitted in cars, or other equipment, for monitoring the safety and location of assets. This is clearly intrusive. It is expected that certain types of LBS would be optional, especially the ones which would require subscription. But there would be some that would be compulsory, e.g., the ID card with magnetic chips and the Radio Frequency ID (RFID) body tags. We may like them or loathe them but an increase in the number of identity fraud cases, and rogue groups, means that governments are likely to enforce them. The opportunities presented by the existence of such tracking devices, and the need to recover costs or increase profits, will result in numerous innovative applications of GI technology.

14.3.4 Potential risks to location based services from over-democratisation of GI Technology

While the application of geospatial data is unlimited, its actual deliverables are fixed. The primary deliverable of GI Technology has been, and is likely to be, the *locational information*. The democratisation will appear in the form of various ways of communicating the locational information and exploiting relationships between the locational information and other types of geographic information. The public at large would be able to collect (e.g., using GPS), analyse (e.g., simple online GIS) and visualise the basic geospatial information by themselves. While this scenario does appear to empower the public, these location based services services are not sustainable especially as they provide little superior margins to big vendors. This prediction is opposite to the one by Wegener and Messer (1996), who predict that the public will lose access to geospatial data.

The overall message of this book is "Integration, Integration, Integration". The signs of such systems have started to appear. According to NSF chief computer scientist, they "..are planning for facilities that a chemist can use this morning, a physicist can use this afternoon, and an earthquake engineer can use tonight" (Colwell 2004). Another expectation is that the new GI technology could also be in the form of "GI inside or GI in disguise" (Lawrence 2005). In any case, the likelihood of such open and interoperable GI technologies is actually fairly realistic. Even today, supercomputer facilities are used around the world by scientists from various disciplines, ranging from weather modelling to studying the gene sequence of a genetically-modified tomato, on a time-share basis. However, while technological advancements can be achieved by focussed financial and intellectual investments, the challenge to uplift the technologically poor nations is not as straightforward. Therefore, GI Technologists have now the challenging and rewarding responsibility to be creative with the proposed technologies, in order to produce far reaching ideas and instill the hope for a better future for all.

References

1. Abowd, G.D., Atkenson, C.G., Hong, J., Long, S., Kooper, R., Pinkerton, M., 1997. Cyberguide: A Mobile Context-Aware Tour Guide, *Wireless Networks*, **3**, 421-433.

2. Adly, N., Steggles, P., and Harter, A., 1997. Spirit: A Resource Database for Mobile Users, In Proceedings *Workshop on Ubiquitous Computing*, Atlanta, GA, USA.

3. Agrawala, M., and Stolte, C., 2001. Rendering Effective Route Maps: Improving Usability through Generalization, In Proceedings *SIGGRAPH 2001*, Los Angeles, CA, USA.

4. Ahlqvist, O., Keukelaar, J., and Oukbir, K., 2000. Rough Classification and Accuracy Assessment, *International Journal of Geographical Information Science*, **14**, 475-496.

5. Ahlqvist, O., Keukelaar, J., and Oukbir, K., 2003. Rough and Fuzzy Geographical Data Integration, *International Journal of Geographical Information Science*, **17**, 223-234.

6. Akamatsu, M., Sato, S., and MacKenzie, I.S., 1994. Multimodal Mouse: A Mouse-Type Device with Tactile and Force Display, *Presence*, **3**, 73-80.

7. Anderson, A., 2001. *Security Engineering: A Guide to Building Dependable Systems*, John Wiley and Sons, NY, USA.

8. Andrews, S.K., 1983. Spatial Cognition through Tactual Maps, In Proceedings 1st *International Symposium on Maps and Graphics for the Visually Handicapped*, Washington DC, USA.

9. Applewhite, A., 2004. The View from the Top, *IEEE Spectum*, **41**, 36-38.

10. Arens, C., Stoter, J.E., and van Oosterom, P.J.M., 2003. Modelling 3D Spatial Objects in a Geodbms using a 3D Primitive, In *Proceedings AGILE*, Lyon, France.

11. Armstrong, M., and Wang, S., 2004. Geomiddleware to Support Interoperability for Grid Computing, Position paper for *Specialist Meeting on Spatial Webs*, Santa Barbara, CA, USA.

12. Arthur, P., and Passini, R., 1992. *Wayfinding: People, Signs, and Architecture*, McGraw-Hill Book Co., NY, USA.

13. ASTM (American Society for Testing and Materials), 1985. *Annual Book of ASTM Standards: Section 4 Construction*, **04.08**, Philadelphia, USA.

14. Atanassov, K.T., 1999. *Intuitionistic Fuzzy Sets*, Springer-Verlag, Heidelberg, Germany, NY, USA.

15. Baer, A.E., 1974. *A Simulation Model of Multidirectional Pedestrian Movement within Physically Bounded Environments*, Institute of Physical Planning Report no. 47, Carnegie-Mellon University, Pittsburgh, PA, USA.

16. Ballas, J.A., and Howard, J.H., 1987. Interpreting the Language of Environmental Sounds, *Environment and Behavior*, **19**, 91-114.

17. Barkai, D., 2001. *Peer-to-Peer Computing: Technologies for Sharing and Collaborating on the Net*, Intel Press, Hillsboro, OR, USA.

18. Baumgart, B.G., 1975. A Polyhedron Representation for Computer Vision, In *National Computer Conference*, Anaheim, CA, USA.

19. Baus, J., Krüger, A., and Wahlster, W., 2002. A Resource-Adaptive Mobile Navigation System, In Proceedings *International Conference on Intelligent User Interfaces*, San Francisco, CA, USA.

20. Begault, D.R., Wenzel, E.M., Shrum, R., and Miller J., 1996. A Virtual Audio Guidance and Alert System for Commercial Aircraft Operations, In Proceedings *3rd International Conference on Auditory Display*, Palo Alto, CA, USA.

21. Behringer, R., 1999. Registration for Outdoor Augmented Reality Applications using Computer Vision Techniques and Hybrid Sensors, In Proceedings *IEEE Virtual Reality*, Houston, TX, USA.

22. Beresford, A.R., 2004a. *Location Privacy in Ubiquitous Computing*, Ph.D. thesis, University of Cambridge, Cambridge, UK.

23. Beresford, A.R., and Stajano, F., 2003. Location Privacy in Pervasive Computing, *IEEE Pervasive Computing*, **3**, 46-55.

24. Beresford, A.R., and Stajano, F., 2004. Mix Zones: User Privacy in Location-Aware Services, In Proceedings *IEEE International Workshop on Pervasive Computing and Communication Security*, Orlando, FL, USA.

25. Berners-Lee,T., Hendler, J., and Lassila, O., 2001. The Semantic Web, *Scientific American*, **284**, 34-43.

26. Bertin, J., 1967. *Semiologie Graphique*, Mouton, Den Hagg.

27. Bishr, Y., 1997. *Semantic Aspects of GIS*, ITC Publications, Enschede, The Netherlands.

28. Blades, M., 1991. Wayfinding Theory and Research: The Need for a New Approach, In: D.M. Mark, and A.U. Frank (Eds.), *Cognitive and Linguistic Aspects of Geographic Space*, Kluwer Academic Publishers, Dordrecht, The Netherlands, 137-165.

29. Blaschke, T., and Tiede, D., 2003. Bridging GIS – Landscape Analysis Modelling and 3D Simulation, Is this Already 4D?, In Proceedings *CORP 2003*, Vienna University of Technology, Austria.

30. Bly, S., 1982. Presenting Information in Sound, In Proceedings *Human Factors in Computer Systems*, Gaithersburg, MD, USA.

31. Bonabeau, E., Dorigo, M., and Theraulaz, G., 1999. *Swarm Intelligence: From Natural to Artificial Systems*, Oxford University Press, NY, USA.

32. Borgers, A., and Timmermans, H., 1986. A Model of Pedestrian Route Choice and Demand for Retail Facilities within Inner-City Shopping Areas, *Geographical Analysis*, **18**, 115-128.

33. Braithwaite, C., and Scott, M., 2004. *UMTS Network Planning and Development: Design and Implementation of the 3G CDMA Infrastructure*, Elsevier, Amsterdam, The Netherlands.

34. Breen, D. E., Rose E., and Whitaker, R.T., 1995. *Interactive Occlusion and Collision of Real and Virtual Objects in Augmented Reality*, Technischer Bericht ECRC-95-02, European Computer-Industry Research Centre GmbH, Munich, Germany.

35. Bresters, P.W., 2003. 3D Visualisations with the Height Model of the Netherlands, In Proceedings *EuroSDR Commission V Workshop on Visualisation and Rendering*, Enschede, The Netherlands.

36. Brewer, I., 2002. Cognitive Systems Engineering and Giscience: Lessons Learned from a Work Domain Analysis for the Design of a Collaborative, Multimodal Emergency Management GIS, In Proceedings *GIScience 2002*, Boulder, CO, USA.

37. Brewster, S.A., Wright, P.C., and Edwards, A.D.N., 1993. An Evaluation of Earcons for use in Auditory Human-Computer Interfaces - Human Factors in Computing Systems, In Proceedings *Conference on Human Factors in Computing Systems,* Amsterdam, The Netherlands.

38. Brown, A., Johnson, S., and Kelly, K., 2002. *Using Service-Oriented Architecture and Component-based Development to Build Web Service Application*, Rational Software Corporation White Paper.

39. Brunsdon, C., and Openshaw, S., 1994. Error Simulation in Vector GIS using Neural Computing Methods, In: M.F. Worboys (Ed.), *Innovation in GIS*, Taylor and Francis Ltd., London, UK.

40. Buehler, K., and McKee, L. (Eds.), 1998. *The OpenGIS Guide: Introduction to Interoperable Geoprocessing and the OpenGIS Specification*, Open GIS Consortium, Wayland, MA, USA.

41. Burrough, P.A., 1999. GIS and GeoStatistics: Essential Partners for Spatial Analysis, In Proceedings *International Symposium on Spatial Data Quality*, Hong Kong, China.

42. Burrough, P.A., and McDonnell, R., 1998. *Principles of Geographical Information Systems*, Oxford University Press, Oxford, UK

43. Burstedde, C., Klauck, K., Schadschneider, A., and Zittarz, J., 2001. Simulation of Pedestrian Dynamics using a Two-Dimensional Cellular Automaton, *Physica A*, **295**, 507-525.

44. Buxton, W., 1985. Communicating with Sound, In Proceedings *Human Factors in Computer Systems*, San Francisco, CA, USA.

45. Camazine, S., Deneubourg, J-L., Franks, N.R., Sneyd, J., Theraulaz, G., and Bonabeau, E., 2001. *Self-Organization in Biological Systems*, Princeton University Press, Princeton, NJ, USA.

46. Cambray, B., 1993. Three-Dimensional Modelling in a Geographical Database, In Proceedings *11th International Conference on Computer Assisted Cartography*, Minneapolis, MN, USA.

47. Carpenter, G.A., and Grossberg, S., 1995. Adaptive Resonance Theory, In: M. Arbib (Ed.), *Handbook of brain theory and neural networks*, MIT Press, Cambridge MA, USA.

48. Catlett, C., and Smarr, L., 1992. Metacomputing, *Communications ACM*, 35, 44-52.

49. Cesarone, R.J., Hastrup, R.C., Bell, D.J., and Lyons, D.T., 1999. Architectural Design for a Mars Communications and Navigation Orbital Infrastructure, In Proceedings *AAS/AIAA Astrodynamics Specialist Conference*, Girwood, AK, USA.

50. Chalmers, M., 1993. Using a Landscape Metaphor to Represent a Corpus of Documents, In: A.U. Frank, and I. Campari (Eds.), *Lecture Notes in Computer Science 716*, Springer-Verlag, Berlin, Germany, 377-390.

51. Chang, L., and Burrough, P.A., 1987. Fuzzy Reasoning: A New Quantitative Aid for Land Evaluation, *Soil Survey and Land Evaluation*, **7**, 69-80.

52. Chen, T., and Shibasaki, R., 1999. A Versatile AR Type 3D Mobile GIS based on Image Navigation Technology, In Proceedings *IEEE International Conference on Systems on Man and Cybernetics*, Tokyo, Japan.

53. Chervenak, A., Foster, I., Kesselman, C., Salisbury, C., and Tuecke, S., 2001. The Data Grid: Towards an Architecture for the Distributed Management and Analysis of Large Scientific Datasets, *Journal of Network and Computer Applications*, **23**, 187-200.

54. Chrisman, N.R., 1991. The Error Component in Spatial Data, In: D.J. Maguire, M.F. Goodchild, and D.W. Rhind (Eds.), *Geographical information systems: Principles and Applications*, John Wiley and Sons, NY, USA.

55. Clark, J., and Clark, D.D., 1994. Creating Tactile Maps for the Blind using a GIS, In Proceedings *ASPRS/ACSM Annual Convention and Exposition*, Reno, NV, USA.

56. Clarke, K.C., 1997. *Getting Started with Geographic Information Systems*, Prentice Hall, Upper Saddle River, NJ, USA.

57. Clarke, K.C., 1999. Visualizing Different Geofutures, In: P. Longley, S. Brooks, B. Macmillan, and R. MacDonnell (Eds.), *Geocomputation*, John Wiley and Sons, Chichester, UK, 119-137.

58. Clarke, K.C., 2001. What is a GIS?, In K.C. Clarke (Ed.), *Getting Started with Geographic Information Systems*, Prentice Hall, Upper Saddle River, NJ, USA, 1-39.

59. Cleveland, W.S., 1993. *Visualizing Data*, Hobart Press, Summit, NJ, USA.

60. Collins, C.C., 1985. On Mobility Aids for the Blind, In: D.H. Warren, and E.R. Strelow (Eds.), *Electronic Spatial Sensing for the Blind: Contributions from Perception, Rehabilitation, and Computer Vision*, Martinus Nijhoff Publications, Dordrecht, The Netherlands, 35-64.

61. Colwell, R., 2004. The New Landscape of Science: A Geographic Portal, *Annals of the American Geographers*, **94**, 703-708.

62. Connor, T.S., 1980. Integrated Noise Model – The Federal Aviation Administration's Computer Program for Predicting Noise Exposure around an Airport, In Proceedings *International Conference on Noise Control Engineering*, Miami, FL, USA.

63. Conry-Murray, A., 2004. Grid Computing's Promises and Perils, *Network Magazine*, **19**, 30-40.

64. Coors, V., 2002. Resource-Adaptive 3D Maps for LBS, In Proceedings *23rd Urban Data Management Symposium "30 years of UDMS – looking back, looking forward"*, Prague, Czech Republic.

65. Coors, V., 2003. 3D GIS in Networking Environments, *Computers, Environments and Urban Systems*, **27**, 345-357.

66. Couclelis, H., 1998. Worlds of Information: The Geographic Metaphor in the Visualization of Complex Information, *Cartography and Geographic Information Systems*, **25,** 209-220.

67. Coulson, M., Riger, M., and Wheate, R., 1991. Progress in Creating Tactile Maps from Geographic Information Systems Output, In Proceedings *15th Conference and 9th General Assembly of the International Cartographic Association*, Bournemouth, UK.

68. Council of Europe, 1950. *The European Convention on Human Rights*.

69. Cross, V., and Firat, A., 2000. Fuzzy Objects for Geographical Information Systems, Fuzzy Sets and Systems, **113**, 19-36.

70. Cuellar, J.R., Morris, J.B., Mulligan, D.K., Peterson, J., and Polk, J., 2004. *RFC 3693 - GeoPriv Requirements*, The Internet Society – Network Working Group.

71. D'Roza, T., and Bilchev, G., 2004. An Overview of Location-based Services, In: D. Ralph, and S. Searby (Eds.), *Location and Personalisation: Delivering Online and Mobility Services*, Institute of Electrical Engineers, London, UK, 15-28.

72. Daratech, 2004. *GIS Markets and Opportunities*, Daratech Inc., Cambridge, MA, USA.

73. Das, H., Bao, X., Bar-Cohen, Y., Bonitz, R., Lindemann, R., Maimone, M., Nesnas, I., and Voorhees, C., 1999. Robot Manipulator Technologies for Planetary Exploration, In Proceedings *Smart Structures and Integrated System Symposium*, Newport Beach, CA, USA.

74. Deng, J., 1982. Control Problems of Grey Systems, *Systems and Control Letters*, **1**, 288-294.

75. Dewey, C., 2001. *Navigationsdienste Für Rollstuhlfahrer*, Diploma thesis, Westfälische Wilhelms-Universität Münster, Germany.

76. Dey, A.K., and Abowd, G.D., 2000. Towards a Better Understanding of Context and Context-Awareness, In Proceedings *Conference on Human Factors in Computing Systems- Workshop "The What, Who, Where, When, Why and How of Context-Awareness"*, The Hague, The Netherlands.

77. Di, L., and McDonald, K., 1999. Next Generation Data and Information Systems for Earth Sciences Research, In Proceedings *1st International Symposium on Digital Earth*, Beijing, China.

78. Di, L., Yang, W., Deng, M., Deng, D., and McDonald, K., 2001. The Prototype NASA HDF-EOS Web GIS Software Suite (NWGISS), In Proceedings *NASA Earth Science Technologies Conference*, Greenbelt, MD, USA.

79. Di, L., Yang, W., Deng, M., Deng, D., and McDonald, K., 2002. Interoperable Access of Remote Sensing Data through NWGISS, In Proceedings *IEEE International Geoscience and Remote Sensing Symposium*, Toronto, Canada.

80. Dijkstra, J., Jessurun, J., and Timmermans, H.J.P., 2002. A Multi-Agent Cellular Automata Model of Pedestrian Movement, In: M. Schreckenberg, and S.D. Sharma (Eds.), *Pedestrian and Evacuation Dynamics*, Springer-Verlag, Berlin, Germany, 173-180.

81. Dodge, M., and Kitchin, R., 2001. *Atlas of Cyberspace*, Addison-Wesley, Harlow, UK.

82. Eberhardt, S.P., Bernstein, L.E., Coulter, D.C., and Hunckler, L.A., 1993. Omar – A Haptic Display for Speech Perception by Deaf and Deaf-Blind Individuals, In Proceedings *IEEE Virtual Reality Annual International Symposium*, Seattle, WA USA.

83. Elias, B., 2003. Extracting Landmarks with Data Mining Methods, In: W. Kuhn, M. Worboys, and S. Timpf (Eds.), *Spatial Information Theory: Foundations of Geographic Information Science*, Springer-Verlag, Berlin, Germany, 375-389.

84. Engelen, V.W.P. van, and Ting-Tian, W. (Eds.), 1995. *Global and National Soils and Terrain Digital Databases – SOTER: Procedures Manual*, Wageningen, The Netherlands.

85. Eriksson, Y., 2001. How to Make Tactile Maps Discriminable to Touch, In Proceedings *20th International Cartographic Conference*, Beijing, China.

86. Ernest, M., Joseph, J., and Fellenstein, C., 2004. Evolution of Grid Computing Architecture and Grid Adoption Models, *IBM Systems Journal*, **43**, 624-645.

87. European Union, 1995. Data Protection Directive (95/46/ec), *Official Journal of the European Communities*, **L. 281**, 31.

88. Faby, H., 2004. Individuelle Reisevorbereitung Mit Internetkarten – Status Quo Und Potenziale, *Kartographische Nachrichten*, **54**, 3-10.

89. Fairley, P., 2004. Lighting up the Andes, *IEEE Spectrum*, **41**, 32-37.

90. Feiner, S.K., 1999. The Importance of Being Mobile: Some Social Consequences of Wearable Augmented Reality Systems, In Proceedings *International Workshop on Augmented Reality*, San Francisco, CA, USA.

91. Fisher, P.F., 1992. First Experiments in Viewshed Uncertainty: Simulating the Fuzzy Viewshed, *Photogrammetric Engineering and Remote Sensing*, **58**, 345-352.

92. Fisher, P.F., 1993. Algorithm and Implementation Uncertainty in the Viewshed Function, *International Journal of Geographical Information Systems*, **7**, 331-347.

93. Fisher, P.F., 1994a. Probable and Fuzzy Models of the Viewshed Operation, In: M.F. Worboys (Ed.), *Innovation in GIS*, Taylor and Francis Ltd., London, UK, 161-175.

94. Fisher, P., 1994b. Animation and Sound for the Visualization of Uncertain Spatial Information, In: H.M. Hearnshaw, and D.J. Unwin (Eds.), *Visualization in Geographic Information Systems*, John Wiley and Sons, NY, USA, 181-185.

95. Fisher, P.F., 1995. An Exploration of Probable Viewsheds in Landscape Planning, *Environment and Planning-B: Planing and Design*, **22**, 527-546.

96. Flowers, J.H., and Hauer, T.A., 1992. The Ear's versus the Eye's Potential to Assess Characteristics of Numeric Data: Are We Too Visuocentric?, *Behavior Research Methods, Instruments, and Computers*, **24**, 258-264.

97. Fonte, C.C., and Lodwick, W., 2004. Areas of Geographical Entities, *International Journal of Geographical Information Science*, **18**, 127-150.

98. Foresman, T.W., 1998. *The History of Geographic Information Systems: Perspectives from the Pioneers*, Prentice Hall, Upper Saddle River, NJ, USA.

99. Forester, J., 1989. *Planning in the Face of Power*, University Press, Berkley, CA, USA.

100. Foster, I., Geisler, J., Nickless, W., Smith, W., and Tuecke, S., 1996. Software Infrastructure for the I-Way High-Performance Distributed Computing Experiment, In Proceedings *5th IEEE Symposium on High Performance Distributed Computing*, Syracuse, NY, USA.

101. Foster, I., and Kesselman, C., 1997. Globus: A Metacomputing Infrastructure Toolkit, *International Journal of Supercomputer Applications*, 11, 115-128.

102. Foster, I., and Kesselman, C., 1999. *The Grid: Blueprint for a Future Computing Infrastructure*, Morgan Kaufmann Publishers, San Francisco, CA.

103. Foster, I., and Kesselman, C., 2004. *The Grid 2: Blueprint for a New Computing Infrastructure*, Morgan Kaufmann Publishers, San Francisco, CA, USA.

104. Foster, I., Kesselman, C., and Tuecke, S., 2001. The Anatomy of the Grid – Enabling Scalable Virtual Organizations, *International Journal of High Performance Computing Applications*, **15**, 200-222.

105. Friedmann, J., 1987. *Planning in the Public Domain*, Princeton University Press, Princeton, NJ, USA.

106. Frisken-Gibson, S.F., Bach-Y-Rita, P., Tompkins, W.J., and Webster, J.G., 1987. A 64-Solenoid, Four-Level Fingertip Search Display for the Blind, *IEEE Transactions on Biomedical Engineering*, **34**, 963-965.

107. Fruin, J.J., 1971. *Pedestrian Planning and Design*, Metropolitan Association of Urban Designers and Environmental Planners Inc., NY, USA.

108. Furnas, G.W., 1986. Generalized Fisheye Views, In Proceedings *Human Factors in Computing Systems*, Boston, MA, USA.

109. Gardner, J.A., and Bulatov, V., 2001. Smart Figures, SVG, and Accessible Web Grahics, In Proceedings *CSUN International Conference on Technology and Persons with Disabilities*, Los Angeles, CA, USA.

110. Gaver, W.W., 1989. The Sonicfinder: An Interface That Uses Auditory Icons, *Human-Computer Interaction*, **4**, 67-94.

111. Gavison, R., 1980. Privacy and the Limits of Law, *Yale Law Journal*, **89**, 421-471.

112. Gewin, V., 2004. Mapping Opportunities, *Nature*, **427**, 376-377.

113. Goldberg, D., 1989. *Genetic Algorithms in Search, Optimisation and Machine Learning*, Addison-Wesley, Reading, MA, USA.

114. Goldberg, S.B., Maimone, M.W., and Matthies, L., 2002. Stereo Vision and Rover Navigation Software for Planetary Exploration, In Proceedings *IEEE Aerospace Conference*, Big Sky, MT, USA.

115. Golledge, R., 1999. Human Wayfinding and Cognitive Maps, In: R. Golledge (Ed.), *Wayfinding Behavior: Cognitive Mapping and Other Spatial Processes*, John Hopkins University Press, Baltimore, MD, USA, 5-45.

116. Golledge, R.G., 2004. Multidisciplinary Opportunities and Challenges in NBIC, In: M.C. Roco, and C.D. Montemagno (Eds.), *The Coevolution of Human Potential and Converging Technologies*, New York Academy of Sciences, NY, USA, 199-211.

117. Golledge, R.G., Klatzky, R.L., Loomis, J.M., Speigle, J., and Tietz, J., 1998. A Geographic Information System for a GPS based Personal Guidance System, *International Journal of Geographical Information Systems*, **12**, 727-749.

118. Golledge, R.G., Loomis, J.M., and Klatzky, R.L., 1994. Auditory Maps as Alternatives to Tactual Maps, In Proceedings *4th International Symposium on Maps and Graphics for the Visually Impaired*, São Paulo, Brazil.

119. Golledge, R.G., Loomis, J.M., and Klatzky, R.L., 1998. Auditory Maps: What Are They and How Can They Be Used to Teach Geography?, In Proceedings *Morphological Eyes and Ears: Learning and Assessment of Geographic Space*, Southwest Texas State University, TX, USA.

120. Golledge, R.G., Loomis, J.M., Klatzky, R.L., Flury, A., and Yang, X.L., 1991. Designing a Personal Guidance System to Aid Navigation Without Sight: Progress on the GIS Component, *International Journal of Geographical Information Systems*, **5**, 373-395.

121. Golledge, R.G., Rice, M., and Jacobson, R.D., 2005. A Commentary on the Use of Touch for Accessing On-Screen Spatial Representations: The Process of Experiencing Haptic Maps and Graphics, *The Professional Geographer*, **57**, 339-349.

122. Golombek, M.P., Cook RA., Moore, H.J., and Parker, T.J., 1997. Selection of the Mars Pathfinder Landing Site, *Journal of Geophysical Research*, **102**, 3967-3988.

123. Goodchild, M.F., 1999a. Future Directions in Geographic Information Science, *Geographic Information Sciences*, **5**, 1-8.

124. Goodchild, M.F., 1999b. Measurement-based GIS, In Proceedings *International Symposium on Spatial data quality*, Hong Kong, China.

125. Goodchild, M.F., 2004. GIScience, Geography, Form, and Process, *Annals of American Geographers*, **94**, 709-714.

126. Goodchild, M.F., Egenhofer, M.J., Fegeas, R., and Koffman, C.A., 1999. *Interoperating Geographic Information Systems*, Kluwer Academic Publishers, Boston, MA, USA.

127. Goodchild, M.F., and Gopal, S., 1989. *The Accuracy of Spatial Databases*, Taylor and Francis, London, UK.

128. Goodchild, M.F., Sun, G., and Yang, S., 1992. Development and Test of an Error Model for Categorical Data, *International Journal of Geographical Information Systems*, **6**, 87-104.

129. Griffin, A., 2001. Feeling it Out: the Use of Haptic Visualization for Exploratory Geographic Analysis, *Cartographic Perspectives*, **39**, 12-29.

130. Gruteser, M., and Grunwald, D., 2003. Anonymous Usage of Location-based Services through Spatial and Temporal Cloaking, In Proceedings *ACM/USENIX International Conference on Mobile Systems, Applications, and Services*, San Francisco, CA, USA.

131. Guhathakurta, S., 2002. Urban Modeling as Storytelling: Using Simulation Models as Narrative, *Environment and Planning-B: Planning ans Design*, **29**, 895-911.

132. Guesgen, H.W., and Albrecht, J., 2000. Imprecise Reasoning in Geographic Information Systems, *Fuzzy Sets and Systems*, **113**, 121-131.

133. Hagen, A., 2003. Fuzzy Set Approach to Assessing Similarity of Categorical Maps, *International Journal of Geographical Information Science*, **17**, 235-249.

134. Hand, D., Mannila, H., and Smyth, P., 2001. *Principles of Data Mining*, MIT Press, Cambridge MA, USA.

135. Hardwick, A., Rush, J., Furner, S., and Seton, J., 1997. Feeling it as Well as Seeing it-Haptic Display within Gestural HCI for Multimedia Telematics Services, In: P.A. Harling, and A.D.N. Edwards (Eds.), *Progress in Gestural Interaction*, Springer-Verlag, Berlin, Germany, 105-116.

136. Hecht-Nielsen, R., 1987. Kolmogorov's Mapping Neural Network Existence Theorem, In Proceedings *IEEE International Conference on Neural Networks*, San Diego, CA, USA.

137. Hecht-Nielson, R., 1990. *Neurocomputing*, Addison-Welsley, Reading, MA, USA.

138. Helbing, D.B., 1991. A Mathematical Model for the Behavior of Pedestrians, *Behavioral Science*, **36**, 298-310.

139. Helbing, D., Farkas I., and Vicsek, T., 2000. Simulating dynamical features of escape panic, *Nature*, **407**, 487-490.

140. Helbing, D., Schweitzer, F., Keltsch, J., and Molnar, P., 1997. Active Walker Model for the Formation of Animal and Trail Systems, *Physical Review E*, **56**, 2527-2539.

141. Heller, D., 1994. *Proposed ECS Core Metadata Standard – Release 2.0. ECS Document # 420-TP-001-005*, Raytheon Company, Upper Marlboro, MD, USA.

142. Henderson, L.F., 1971. The Statistics of Crowd Fluids, *Nature*, **229**, 381-383.

143. Heuvelink, G.B., 1993. *Error Propagation in Quantitative Spatial Modelling*, Ph.D. Thesis, University of Utrecht, The Netherlands.

144. Heuvelink, G.B., and Burrough, P.A., 1993. Error Propagation in Cartographic Modelling using Boolean Logic and Continuous Classification, *International Journal of Geographical Information Systems*, **7**, 231-246.

145. Heye, C., 2002. *Deskriptive Beschreibung Der Ergebnisse Der Internetbefragung: Umsteigen an Haltestellen Der Vbz*, University of Zurich, Switzerland.

146. Heye, C., Rüetschi, U.J., and Timpf, S., 2003. Komplexität Von Routen in Öffentlichen Verkehrssystemen, In Proceedings *Angewandte Geographische Informationsverarbeitung XV*, Salzburg, Austria.

147. Hightower, J., and Borriello, G., 2001a. *A Survey and Taxonomy of Location Systems for Ubiquitous Computing*, University of Washington, WA, USA.

148. Hightower, J., and Borriello, G., 2001b. Location Systems for Ubiquitous Computing, *IEEE Computer*, **34,** 57-66.

149. Hjelm, J., 2002. *Creating Location Services for the Wireless Web*, John Wiley and Sons, NY, USA.

150. Hoel, E., Menon, S., and Morehouse, S., 2003. Building a Robust Relational Implementation of Topology, In Proceedings *8th International Symposium on Spatial and Temporal Databases*, Santorini, Greece.

151. Höllerer, T., Feiner, S., Terauchi, T., Rashid, D., and Hallaway, D., 1999, Exploring Mars: Developing Indoor and Outdoor User Interfaces to a Mobile Augmented Reality System, *Computers and Graphics*, **23**, 779-785.

152. Hopfield, J.J., 1984. Neurons with Graded Response have Collective Computational Properties like those of Two State Neurons, In *US National Academy of Science Proceedings*, **81**, 3088-3092.

153. Hopper, A., 2000. Sentient Computing, *Philosophical Transactions*, Royal Society, London, **3581**, 2349-2358.

154. Howard, A., and Seraji, H., 2001. Vision-based Terrain Characterization and Traversability Assessment, *Journal of Robotic Systems*, **18**, 577-587.

155. Howard, A., and Seraji, H., 2004. Multi-Sensor Terrain Classification for Safe Spacecraft Landing, *IEEE Transactions on Aerospace and Electronic Systems*, **40**, 1122-1131.

156. Huntsberger, T., Aghazarian, H., Cheng, Y., Baumgartner, E.T., Tunstel, E., Leger, C., Trebi-Ollennu, A., and Schenker, P.S., 2002. Rover Autonomy for Long Range Navigation and Science Data Acquisition on Planetary Surfaces, In Proceedings *IEEE Conference on Robotics and Automation*, Washington DC, USA.

157. IBM, 2000. *IBM DB2 Spatial Extender User's Guide and Reference.*

158. Ilfand, P., 1999. *Taking the Stars: Celestial Navigation from Argonauts to Astronauts*, Krieger Publishing company, FL, USA.

159. Infopolis2, 1999. *Needs of Travellers: An Analysis based on the Study of Their Tasks and Activities*, Commission of the European Communities – DG XIII. WP3.2/DEL3/1999, Brussels, Belgium.

160. Informix, 2000. *Informix Spatial DataBlade Module User's Guide Part no. 000-6868.*

161. Ingres, 1994. INGRES/*Object Management Extension User's Guide, Release 6.5.*

162. Intelligent Space Partnership, 2002. *Carnival Public Safety Project – Assessment of Route Design for the Notting Hill Carnival*, Greater London Authority, London, UK.

163. ISO, 2001. *ISO/DIS 19107: Geographic information and Geomatics – Spatial schema*, International Organization for Standardization.

164. Jacobs, J., 1969. *The Economy of Cities*, Random House, NY, USA.

165. Jacobson, R.D., 1998. Navigating Maps with Little or No Sight: A Novel Audio-Tactile Approach, In Proceedings *Content, Visualisation and Intermedia Representation*, University of Montreal, Canada.

166. Jacobson, R.D., 1999. *Exploring Geographies of Blindness: Learning, Reading, and Communicating Geographic Space*, Ph.D. Thesis, Queen's University of Belfast, Belfast, Northern Ireland.

167. Jansson, G., 2001. Perceiving Complex Virtual Scenes with a Phantom Without Visual Guidance, In Proceedings *Touch in Virtual Environments Workshop*, University of Southern California, Los Angeles, CA, USA.

168. Jeng, J., 2001. Towards a Universal Service-Computing Platform Via Virtual Service Machine, In Proceedings *ACM Symposium on Applied Computing*, Las Vegas, NV, USA, 663-667.

169. Johnson, A., Klumpp, A., Collier, J., and Wolf, A., 2001. LIDAR-based Hazard Avoidance for Safe Landing on Mars, In Proceedings *AAS/AIAA Space Flight Mechanics Meeting*, Santa Barbara, CA, USA.

170. Johnston, W.E., Gannon, D., and Nitzberg, B., 1999. Grids as Production Computing Environments: The Engineering Aspects of NASA's Information Power Grid, In Proceedings *8th IEEE Symposium on High Performance Distributed Computing*, Redondo Beach, CA, USA.

171. Jones, D., 2004. Interview with Digby Jones By Adrian Chiles on the BBC Programme "Working Lunch", November 5.

172. Kaptelinin, V., Nardi, B., and Macaulay, C., 1999. The Activity Checklist: A Tool for Representing the "Space" of Context, *Interactions*, 29-39.

173. Kasabov, N., 1996. Learning Fuzzy Rules and Approximate Reasoning in Fuzzy Neural Networks and Hybrid Systems, *Fuzzy Sets and Systems*, **82**, 2-20.

174. Kay, L., 1964. An Ultrasonic Sensing Probe as a Mobility Aid for the Blind, *Ultrasonics*, **2**, 53.

175. Kelly, A.J., and Salcudean, S.E., 1994. On the Development of a Force-Feedback Mouse and it's Integration into a Graphical User Interface, In Proceedings *ASME Winter Annual Meeting*, New Orleans, LA, USA.

176. Kim, W., and Seo, J., 1991. Classifying Schematic and Data Heterogeneity In Multidatabase Systems, *IEEE Computer*, **24**, 12-18.

177. Klein, M., and Fensel, D., 2004. Ontology Versioning on the Semantic Web, In Proceedings *International Semantic Web Working Symposium*, Stanford University, CA, USA.

178. Klinker, G., 2000. Augmented Reality: A Problem in Need of Many Computer Vision-based Solutions, In: A. Leonardis, F. Solina, and R. Bajcsy (Eds.), *Confluence of Computer Vision and Computer Graphics*, Kluwer Academic Publishers, Dordrecht, The Netherlands, 267-284.

179. Klinker, G., Reiners, D., Stricker, D., and Müller, S., 1998. Augmented Reality for Construction Tasks: Doorlock Assembly Applications, In Proceedings *International Workshop on Augmented Reality*, San Francisco, CA, USA.

180. Kluijver, H., de and J.E. Stoter, 2003. Noise Mapping and GIS: Optimising Quality and Efficiency of Noise Effect Studies, *Computers, Environment and Urban Systems*, **27**, 85-102.

181. Kofler, M., 1998. *R-Trees for Visualizing and Organizing Large 3D GIS Databases*, Ph.D. thesis, Technical University Graz, Austria.

182. Kramer, G. (Ed.), 1994. *Auditory Display: Sonification, Audification, and Auditory Interfaces*, Addison-Wesley, Reading, MA, USA.

183. Krauze-Tomczyk, I., and Korycka-Skorupa, J., 2001. Maps for the Blind and Visually Handicapped, In Proceedings *20th International Cartographic Conference*, Beijing, China.

184. Kretschmer, U., Coors, V., Spierling, U, Grasbon, D., Schneider, K., Rojas, I., and Malaka, R., 2001. Meeting the Spirit of History, In Proceedings *Virtual Reality, Archaeology, and Cultural Heritage*, Greece.

185. Krüger, A., Butz, A., Müller, C., Stahl, C., Wasinger, R., Steinberg, K., and Dirschl, A., 2004. The Connected User Interface: Realizing a Personal Situated Navigation Service, In Proceedings *International Conference on Intelligent User Interfaces*, Funchal, Madeira, Portugal.

186. Krygier, J.B., 1994. Sound and Geographic Visualisation, In: A.M. MacEachren, and D.R. Fraser-Taylor (Eds.), *Visualisation in Modern Cartography*, Elsevier Science, NY, USA, 149-166.

187. Kuhn, W., 2001. Ontologies in Support of Activities in Geographical Space, *International Journal of Geographic Information Science*, **15**, 613-631.

188. Kuipers, B., 1982. The “Map in the Head” Metaphor, *Environment and Behaviour*, **14,** 202-220.

189. Kulkarni, A., 2002. Design Principles of a Reactive Behavioural System for the Intelligent Room, *Bitstream*, 22-26.

190. Kurze, M., 1996. TDraw: A Computer-based Tactile Drawing Tool for Blind People, In Proceedings *2nd Annual ACM Conference on Assistive Technologies*, Vancouver, British Columbia, Canada.

191. Lacroix, S., Mallet, A., Bonnafous, D., Bauzil, G., Fleury, S., Herrb, M., and Chatila, R., 2002. Autonomous Rover Navigation on Unknown Terrains: Functions and Integration, *International Journal of Robotics Research*, **21**, 917-942

192. Lambert, J.M.N., and Widya, I., 2001. Quality of Service and Service Provisioning on a Competitive Market, *Lecture Notes in Computer Science 1890*, Springer-Verlag, London, UK, 17-25.

193. Langheinrich, M., 2001. Privacy By Design – Principles of Privacy-Aware Ubiquitous Systems, In Proceedings *3rd International Conference on Ubiquitous Computing*, Atlanta, GA, USA.

194. Lankenau, A., and Röfer, T., 2000. Smart Wheelchairs – State of the Art in an Emerging Market, *Künstliche Intelligenz*, 37-39.

195. Lawrence, V., 2005. GI Takes it's Place in the Wider World, *GeoInformatics*, **8**, 26-29.

196. Lee, M.S., 2001. *Experiments with a Computerized Self-Administrative Activity Survey*, Ph.D. Thesis, University of California Irvine, Irvine, CA, USA.

197. Lessig, L., 1999. *Code and Other Laws of Cyberspace*, Basic Books, NY, USA.

198. Leung, Y., 1987. On the Imprecision of Boundaries, *Geographical Analysis*, **19**, 125-151.

199. Leung, Y., 1997. *Intelligent Spatial Decision Support Systems*, Springer-Verlag, Berlin/Heidelberg, Germany.

200. Lewis, M.J.T., 2001. *Surveying Instruments of Greece and Rome*, Cambridge University Press, Cambridge, UK.

201. Li, B., 2003. Peer-to-Peer Sharing of Cartographic Data and Software, In: M. Peterson (Ed.), *Maps and the Internet*, Elsevier, Cambridge, UK, 157-168.

202. Li, R., Di, K., Matthies, L., Arvidson, R.E., Folkner, W., and Archinal, B., 2003. Rover Localization and Landing Site Mapping Technology for the 2003 Mars Exploration Rover Mission, *Photogrammetric Engineering and Remote Sensing*, **70**, 77-90.

203. Li, R., Ma, F., Xu, F., Matthies, L.H., Olson, C.F., and Arvidson, R.E., 2002. Localization of Mars Rovers using Descent and Surface-based Image Data, *Journal of Geophysical Research - Planets*, **107 (E11)**, 8004.

204. Lindstrom, P., Koller, D., Ribarsky, W., Hodges, L.F., and Faust, N., 1996. Real-Time, Continuous Level of Detail Rendering of Height Fields, In Proceedings *ACM SIGGRAPH*, New Orleans, LA, USA.

205. Linvill, J., 1973. *Research and Development of Tactile Facsimile Reading Aid for the Blind: The Optacon*, U.S. Department of Health, Education, and Welfare, Office of Education, Bureau of Education for the Handicapped, Washington DC, USA.

206. Liu, S., Guo, T., and Dang, Y., 2000. *Grey System Theory and it's Application*, Science Press of China, Beijing, China.

207. Lloyd, L.L., and Schleppenbach, D.A., 1997. *An Audiotactile General Chemistry Course for the Visually Impaired*, National Science Foundation Award No. 9722030.

208. Lodwick, W.A., Monson, W., and Svoboda, L., 1990. Attribute Error and Sensitivity Analysis of Map Operations in Geographical Information Systems, *International Journal of Geographical Information Systems*, **4**, 413-427.

209. Longley, P.A., Brooks, S., McDonnell, R., and MacMillan, B. (Eds.), 1998. *GeoComputation – A Primer*, John Wiley and Sons, Chichester, UK.

210. Longley, P.A., Goodchild, M.F., Maguire, and D.J., Rhind, D.W., 1999. Intoduction, In: P.A. Longley, M.F. Goodchild, D.J. Maguie, D.W. Rhind (Eds.), *Principles of Geographical Information Systems – Principles and Applications*, John Wiley and Sons, Chichester, UK, 23-27.

211. Loomis, J.M., Golledge, R.G., and Klatzky, R.L., 1998. Navigation System for The Blind: Auditory Display Modes and Guidance, *Presence: Teleoperators and Virtual Environments*, **7**, 193-203.

212. Loomis, J.M., Golledge, R.G., and Klatzky, R.L., 2001. GPS-based Navigation Systems for the Visually Impaired, In: W. Barfield, and T. Caudell (Eds.), *Fundamentals of Wearable Computers and Augmented Reality*, Lawrence Erlbaum Associates, Mahwah, NJ, USA, 429-446.

213. Loomis, J.M., Hebert, C., and Cicinelli, J.G., 1990. Active Localization of Virtual Sounds, *Journal of the Acoustical Society of America*, **88,** 1757-1764.

214. Loomis, J.M., and Soule, J.I., 1996. Virtual Acoustic Displays for Real and Synthetic Environments, In Proceedings *Society for Information Display International Symposium*, San Diego, CA, USA.

215. López de Ipiña, D., 2003. *Visual Sensing and Middleware Support for Sentient Computing*, Ph.D. Thesis, University of Cambridge, Cambridge, UK.

216. Loubal, P.S., 1992. Fingertip Map Tracing Devices for the Blind, In Proceedings *Conference on Technology and Persons with Disabilities*, Los Angeles, CA, USA.

217. Louwsma, J., Tijssen, T., and van Oosterom, P., 2003. A Comparison Between Topologically Structured and "Plain" Spatial Data, *GEO:connexion*, **June**.

218. Lovas, G.G., 1994. Modeling and Simulation of Pedestrian Flow Traffic, *Transportation Research B*, **28B**, 429-443.

219. Lunney, D., 1983. A Microcomputer-based Laboratory Aid for Visually Impaired Students, *IEEE Micro*, **3,** 19-31.

220. Luo, Y., 1999. *Study on Agent-based Distributed GIS*, Ph.D Thesis, Peking University, Beijing, China.

221. Luo, Y., Wang, X., and Xu, Z., 2001. Design of a Framework for Multi-User/Application Oriented Webgis Services, In Proceedings *International Conference on Computer Networks and Mobile Computing*, Beijing, China.

222. Lynch, K., 1960. *The Image of the City*, MIT Press, Cambridge, MA, USA.

223. MacEachren, A.M., 1994a. *Some Truth with Maps: A Primer on Symbolization and Design*, Association of American Geographers, Washington DC, USA.

224. MacEachren, A.M., 1994b. Visualization in Modern Cartography: Setting the Agenda, In: A.M. MacEachren (Ed.), *Visualization in Modern Cartography*, Pergamon Press, Oxford, UK.

225. MacEachren, A.M., 1995. *How Maps Work: Representation, Visualization, and Design*, Guildford Press, NY, USA.

226. MacEachren, A.M., Brewer, I., and Steiner, E., 2001. Geovisualization to Mediate Collaborative Work: Tools to Support Different-Place Knowledge Construction and Decision-Making, In Proceedings *20th International Cartographic Conference*, Beijing, China.

227. Madsen, W., 1994. Protecting Indigenous Peoples' Privacy from Eyes in the Sky, In Proceedings *Conference on Law and Information Policy for Spatial*

Databases, National Center for the Study of Law, Science and Technology, AZ, USA.

228. Maguire, D.J., 1989. *Computers in Geography*, Longman Scientific and Technical, Essex, UK.

229. Makino, H., Ishii, I., and Nakashizuka, M., 1996. Development of Navigation System for the Blind using GPS and Mobile Phone Connection, In Proceedings *18th Annual Meeting of the IEEE Engineering in Medicine and Biology Society*, Amsterdam, The Netherlands.

230. Malaka, R., and Zipf, A., 2000. Deep Map – Challenging IT Research in the Framwork of a Tourist Information System, In Proceedings *Information and Communication Technologies in Tourism*, Barcelona, Spain.

231. Malczewski, J., 1999. *GIS and Multicriteria Decision Analysis*, John Wiley and Sons, NY, USA.

232. Maor, E., 1998. *Trigonometric Delights*, Princeton University Press, Princeton, NJ, USA.

233. Mark, D.M., 2003. Geographic Information Science: Defining the Field, In: M. Duckham, M.F. Goodchild, and M.F.Worboys (Eds.), *Foundations of Geographic Information Science*, Taylor and Francis, NY, USA, 3-18.

234. Massie, T.H., and Salisbury, K., 1994. The Phantom Haptic Interface: A Device for Probing Virtual Objects, In Proceedings *ASME Winter Annual Meeting - Symposium on Haptic Interfaces for Virtual Environment and Teleoperator Systems*, Chicago, IL, USA.

235. McCall, M.K., 2002. Seeking Good Governance in Participatory GIS: A Review of Indigenous Spatial Knowledge, Participatory Processes, and Governance Dimensions in Local Level GIS for Development, In *Proceedings Governance and the use of GIS in developing countries*, ITC, Enschede, The Netherlands.

236. Mendel, J., and John, R., 2002. Type-2 Fuzzy Sets Made Simple, *IEEE Transactions on Fuzzy Systems*, **10**, 17–127.

237. Mestel, A., and Albus, J., 2002. *Intelligent Systems, Architecture, Design, and Control*, John Wiley and Sons, NY, USA.

238. Moeslund, T.B., Stoerring, M., Liu, Y., Broll, W., Lindt, I., Yuan, C., and Wittkaemper, M., 2004. Towards Natural, Intuitive and Non-Intrusive HCI Devices for Roundtable Meetings, In Proceedings *Workshop on Multi-User and Ubiquitous User Interfaces*, Funchal, Madeira, Portugal.

239. Molenaar, M., and Cheng, T., 2000. Fuzzy Spatial Objects and Their Dynamics, *ISPRS Journal of Photogrammetry and Remote Sensing*, 55, 164-175.

240. Monmonier, M.S., 1982. *Computer-Assisted Cartography: Principles and Prospects*, Prentice Hall, Englewood Cliffs, NJ, USA.

241. Monmonier, M., 1996. *How to Lie with Maps*, University of Chicago Press, Chicago, IL, USA..

242. Montello, D.R., 1993. Scale and Multiple Psychologies of Space, In Proceedings *Conference on Spatial Information Theory*, Marciana Marina, Elba Island, Italy.

243. Montello, D.R., 2005. Navigation, In: P. Shah, and A. Miyake (Eds.), *Handbook of visuospatial cognition*, Cambridge University Press, Cambridge, UK.

244. Moore, G., 1965. Cramming more Components Onto Integrated Circuits, *Electronics*, **38**, 56-59.

245. Nardi, B. (Ed.), 1996. *Context and Consciousness – Activity Theory and Human-Computer Interaction*, MIT Press, Cambridge, MA, USA.

246. National Research Council, 1997. *The Future of Spatial Data and Society*, National Academic Press, Washington DC, USA.

247. National Research Council, 2003. *IT Roadmap to a Geospatial Future*, National Academic Press, Washington DC, USA.

248. Nebiker S, 2003. Support for Visualization and Animation in a Scalable 3D GIS Environment: Motivation, Concepts and Implementation, *International Archives of Photogrammetry, Remote Sensing and Spatial Information Science*, **XXXIV-5/W10**.

249. Negroponte, N., 1995. *Being Digital*, Alfred A. Knopf, NY, USA.

250. Netzel, H., and Kaalberg, F., 1999. Settlement Risk Management with GIS for the North/South Metroline in Amsterdam, In Proceedings *World Tunnel Congress*, Oslo, Norway.

251. Newbould, R., and Collingridge, R., 2004. Profiling – Technology, In: R.D. Ralph, and S. Searby (Eds.), *Location and Personalisation: Delivering Online and Mobility Services*, Institute of Electrical Engineers, London, UK, 59-80.

252. Nissanke, N., 1997. *Realtime Systems*, Prentice Hall, Upper Saddle River, NJ, USA.

253. Norman, D.A., 1988. *The Design of Everyday Things*, Doubleday Press, NY, USA.

254. Odlyzko, A.M., 2003. Privacy, Economics, and Price Discrimination on the Internet, In Proceedings *5th International Conference on Electronic Commerce*, Pittsburgh, PA, USA.

255. OGC, 1999. *OpenGIS Simple Features Specification for SQL Revision 1.1*, OpenGIS Project Document 99-049.

256. OGC, 2003. *OpenGIS Web Services Architecture*, OpenGIS Interoperability Program Report OGC 03-025.

257. Olson, C.F., Matthies, L.H., Schoppers, M., and Maimone, M.W., 2000. Robust Stereo Ego-Motion for Long Distance Navigation, In Proceedings *IEEE Computer Vision and Pattern Recognition*, Hilton Head, SC, USA.

258. Oosterom, P.J.M. van, and Lemmen, C.H.J., 2001. Spatial Data Management on a Very Large Cadastral Database, *Computers, Environments and Urban Systems*, **25**, 509-528.

259. Oosterom, P.J.M. van, and Lemmen, C., 2003. 3D Cadastres, *Computers, Environment and Urban Systems*, 27, Pages 337-343.

260. Oosterom, P.J.M. van, Stoter, J.E., Quak, C.W., and Zlatanova, S., 2002a. The Balance Between Geometry and Topology, In Proceedings *10th International Symposium on Spatial Data Handling*, University of Leicester, Leicester, UK.

261. Oosterom, P.J.M. van, Stoter, J.E., Verbree, E., and Zlatanova, S., 2002b. 3D GIS komt er wel, maar 't zal wel even duren, *VI Matrix*, **10**, 0-23.

262. Oosterom, P.J.M. van, Stoter, J.E., Verbree, E., and Zlatanova, S., 2002c. Onderzoek brengt 3D GIS in gangbare geo-informatie naderbij, *VI Matrix*, **10**, 26-29.

263. Oosterom, P.J.M. van, Quak, C.W., and Tijssen, T.P.M., 2003. Polygons: The Unstable Foundation of Spatial Modeling, In Proceedings *ISPRS Joint Workshop on Spatial, Temporal and Multi-Dimensional Data Modelling and Analysis*, Québec, Canada.

264. Oosterom, P.J.M. van, Vertegaal, W., van Hekken, M., and Vijlbrief, T., 1994. Integrated 3D modelling within a GIS, In Proceedings *Advanced Geographic Data Modelling: spatial data modelling and query languages for 2D and 3D applications*, Delft, The Netherlands.

265. Openshaw, S., 1999. GeoComputation, In: S. Openshaw, and R.J. Abrahart (Eds.), *GeoComputation*, Taylor and Francis, London, UK, 1-32.

266. Openshaw, S., Fischer, M.M., Benwell, G., and Macmillan, B., 1999. Geocomputation Research Agendas and Futures, In: S. Openshaw, and R.J. Abrahart (Eds.), *GeoComputation*, Taylor and Francis, London, UK, 379-400.

267. Oppermann, R., and Specht, M., 1999. A Nomadic Information System for Adaptive Exhibition Guidance, *Archives and Museum Informatics*, **132**, 27-138.

268. Oracle, 2001. *Oracle Spatial User's Guide and Reference Release 9.0.1*, Part Number A88805-01.

269. Orwell, G., 1949. *Nineteen Eighty-Four*, Buccaneer Books, NY, USA.

270. Park, J., and Sandberg, J.W., 1991. Universal Approximation Function using Radial Basis Functions Network, *Neural Computation*, **3**, 246-257.

271. Parkes, D., and Dear, R., 1989. *Nomad: An Audio-Tactile Interactive Graphics Processor for the Visually Impaired and Blind - Reference Manual (Nomad Manual Version 1)*, Quantum Technology Pvt. Ltd., Sydney, Australia.

272. Pascarelli, E.F., 1994. *Repetitive Strain Injury: A Computer User's Guide*, John Wiley and Sons, NY, USA.

273. Pasman, W., and Jansen, F.W., 2002. Scheduling Level of Detail with Guaranteed Quality and Cost, In Proceedings *Web3D Conference*, Tempe, AZ, USA.

274. Pawlak, Z., 1982. Rough Sets, *International Journal of Computer and Information Sciences*, **11**, 341-356.

275. Peng, Z.-R., and Tsou, M-H., 2003. *Internet GIS – Distributed Geographic Information Services for the Internet and Wireless Networks*, John Wiley and Sons, NY, USA.

276. Penninga, F., 2004. *Oracle 10g Topology: Testing Oracle 10g Topology using Cadastral Data*, GISt report No 26, University of Delft, Delft, The Netherlands.

277. Pennock, K.A., and Lantrip, D.B., 1995. Themescapes: A Landscape Representation of Themes in Text, In Proceedings *Symposium on Advanced Intelligence Processing and Analysis*, Tysons Corner, VA, USA.

278. Perkins, C., 2001. Tactile Campus Mapping: Evaluating Design and Production Technologies, In Proceedings *20th International Cartographic Conference*, Beijing, China.

279. Peterson, M. (Ed.), 2003. *Maps and the Internet*, Elsevier, Cambridge, UK.

280. Petrie, H., Johnson, V., Strothotte, T., Raab, A., Fritz, S., and Michel, R., 1996. Mobic: Designing A Travel Aid for Blind and Elderly People, *Journal of Navigation*, **49**, 45-52.

281. Petrucci, L.S., Roth, P., Harth, E., Assimicoupolos, A., and Pun, T., 2000. Websound: A Generic Sonification Tool and it's Application to an Auditory Web Browser for Blind and Vision Impaired Users, In Proceedings *6th International Conference on Auditory Display*, Atlanta, GA, USA.

282. Petry, F.E., Cobb, M.A., Wen, L., and Yang, H., 2003. Design of System for Managing Fuzzy Relationships for Integration of Spatial Data in Querying, *Fuzzy Sets and Systems*, **140**, 51-73.

283. Pfitzmann, A., and Köhntopp, M., 2000. Anonymity, Unobservability and Pseudonymity–A Proposal for Terminology, In Proceedings *International Workshop on the Design Issues in Anonymity and UnObservability*, Berkeley, CA, USA.

284. Phillips, D., and Curry, M., 2003. Privacy and the Phenetic Urge: Geodemographics and the Changing Spatiality of Local Practice, In: D. Lyon (Ed.), *Surveillance as Social Sorting: Privacy, Risk, and Automated Discrimination*, Routledge, London, 137-152.

285. Piekarski, W., Gunther, B., and Thomas, B., 1999. Integrating Virtual and Augmented Realities in an Outdoor Application, In Proceedings *IEEE International Workshop on Augmented Reality*, Cathedral Hill Hotel, San Francisco, CA, USA.

286. Pitoura, E., Abiteboul, S., Pfoser, D., Smaras, G., and Vazirgiannis, M., 2003. Dbglobe: A Service-Oriented P2P System for Global Computing, *SIGMOD Record*, **32**, 77-82.

287. Pospischil, G., Umlauft, M., and Michlmayr, E., 2002. Designing LoL@, a Mobile Tourist Guide for UMTS, In: F. Paternò (Ed.), *Mobile HCI – Lecture Notes in Computer Science 2411*, Springer-Verlag, Berlin, Germany, 140-154.

288. Quak, C.W., Stoter J.E., and Tijssen, T.P.M., 2003. Topology in Spatial DBMSs, In Proceedings *Digital Earth*, Brno, Czech Republic.

289. Rahimi, S., Cobb, M., Ali, D., Paprzycki, M., and Petry, F., 2002. A Knowledge-based multi-agent system for Geospatial Data Conflation, *Journal of Geographic Information and Decision Analysis*, **62**, 67-81.

290. Ralph, D., and Searby, S., 2004. *Location and Personalisation: Delivering Online and Mobility Services*, Institute of Electrical Engineers, London.

291. Ramakrishnan, R., and Gehrke, J., 2003. *Database Management Systems*, McGraw-Hill Higher Education, NY, USA.

292. Ramloll, R., Yu, W., Brewster, S., Riedel, B., Burton, M., and Dimigen, G., 2000. Constructing Sonified Haptic Line Graphs for the Blind Student: First Steps, In Proceedings *4th International ACM Conference on Assistive Technologies*, Arlington, VA, USA.

293. Ramstein, C., and Hayward, V., 1994. The Pantograph: A Large Workspace Haptic Device for a Multi-Modal Human Computer Interaction, In Proceedings *Conference on Human Factors in Computing Systems*, Boston, MA, USA.

294. Randell, C., and Muller, H.L., 2002. The ESleeve: A Novel Wearable Computer Configuration for the Discovery of Situated Information, In Proceedings *22nd International Conference on Distributed Computing Systems*, Vienna, Austria.

295. Raper, J., 2000. *Multidimensional Geographic Information Science*, Taylor and Francis, London.

296. Reynolds, C.W., 1987. Flocks, Herds, and Schools: A Distributed Behavioral model, *Computer Graphics*, **21**, 25-34.

297. Ribo, M., Lang, P., Ganster, H., Brandner, M., Stock, C., and Pinz, A., 2002. Hybrid Tracking for Outdoor Augmented Reality Applications, *IEEE Computer Graphics and Applications,* **22**, 54-63.

298. Rice, M., Jacobson, R.D., Golledge, R.G., and Hegarty, M., Submitted. Non-visual Interfaces for Interpreting Spatial Information: Force Feedback and Auditory Interfaces for Interpreting On-Screen Shapes, *Cartography and Geographic Information Science.*

299. Rigol, J.P., Jarvis, C.H., and Stuart, N., 2001. Artificial Neural Networks as a Spatial Interpolation, *International Journal of Geographical Information Science*, **15**, 323-343.

300. Rikkers, R., Molenaar, M., and Stuiver, J., 1993. A query oriented implementation of a 3D topological data structure, In Proceedings *4th European Conference and Exhibition on Geographical Information Systems*, Genoa, Italy.

301. Roberts, G., Evans., A., Dobson, A., Denby, B., Cooper, S., and Hollands, R., 2002. Look Beneath the Surface with Augmented Reality, *GPS World*, **February**, 14-20.

302. Robinson, A.H., 1985. Cartography as a Visual Technique, In: A.H. Robinson, *The Look of Maps: An Examination of Cartographic Design*, University of Wisconsin Press, Madison, WI, USA.

303. Robinson, A.H., Morrison, J.L., Muehrcke, P.C., Kimerlink, A.J., and Guptill, S.C., 1995. Elements of Cartography, John Wiley and Sons, NY, USA.

304. Rotenberg, M., and Laurant, C., 2004. *Privacy and Human Rights: An International Survey of Privacy Laws and Developments*, Privacy International and the Electronic Privacy Information Center, London, UK.

305. Rumelhart, D.E., and McClelland, J.L., 1986. *Parallel Distributed Processing: Explorations in the Microstructure of Cognition*, MIT Press, Cambridge, MA, USA.

306. Sager, T., 1992. Why Plan? – A Multi-Rationality Foundation for Planning, In *Scandinavian Housing and Planning Research*, **9**, 129-147.

307. Salisbury, J.K., and Srinivasan, M.A., 1992. Haptic Interfaces, In: V. MIT (Ed.), *BBN Report No. 7661- Virtual Environment Technology for Training*, BBN.

308. Sambuc, R., 1975. *Fonctions φ-floues. Application l'aide au diagnostic en pathologie thyroidienne*, Ph.D. Thesis, University of Marseille, France.

309. Sandhu, R., Ferraiolo, D., and Kuhn, R., 2000. The NIST Model for Role-based Access Control: Towards a Unified Standard, In Proceedings *5th ACM Workshop on Role-based Access Control*, Berlin, Germany.

310. Scaletti, C., and Craig, A.B., 1991. Using Sound to Extract Meaning from Complex Data. In Proceedings *Society for Optical Engineering Conference on Extracting Meaning from Complex Data*, Bellingham,WA, USA.

311. Schilit, B., Adams, N., and Want, R., 1994. Context-Aware Computing Applications, In Proceedings *Workshop on Mobile Computing Systems and Applications*, Santa Cruz, CA, USA.

312. Schoggen, P., 1989. *Behavior Settings*, Stanford University Press, Stanford, CA, USA.

313. Schuurman, N., 2004. *GIS: A Short Introduction*, Blackwell, Malden, MA, USA.

314. Schwald, B., Seibert, H., Weller, T., and Flexible, A., 2002. Tracking Concept Applied to Medical Scenarios using an AR Window, In Proceedings *International Symposium on Mixed and Augmented Reality*, Darmstadt, Germany.

315. Serjantov, A., and Danezis, G., 2002. Towards an Information Theoretic Metric for Anonymity, In Proceedings *Workshop on Privacy Enhancing Technologies*, San Francisco, CA, USA.

316. Shannon, C. E., 1948. A Mathematical Theory of Communication, *Bell System Technical Journal*, **27**, 379-423 and 623-656.

317. Shapiro, C., and Varian, H.R., 1999. *Information Rules: A Strategic Guide to The Network Economy*, Harvard Business School Press, Boston, MA, USA.

318. Sharma, R., Yeasin, M., Krahnstöver, N., Rauschert, I., Cai, G., Brewer, I., Maceachren, A.M., and Sengupta, K., 2003. Speech-Gesture Driven Multimodal Interfaces for Crisis Management, *Proceedings of IEEE - Special Issue on Multimodal Human-Computer Interface*, **91**, 1327-1354.

319. Shi, W. Z., Yang, B.S., and Li, Q.Q., 2003. An Object-Oriented Data Model for Complex Objects in Three-Dimensional Geographic Information Systems, *International Journal of Geographic Information Science*, **17**, 411-430.

320. Shiffer, M.J., 1995. Environmental Review with Hypermedia Systems, *Environment and Planning-B: Planning and Design*, **22**, 359-372.

321. Shiffer, M.J., 1999. Managing Public Discourse: Towards the Augmentaion of GIS with Multimedia, In: P.A Longley, M.F. Goodchild, D.J. Maguire, and D.W. Rhind (Eds.), *Geographical Information Systems: Management Issues and Applications*, John Wiley and Sons, NY, USA, 723-732.

322. Shneiderman, B., 2002. *Leonardo's Laptop: Human Needs and the New Computing Technologies*, MIT Press, Cambridge, MA, USA.

323. Siegel, A.W., and White, S.H., 1975. The Development of Spatial Representations of Large-Scale Environments, In: H.W. Reese (Ed.), *Advances in Child Development and Behavior*, Academic Press, NY, USA, 9-55.

324. Siekierska, E., Labelle, R., and O'Neil, L., 2001. Tactile Mapping Project at Canadian Federal Mapping Agency – Mapping Services Branch, In Proceedings *20th International Cartographic Conference*, Beijing, China.

325. Sillitti, A., Vernazza, T., and Succi, G., 2002. Service Oriented Programming: A New Paradigm of Software Reuse, In: C. Gacek (Ed.), *Lecture Notes in Computer Science 2319*, Springer-Verlag, Berling, Germnay, 269-280.

326. Sims, R.A., Corns, I.G.W., and Klinka, K. (Eds.), 1996. *Global to Local, Ecological Land Classification*, Kluwer Academic Publishers, Dordrecht, The Netherlands.

327. Sommerville, I., 2000. *Software Engineering*, Addison Wesley Professional, Boston, MA, USA.

328. Stell, J.G., and Worboys, M.F., 1999. Generalizing Graphs using Amalgamation and Selection, In: R.H. Gueting, D. Papadias, and F. Lochovsky (Eds.), *Lectue Notes in Computer Science 1651*, Springer-Verlag, London, UK, 19-32.

329. Stevens, R.L., 1999. H-Sense – Sediment Perspectives upon Harbour Sustainability, In Proceedings *4th International Congress on Characterisation and Treatment of Sediments*, Antverp, Belgium.

330. Still, G.K., 2001. *Crowd Dynamics*, Ph.D. Thesis, University of Warwick, Warwick, UK.

331. Stone, M.C., Fishkin, K., and Bier, E.A., 1994. The Movable Filter as a User Interface Tool, In Proceedings *Human Factors in Computing Systems*, Boston, MA, USA.

332. Stoter, J.E., and Zlatanova, S., 2003. Visualisation and Editing of 3D Objects organised in a DBMS, In *Proceedings EUROSDR workshop: Rendering and visualisation*, Enschede, The Netherlands.

333. Stricker, D., 2001. Tracking with Reference Images: A Real-Time and Markerless Tracking Solution for Out-Door Augmented Reality Applications, In Proceedings *Conference on Virtual Reality, Archaeology, and Cultural Heritage*, Greece.

334. Sui, D. Z., 2004. GIS, Cartography and the "Third Culture": Geographic Imaginations in the Computer Age, *The Professional Geographer*, **561**, 62-72.

335. Sui, D.Z., and Goodchild M.F. 2001. GIS as Media?, *International Journal of Geographical Information Science*, **15**, 387-390.

336. Sullivan, A., 2005. 3-Deep, *IEEE Spectrum*, **42**, 22-27.

337. Tao, V., 2001. Online GIServicess, *Journal of Geospatial Engineering*, **32**, 135-143.

338. Tao, V., and Yuan, S., 2000. GeoServNet: Renting Geotools Over Internet, *Geo-Informatics*, **1412**, 12-15.

339. Tarafdar, A.K., and Krishna, B., 2002. Community GIS vs GIS Community, *GIS@development*, **6**.

340. Tarafdar, A. K., Krishna, B., and Gupta, R., 2002. In the Maze of Disaster Management, *GIS@development*, **6**.

341. Tatham, A.F., 1991. The Design of Tactile Maps: Theorectical and Practical Considerations, In Proceedings *15th Conference and 9th General Assembly of the International Cartographic Association*, Bournemouth, UK.

342. Tatham, A.F., 2001. Achievable Goal or Chimaera? Standardisation in Tactile Mapping, In Proceedings *20th International Cartographic Conference*, Beijing, China.

343. Timpf, S., 2002. Ontologies of Wayfinding: A Traveler's Perspective, *Networks and Spatial Economics*, **2**, 9-33.

344. Timpf, S., 2003. Geographic Activity Models, In: M. Duckham, M.F. Goodchild, and M.F. Worboys (Eds.), *Foundations of Geographic Information Science*, Taylor and Francis, London (UK) and NY (USA), 241-254.

345. Timpf, S., Volta, G.S., Pollock, D.W., and Egenhofer, M.J., 1992. A Conceptual Model of Wayfinding using Multiple Levels of Abstractions, In: A.U. Frank, I. Campari, and U. Formentini (Eds.), *Theories and Methods of Spatio-Temporal Reasoning in Geographic Space*, Springer-Verlag, Heidelberg-Berlin, Germany, 348-367.

346. Tomlinson, R., 1998. The Canada Geographic Information System, In: T.W. Foresman (Ed.), *The History of Geographic Information Systems*, Prentice Hall, Upper Saddle River, NJ, USA, 21-32.

347. Tsichritzis, D., and Lochovsky, F., 1982. *Data Models*, Prentice-Hall, Englewood Cliffs, NJ, USA.

348. Tsou, M., 2002. An Operational Metadata Framework for Searching, Indexing, and Retrieving Distributed Geographic Information Services on the Internet, In: M. Egenhofer, and D. Mark (Eds.), *Lecture Notes in Computer Science 2478*, Springer-Verlag, Berlin, 313-332.

349. Tsou, M., and Buttenfield, B., 1998. Client/Server Components and Metadata Objects for Distributed Geographic Information Services, In Proceedings *GIS/LIS Conference*, Fort Worth, TX, USA.

350. Tsou, M., and Buttenfield, B., 2002. A Dynamic Architecture for Distributing Geographic Information Services, *Transactions in GIS*, **6,** 355-381.

351. Tufte, E.R., 1983. *The Visual Display of Quantitative Information*, Graphics Press, Cheshire, CT, USA.

352. Tufte, E.R., 1997. *Visual and Statistical Thinking: Displays of Evidence for Making Decisions*, Graphics Press, Cheshire, CT, USA.

353. Tunstel, E., 1995. Fuzzy Spatial Map Representation for Mobile Robot Navigation, In Proceedings *10th ACM Annual Symposium on Applied Computing*, Nashville, TN, USA.

354. Tunstel, E., and Howard, A., 2003. Approximate Reasoning for Safety and Survivability of Planetary Rovers, *Fuzzy Sets and Systems*, **134**, 27-46.

355. Tukey, J.W., 1977. *Exploratory Data Analysis*, Reading, MA, USA.

356. Tversky, B., 1993. Cognitive Maps, Cognitive Collages, and Spatial Mental Model, In: A.U. Frank, and I. Campari (Eds.), *Spatial Information Theory: Theoretical Basis for GIS*, Springer-Verlag, Heidelberg/Berlin, Germany, 14-24.

357. Upham, P., Thomas, C., Gillingwater, D., and Raper, D., 2003. Environmental Capacity and Airport Operations, *Journal of Air Transport Management*, **9**, 145-151.

358. Upham, P., Yang, Y., Raper, D., Thomas, C., Gillingwater, D., and Hinde, C., 2004. Mitigating Environmental Constraints at Airports through Long Term Planning, *Air Traffic Control Quarterly*, **12**, 107-124.

359. United Nations, 1948. *Universal Declaration of Human Rights: General Assembly Resolution 217a (Iii).*

360. United Nations, 1966. *International Covenant on Civil and Political Rights.*

361. Unwin, D., 1998. Book Review: The History of Geographic Information Systems: Perspectives from the Pioneers, *URISA Journal*, **10**.

362. Vijlbrief, T., and Oosterom, P.J.M. van, 1992. GEO++: An Extensible GIS, In Proceedings *5th International Symposium on Spatial Data Handling*, Charleston, SC, USA.

363. Vckovski, A., 1998. *Interoperable and Distributed Processing in GIS*, Taylor and Francis, London, UK.

364. Vicsek, T., Czirok, A. B-J. E., Cohen, I., and Shochet, O., 1995. Novel Type of Phase Transition in a Systems of Self-Driven Particle, *Physical Review Letters*, **75**, 1226-1229.

365. Waddell, C.D., 1999. The Growing Digital Divide in Access for People with Disabilities: Overcoming Barriers to Participation in the Digital Economy, In Presentation *Understanding the Digital Economy: Data, Tools and Research conference*, US Department of Commerce, Washington DC, USA.

366. Wagner, M.G., and Ludi, S., 1999. Animated Icons as an Assistant During Interaction Between a Graphical User Interface and the Legally Blind/Partially Sighted, *National Science Foundation award 9978183*.

367. Wang, F., Hall, G.B., and Subaryono, 1990. Fuzzy Information Representation and Processing in Conventional GIS Software: Database Design and Application, *International Journal of Geographical Information Systems*, **4**, 261-283.

368. Wang, X., and Gruen, A., 2000. A Hybrid GIS for 3-D City Models, *International Archives of Photogrammetry and Remote Sensing*, **33**, 1165-1172.

369. Ward, A., Jones, A., and Hopper, A., 1997. A New Location Technique for the Active Office, *IEEE Personal Communications*, **4**, 42-47.

370. Warren, S., and Brandeis, L., 1890. The Right to Privacy, *Harvard Law Review*, **4**, 193-200.

371. Wasinger, R., Stahl, C., and Krüger, A., 2003. M3i in a Pedestrian Navigation and Exploration System, In: L. Chittaro (Ed.), *Mobile HCI*, Springer-Verlag, Berlin/Heidelberg, Germany, 481-485.

372. Wees, J.D., van, Versseput, R.W., Simmelink, H.J., Allard, R.R.L., and Pagnier, H.J.M., 2002. *Shared Earth System Models for the Dutch Subsurface*, Netherlands Institute of Applied Geoscience TNO - National Geological Survey, The Netherlands.

373. Wegener, M., and Masser, I., 1996. Brave New GIS Worlds, In: I Masser, H. Campbell, and M. Craglia (Eds.), *GIS Diffusion: The Adoption and Use of Geographical Information Systems in Local Government in Europe*, Taylor and Francis, London, UK.

374. Weiser, M., 1991. The Computer for the 21st Century, *Scientific American*, **365**, 94-104.

375. Wenzel, E.M., Fisher, S.S., Stone, P.K., and Foster, S.H., 1990. A System for Three-Dimensional Acoustic Visualization in a Virtual Environment Workstation, In Proceedings *IEEE Visualization*, Los Alamitos, CA, USA.

376. Wenzel, E.M., and Foster, S.H., 1990. Digital Synthesis of Virtual Acoustic Environments, *Computer Graphics*, **24**, 139-140.

377. Werner, M., 2002. Integrating GIS with Inundation Models for Flood Extent Mapping, In Proceedings *Urban Data Management Symposium*, Prague, Czech Republic.

378. Westin, A.F., 1970. *Privacy and Freedom*, Bodley Head, London, UK.

379. Wickens, C.D., and Baker, P., 1995. Cognitive Issues in Virtual Reality, In: W. Barfield, and T.A. Furness (Eds.), *Virtual Environments and Advanced Interface Design*, Oxford University Press, NY, USA, 514-541.

380. Willenborg, L., and de Waal, T., 2001. *Elements of Statistical Disclosure Control, Lecture Notes in Statistics 155*, Springer-Verlag, NY, USA.

381. Wilson, I.D., Ware, J.M., and Ware, J.A., 2003. A Genetic Algorithm Approach to Cartographic Map Generalisation, *Computers in Industry*, **52**, 291-304.

382. Winter, S., 2003. Route Adaptive Selection of Salient Features, In: W. Kuhn, M. Worboys, and S. Timpf (Eds.), *Spatial Information Theory: Foundations of Geographic Information Science*, Springer-Verlag, Berlin, Germany, 349-361.

383. Wood, M., 1993. Interacting with Maps, In: D. Medyckyj-Scott, and H.M. Hearnshaw (Eds.), *Human Factors in Geographical Information Systems*, Belhaven Press, London, UK, 111-123.

384. Yang, C., 2000. *Theory, Techniques and Implementation Methods of WebGIS*, Ph.D Thesis, Peking University, Beijing, China.

385. Yang, C., Kafatos, M., Yang, R., Wong, D., and Qu, J., 2002. Geographic Information Service Environment, In Proceedings *GeoInformatics*, Nanjing, China.

386. Yang, C, Kafatos, M., Wong, D., Yang, R., and Cao, Y., 2004a. Gridgis: A Next Generation GIS, In Proceedings *International Conference on Cybernetics and Information Technologies, Systems and Applications*, Orlando, FL, USA.

387. Yang, C., Wong, D., Yang, R.X., Kafatos, M., and Li, Q., 2004b. Performance Improving Techniques in WebGIS, *International Journal of Geographical Information Science*, **319**, 319-341.

388. Yang, Y., Hinde, C., and Gillingwater, D., 2001a. A New Method to Evaluate a Trained Artificial Neural Network, In Proceedings *International Joint Conference on Neural Networks*, Washington DC, USA.

389. Yang, Y., Gillingwater, D., and Hinde, C., 2001b. An Intelligent System for the Sustainable Development of Airports, In Proceedings *9th World Conference on Transportation Research*, Seoul, Korea.

390. Yang, Y., Gillingwater, D., Hinde, C., and Upham, P., 2001d. A Scaled Approach to Developing a Decision Support System for Sustainable Airport

Development, In Proceedings *UK Sustainable Cities and Aviation Network Conference*, Manchester, UK.

391. Yang, Y., Hinde, C., and Gillingwater, D., 2003a. A New Method in Explaining Neural Network Reasoning, In Proceedings *International Joint Conference on Neural Networks*, Portland, OR,USA.

392. Yang, Y, Hinde C., and Gillingwater D 2003b. Improved Neural Network Training using Redundant Structure, In Proceedings *International Joint Conference on Neural Networks*, Portland, OR,USA.

393. Yang, Y., and Rosenbaum, M., 1999. Spatial Data Analysis with ANN: Modelling to Manage Geoenvironmental Problems Caused By Harbour Siltation, In Proceedings *International Symposium on Spatial Data Quality*, Hong Kong, China.

394. Yang, Y., and Rosenbaum, M., 2001. Artificial Neural Networks Linked to GIS for Determining Sedimentology in Harbours, *Journal of Petroleum Science and Engineering*, **29**, 213-220.

395. Yang, Y., Rosenbaum, M., 2003. Artificial Neural Networks Linked to GIS, In: M. Nikravesh, F. Aminzadeh, and L.A. Zadeh (Eds.), *Developments in Petroleum Science:Soft Computing and Intelligent Data Analysis in Oil Exploration*, **51**, Elsevier Science, Amsterdam, The Netherlands.

396. Yang, Y., Rosenbaum, M., and Burton, C., 2001c. An Intelligent Database for Managing Geoenvironmental Change within Harbours, *Environmental Geology*, **40**, 1224-1231.

397. York, B.W., 1989. Tools to Support Blind Programmers, In Proceedings *17th Annual ACM Computer Science Conference*, Louisville, KY, USA.

398. Zadeh, L.A., 1965. Fuzzy Sets, *Information and Control*, **8**, 338–353.

399. Zadeh, L.A., 1994. Fuzzy Logic, Neural Networks and Soft Computing, *ACM Communications*, **37**, 77-84.

400. Zhou, J., and Golledge, R.G., 2004. *Real-Time Tracking of Activity Scheduling/Schedule Execution within Unified Data Collection Framework*, University of California Berkeley Transport Centre, **Working Paper No. 720**.

401. Zipf, A., 2004. Mobile Internet GIS based Flood Warning and Information Systems, In Proceedings *Symposium on Location based Services and TeleCartography*, Vienna, Austria.

402. Zipf, A., and Aras, H., 2002. Proactive Exploitation of the Spatial Context in Lbs through Interoperable Integration of GIS-Services with a Multi-Agent System, In Proceedings *5th AGILE Conference in Geographic Information Science*, Palma di Mallorca, Spain.

403. Zlatanova, S., 2000. Toward a 3D GIS for Local Governing: A Web-Oriented Approach, In Proceedings *22nd Symposium on Urban and Regional Data Management*, Delft, The Netherlands.

404. Zlatanova, S., 2001. 3D Modelling for Augmented Reality, In Proceedings *3rd ISPRS Workshop on Dynamic and Multi-Dimensional GIS*, Bangkok, Thailand.

405. Zlatanova, S., Rahman, A. A., and Shi, W., 2004. Topological Models and Frameworks for 3D Spatial Objects, *Computers and Geosciences*, **30**, 419-428.

Internet References

1. http://www.magic.gov.uk
2. http://www-groups.dcs.st-and.ac.uk/~history/HistTopics/Measurement.html
3. http://www.egyptologyonline.com/astronomy.htm
4. http://www.cartage.org.lb/en/themes/Sciences/Astronomy/Equipementdevices/Telescope/Historyoftelescope/beforetelescope/beforetelescope.htm
5. http://www-gap.dcs.st-and.ac.uk/~history/HistTopics/Chinese_overview.html
6. http://www.westsea.com/tsg3/octlocker/octchart.htm
7. http://pwifland.tripod.com/historysextant/index.htm
8. http://en.wikipedia.org/wiki/Astrolabe
9. http://www.rleggat.com/photohistory/
10. http://www.cs.wisc.edu/condor/
11. http://setiathome.ssl.berkeley.edu/
12. http://data.geocomm.com/helpdesk/formats.html#types
13 http://edc.usgs.gov/guides/dlg.html#dlg1
14. http://www.imdb.com/title/tt0088846/
15. http://www2.jpl.nasa.gov/basics/
16. http://marsnet.jpl.nasa.gov
17. http://angel.elte.hu/~panic/
18. http://www.helbing.com/
19. http://www.crowddynamics.com/
20. http://fseg.gre.ac.uk/
21. http://fseg.gre.ac.uk/fire/Gothenburg_fire_proj.html
22. http://vwisb7.vkw.tu-dresden.de/TrafficForum/
23. http://www.london.gov.uk/mayor/carnival/interim_report/interim_review_report.pdf
24. http://www-106.ibm.com/developerworks/webservices/library/ws-soa-design1/
25. http://www.opengis.org/
26. http://www.fgdc.gov/
27. http://nationalmap.usgs.gov/
28. http://www.geographynetwork.com/
29. http://www.GeoConnections.org
30. http://www.geo-one-stop.gov/
31. http://www.alexandria.ucsb.edu/
32. http://www-106.ibm.com/developerworks/webservices/library/ws-soad1/
33. http://imgs.intergraph.com/mge/
34. http://www.esri.com/software/arcview/index.html
35. http://www.esri.com/software/arcgis/
36. http://imgs.intergraph.com/geomedia/
37. http://portal.opengis.org/files/?artifact_id=5316
38. https://portal.opengeospatial.org/files/?artifact_id=7176
39. https://portal.opengeospatial.org/files/?artifact_id=3837
40. http://www.opengis.org/docs/01-024r1.pdf
41. http://agents.umbc.edu/
42. http://www.openp2p.com

43. http://www.ggf.org/
44. http://www.service-architecture.com/web-services/articles/service-oriented_architecture_soa_definition.html
45. http://www.service-architecture.com/
46. http://www.edge.org/3rd_culture/gelernter/gelernter_p1.html
47. http://www.bittorrent.com/
48. http://www.geospatial-online.com/geospatialsolutions/article/articleDetail.jsp?id=8141
49. http://dis.cs.umass.edu/
50. http://citeseer.ist.psu.edu/533959.html
51. http://www.casa.ucl.ac.uk/research/agent.htm
52. http://www.sc2000.org/techpapr/papers/pap.pap253.pdf
53. http://cswww.essex.ac.uk/staff/doran/Indexbar.htm
54. http://www.nsf-middleware.org/
55. http://www.globus.org/
56. http://collab.nsf-middleware.org/Lists/NMIR6/AllItems.aspx
57. http://www.w3.org/TR/wsdl
58. http://www.microsoft.com/com/default.mspx
59. http://www.corba.org/
60. http://www.opengroup.org/dce/
61. http://www.w3.org/Protocols/
62. http://www.w3.org/2002/ws/
63. http://www.w3.org/TR/soap/
64. http://www.uddi.org/specification.html
65. http://www.w3.org/XML/
66. http://www.globus.org/ogsa/
67. http://www.ncgia.ucsb.edu/projects/nga/
68. http://citeseer.ist.psu.edu/foster02physiology.html
69. http://www-fp.mcs.anl.gov/~foster/Articles/WhatIsTheGrid.pdf
70. http://www.cacr.caltech.edu/publications/index.cfm
71. ftp://ftp.globus.org/pub/globus/papers/globus-hcw98.pdf
72. http://www.globus.org/alliance/publications/papers.php
73. http://www.ggf.org
74. http://www.globus.org
75. https://portal.opengeospatial.org/files/?artifact_id=3837
76. http://www.opengeospatial.org
77. http://www.opengeospatial.org/docs/02-058.pdf
78. http://www.opengeospatial.org/docs/01-068r2.pdf
79. http://www.opengeospatial.org/docs/01-024r1.pdf
80. http://lennier.gsfc.nasa.gov/grid/current_work/grid_task_plan_V1_1.doc
81. http://www.ceos.org
82. http://www.opengeospatial.org/docs/02-087r3.pdf
83. http://lennier.gsfc.nasa.gov/grid/
84. http://www.isi.edu/~deelman/MCS/
85. http://www.globus.org/rls/
86. http://www.opengeospatial.org/docs/01-111.pdf

87. http://www.fgdc.gov/standards/status/csdgm_rs_ex.html
88. http://www.laits.gmu.edu
89. http://www.globus.org/wsrf/foster_wsrf.pdf
90. http://www.daml.org/services/owl-s/1.1/
91. http://www.w3.org/RDF/
92. http://www.w3.org/2004/OWL/
93. http://www.ontoknowledge.org/oil/
94. http://www.daml.org/
95. http://www.ruleml.org/
96. http://www.w3.org/Submission/2004/SUBM-SWRL-20040521/
97. http://www.sts.tu-harburg.de/~r.f.moeller/racer/
98. http://herzberg.ca.sandia.gov/jess/index.shtml
99. http://www.opengis.org/techno/specs.htm
100. http://www.radius.laser-scan.com/
101. http://www.oracle.com/pls/db10g/
102. http://www.serc.nl/resources/publicaties/artikelen/VirtualReality.pdf
103. http://www.ccic.gov/pubs/blue00/pitac.html (Unstable link)
104. http://ccrma-www.stanford.edu/~sile/papers/www6-paper.html
105. http://www.icad.org/websiteV2.0/References/nsf.html
106. http://www.research.att.com/conf/hfweb/proceedings/tannen/
107. http://www.wapforum.org/what/WAPForumBrochureOCT00.pdf
108. http://www.esri.com/news/arcnews/fall00articles/arcimslets.html
109. http://www.pocketpcmag.com/_archives/nov02/
110. http://soundscapes.geog.ucsb.edu/
111. http://www.blindsight.com/
112. http://www.ski.org/tmap/
113. http://www.nctd.org.uk
114. http://www.multireader.org/design.htm
115. http://www2.cmp.uea.ac.uk/Research/researchareas/CGP/vis96/hughesforrest.html
116. http://www.hitl.washington.edu/research/multimodal/HMRS.html
117. http://pollux.geog.ucsb.edu/%7Egolledge/haptics/haptics.html (Unstable link)
118. http://www.geog.ucsb.edu/pgs/papers/loomis_1985.pdf
119. http://www.senderogroup.com
120. http://www.pulsedata.com/
121. http://www.hitl.washington.edu/research/lowvision/
122. http://www.jlocationservices.com/LBSArticles/
123. http://zdnet.com.com/2100-11-523742.html
124. http://www.pcworld.com/news/article/0,aid,118310,00.asp
125. http://www.intel.com/technology/mooreslaw/index.htm
126. http://news.bbc.co.uk/1/shared/bsp/hi/pdfs/dyson_10_12_04.pdf

Subject Index

Printing: Krips bv, Meppel
Binding: Stürtz, Würzburg